Science Networks · Historical Studies
Volume 15

Edited by Erwin Hiebert and Hans Wussing

Birkhäuser Verlag
Basel · Boston · Berlin

The Intersection of History and Mathematics

Editors:
Sasaki Chikara
Sugiura Mitsuo
Joseph W. Dauben

1994

Birkhäuser Verlag
Basel · Boston · Berlin

Editors:

Professor Ch. Sasaki
The University of Tokyo
Department of History & Philosophy
of Science
3–8–1, Komaba, Meguro-ku,
Tokyo 153, Japan

Professor M. Sugiura
Vice-Chairman
Tsuda College
2–1–1, Tsuda-cho, Kodaira-shi,
Tokyo 187, Japan

Professor J. W. Dauben,
Chairman
Program in History
The Graduate Center, CUNY
33 West 42nd Street
New York, NY 10036–8099, USA

A CIP catalogue record for this book is available from the Library of Congress,
Washington D.C., USA

Deutsche Bibliothek Cataloging-in-Publication Data

The **intersection of history and mathematics** / ed.: Chikara
Sasaki ... – Basel ; Boston ; Berlin : Birkhäuser, 1994
　(Science networks ; Vol. 15)

NE: Sasaki, Chikara [Hrsg.]; GT

© 1994 Birkhäuser Verlag, P.O. Box 133, CH–4010 Basel, Switzerland
Softcover reprint of the hardcover 1st edition 1994
Camera-ready copy prepared by the editors
Printed on acid-free paper produced from chlorine-free pulp

ISBN 978-3-0348-7523-3　　ISBN 978-3-0348-7521-9 (eBook)
DOI 10.1007/978-3-0348-7521-9
9 8 7 6 5 4 3 2 1

Table of Contents

Preface

The Twenty-First International Congress of Mathematicians (ICM) was held in Kyoto, Japan, from August 21 through 29, 1990, the first congress that has taken place in the Eastern hemisphere. On this occasion, Japanese historians of mathematics organized the History of Mathematics Symposium which was held at the Sanjo Conference Hall of the University of Tokyo on August 31 and September 1, as one of the related conferences of the Congress. The symposium was officially sponsored by the Executive Committee of the ICM 90, the History of Science Society of Japan, and the International Commission on the History of Mathematics. The Executive Committee consisted of Murata Tamotsu (Chairperson, Momoyama Gakuin University), Sugiura Mitsuo (Vice-Chairperson, Tsuda College), Sasaki Chikara (Secretary, The University of Tokyo), Adachi Norio (Waseda University), Nagaoka Ryosuke (Tsuda College until 1990, now Daito Bunka University), and Hirano Yoichi (Treasurer, Tokai University).

The symposium emphasized the following three fields of study: (1) mathematical traditions in the East, (2) the history of modern European mathematics, and (3) interaction between mathematical research and the history of mathematics. These fields were chosen mainly because, first, the symposium was related to the ICM, the most important congress of working mathematicians, and, second, the Kyoto ICM was held in a non-Western country for the first time. The symposium consisted of the two Sessions: Session A for invited speakers and Session B for short communications. Speakers of Session A were deliberately nominated through discussions in the Executive Committee (as to the institutional affiliations or mailing addresses of the 22 invited speakers, see the list attached at the end of the volume). Each speaker in Session A and Session B delivered a lecture of 30 minutes and 15 minutes, respectively.

The symposium was carried out in the following sequence. In the morning of August 31, the Opening Ceremony was held, where the chairperson Murata gave the opening address, and the president of the University of Tokyo, Arima Akito, gave welcoming greetings. Professor Arima is a theoretical physicist, a *haiku* poet, and noteworthily a great-grand son of the well-known *wasan* mathematician Hasegawa Hiroshi of the Bakumatsu period. He also showed a map of a castle which his family inherited and in which the art of surveying in *wasan* was used, and insisted on the importance of the role of pre-modern mathematics in the successful introduction of Western science into modern Japan. At the end of the

Opening Ceremony, Joseph W. Dauben, chairman of the International Commission on the History of Mathematics, delivered a lecture under the title "Mathematics: An Historian's Perspective," claiming that an independent historian's research, as distinguished from historical studies by a working mathematician, can make an indispensable contribution to the fruitful development of mathematics.

The ceremony was followed by Session A1 "Mathematics from the 17th through the 19th Century," chaired by Erwin Neuenschwander and Miyake Katsuya. The first speaker Hara Kokiti talked about "Une Méthode de Restitution — quelques examples dans le cas de Pascal," discussing the historiography of the critical historical reconstruction of the mathematical works of Blaise Pascal. Craig G. Fraser's lecture "The Interaction of Analysis and Mechanics in Euler's Science" emphasized that mechanical research functioned as an heuristic device in Euler's mathematical treatises such as that on the calculus of variations. Nagaoka Ryosuke in his lecture "Algebraic Formalism as a Method of Discovery and Proof," on the other hand, pointed out that algebraic analysis beyond the theory of equations played an important role as a tool of both mathematical discovery and demonstration by mathematicians, especially of the eighteenth century, an heroic age of mathematical discoveries.

The afternoon of August 31 was given to Session A2 "Mathematics in the 19th and 20th Centuries," of which the chairpersons were Christian Houzel and Adachi Norio. Jeremy Gray, first of all, delivered a lecture entitled "Complex Curves in the Nineteenth Century," asserting that complex curves were the outcome of attempts to geometrize complex function theory, notably the newly-discovered elliptic function. Eberhard Knobloch's talk "From Gauß to Weierstraß: Determinant Theory and Its Historical Evaluation" argued how linear algebra, especially the theory of determinant, was developed in the nineteenth century. Umberto Bottazzini talked about "The History of Elliptic Functions According to Hermite and Weierstraß" through a careful examination of the Swedish mathematician Gösta Mittag-Leffler's lecture notes for the courses on the theory of elliptic functions by Hermite and Weierstraß. Jesper Lützen's lecture "Julius Petersen's Work on Geometric Construction and Galois Theory" shed light on the early reception of the Galois theory in Denmark. The following three lectures argued about how the modern theory of numbers was formed: Günther Frei's "The Reciprocity Law from Euler to Artin," Takase Masahito's "Three Aspects of the Theory of Complex Multiplication," and Miyake Katsuya's "The Establishment of the Takagi-Artin Class Field Theory." The last speaker of Session A2, Chandler Davis, gave an 'agitational' speech under the title "Where Did Twentieth-Century Mathematics Go Wrong?", pointing out that the pervasion of 'pure mathematics' was one of the factors which made the mathematics of the 20th century uninteresting in general.

Session A3 on "Mathematical Traditions in the East," chaired by Joseph W. Dauben and Sugiura Mitsuo, discussed various aspects of the history of Asian mathematics. Roshdi Rashed's lecture "Indian Mathematics in Arabic" showed how Indian Mathematics was introduced into the Islamic world in the Middle Ages through studying a hitherto unknown text of al-Bīrūnī with a commentary of al-Samaw'al. Annick Horiuchi examined mathematical methods used by Takebe Katahiro, the most talented disciple of Seki Takakazu, the founder of *wasan*, in her talk "Takebe Katahiro (1664–1739)'s Conception of Mathematics and Its Historical Background." Sasaki Chikara's "The Adoption of Western Mathematics in Meiji Japan, 1853–1903" argued that institutional factors played a crucial role in Japan's successful introduction of Western mathematics.

Session A4 "Contemporary Mathematics and Historiographical Problems" was held in the afternoon of September 1 and chaired by Eberhard Knobloch and Hirano Yoichi. The first lecture was David E. Rowe's "The Philosophical Views of Klein and Hilbert," in which the speaker explicated the mathematico-philosophical points of view of the two eminent mathematicians at Göttingen at the beginning period of the 20th century and pointed out the two shared many of the same fundamental ideas about mathematical knowledge. The next lecture by Erhard Scholz was "Hermann Weyl's Contribution to Geometry in the Years 1918 to 1923," in which Weyl's lectures in Zürich on differential geometry related to Einstein's general theory of relativity were analyzed in detail. Sugiura Mitsuo related the development of the theory of Lie groups to that of quantum mechanics in his talk "The Origins of Infinite Dimensional Unitary Representations of Lie Groups." Christian Houzel's "The Beginnings of Sheaf Theory: Works by Leray, Cartan, and Grothendieck, 1945–1960" traced the origins of the abstract sheaf theory by A. Grothendieck to works by Jean Leray, Henri Cartan, and Oka Kiyoshi. Liliane Beaulieu described the early activities and initial goals of Nicholas Bourbaki in her lecture "Dispelling a Myth: Questions and Answers about Bourbaki's Early Works, 1934–1944." The very last lecture by Erwin Neuenschwander under the title "Questions in the Historiography of Modern Mathematics: Documentation and the Use of Primary Sources" argued how historians of mathematics search for new historical materials and use them for their historical works by taking as examples the manuscript sources of Joseph Liouville, Bernhard Riemann, and Felice Casarati.

After Session A4, the participants attended the Closing Ceremony, where Iyanaga Shokichi, professor emeritus of the University of Tokyo and the former president of the Japan Mathematical Society, gave farewell remarks emphasizing the interpenetration of the history of mathematics and mathematical research.

As speakers in Sessions A1 and A3, William J. Ellison and Du Shiran were invited and intended to deliver lectures "The History of Maxwell's Equations and the Formulation of Ideas about Electromagnetic Fields in Mathematical Terms" and "The Basic Features of Traditional Chinese Mathematics," respectively, but both were prevented from visiting Japan because of illness. Fortunately, however, Professor Ellison submitted his paper afterwards. In parallel with Session A, Session B was organized by Nagaoka Ryosuke. The names of speakers and the titles of their lectures can be seen in the list attached at the end of the book.

This book is not literally the proceedings of the symposium, but collected papers based on it. Space has not permitted us to insert the papers delivered at Session B. Some invited speakers have refrained from presenting their full papers

for this volume for various reasons. Some of the titles of the papers published in
the book have been slightly changed from the original. Throughout the volume in
principle, the names of Japanese appear in the normal Japanese order, family name
first and given name second, which has become the ordinary practice nowadays.

Finally, I should like to extend appreciation to Mr. Maeda Yasushige, who
devoted himself to the work of converting the present texts into the TEX format,
and to express our gratitude to mathematics editor Dr. Benno Zimmermann and
Ms. Doris Wörner of Birkhäuser Verlag, who endeavored to make this publication
possible in this form.

As stated above, the symposium emphasized the mathematical tradition of
the non-Western world and the interaction between mathematics proper and its
historical studies. May this volume contribute to the flourishing of mathematical
culture, especially in non-Western countries, and to the further acknowledgement
of the importance of the history of mathematics for mathematical culture as a
whole!

November 12, 1993, Tokyo

SASAKI Chikara
Secretary of the Executive Committee

Mathematics: an Historian's Perspective

Joseph W. Dauben

> If the history of science is a secret history,
> then the history of mathematics is doubly secret,
> a secret within a secret.
>
> — G. Sarton
>
> The craft of mathematical history
> can best be practiced by those of us
> who are or have been active mathematicians...
>
> — A. Weil
>
> History of Mathematics...
> too mathematical for historians
> and too historical for mathematicians.
>
> — I. Grattan-Guinness

Abstract

This article is based upon remarks originally prepared for the Tokyo History of Mathematics Symposium held at the University of Tokyo, August 31 – September 1, 1990, in conjunction with the International Congress of Mathematicians held in Kyoto the week before. Considering the interest the International Mathematical Union has shown in history of mathematics by virtue of its recent unanimous vote to recognize the International Commission on History of Mathematics as a joint IMU commission with the International Union of the History and Philosophy of Science, it seemed appropriate to consider a question that is by no means new, but one which has often stimulated considerable controversy among mathematicians and historians of mathematics alike — namely, on the subject of history of mathematics, what should the discipline include, and who should be included in defining the discipline?

This is an especially pertinent question with respect to the position advocated not long ago (at the 1978 International Congress of Mathematicians in Helsinki) by the eminent mathematician André Weil, who made this the subject of his invited plenary lecture.[1] Weil, speaking as a mathematician, used the occasion to argue forcefully that only mathematicians like himself

---

[1] André Weil, "History of Mathematics: Why and How," *Proceedings of the International Congress of Mathematicians, Helsinki, 1978*, O. Lehto, ed., Helsinki: Academia Scientiarum Fennica, 1980, vol. 1, pp. 227–236.

were qualified to write history of mathematics, and the better the mathematician, the better the history was likely to be. This may at first seem so obvious as to hardly bear discussion, let alone much room for doubt. But in the following, I shall suggest some reasons why I believe the view put forward by Weil to be mistaken, not only from the general point of view of the history of mathematics, but equally pernicious strategically from the more sublime perspective of mathematics itself.

Encouraging History of Mathematics:

Although the subject can trace its roots back to antiquity, when the Greek writer Eudemos of Rhodes wrote the first "history" of mathematics, as a professional discipline it may be said to have originated in the 19th century. Even today, despite increasing interest in history of mathematics, it remains the province of a relatively small number of specialists. As Judith Grabiner put it at a workshop on the evolution of modern mathematics sponsored by the American Academy of Arts and Sciences in Boston in 1974:

> There are now too few historians of mathematics. The path for the historian of mathematics is difficult; he needs the historian's training, but also needs to know a great deal of mathematics. The history of science is itself a young and relatively small profession; the number of historians of mathematics, because of the types of knowledge needed, is even smaller. Still the need for such people is apparent.[2]

Clearly, mathematicians and historians of mathematics alike need to promote rather than limit the number of scholars interested in the subject. But mathematicians, who bring special insights to historical questions because of their technical expertise, usually bring very special interests as well. As Grabiner says, the mathematician "is oriented to the present, and toward past mathematics chiefly insofar as it led to important present mathematics.... The history of mathematics as written by mathematicians tends to be technical, to focus on the content of specific papers." [3]

Weil sees history of mathematics primarily from this limited perspective. It is a subject meant above all for mathematicians, only the best of whom are really in a good position, he maintains, to write the history of this admittedly demanding subject. He begins his argument by quoting Leibniz, one of the earliest mathematicians to give a rationale for interest in history of the subject:

> Its use is not just that History may give everyone his due and that others may look forward to similar praise, but also that the art

---

[2]Judith Grabiner, "The Mathematician, the Historian, and the History of Mathematics," *Historia Mathematica*, **2** (1975), pp. 439–447; esp. p. 443.

[3]Grabiner 1975, p. 439.

of discovery be promoted and its methods known through illustrious examples.[4]

Weil interprets this to mean that Leibniz wanted the historian of science to write in the first place for creative or would-be creative scientists. This was the audience Leibniz had in mind, according to Weil, as he wrote in retrospect about his "most noble invention" — the calculus. But this is a very curious example to rationalize interest in history of mathematics, especially with Weil's underlying theme in mind that mathematicians are best prepared to undertake this task. For there are very good reasons to believe that, at the time he was writing, the significance Leibniz attached to his words about history of mathematics was not as straightforward as Weil would have us believe.

Who Should Write History of Mathematics?
The Case of Newton and Leibniz

Weil responds to the question of who should write history of mathematics with a very restrictive answer: "the craft of mathematical history can best be practiced by those of us who are or have been active mathematicians." If this be true, then what better examples might we take of this principle than mathematicians of the stature of Leibniz — or Newton? In fact, we are even in a position to assess Weil's assertion, because Newton wrote, and Leibniz drafted, what they each called "histories" of their famous co-discovery, namely the infinitesimal or differential calculus.

Briefly, claims over their respective priorities in discovering the calculus (which at first Leibniz felt he was entitled to share with Newton; later he claimed it directly by right of his priority in publication) led to acrimonious debate. This eventually prompted Leibniz to ask that the Royal Society investigate his claims vis-à-vis Newton's. The Royal Society obliged, and a year later produced a "history" of the calculus — little more than a set of documents (assembled secretly with comments by Newton himself). The resulting *Commercium Epistolicum* of 1712, not surprisingly, found unequivocally in Newton's favor.

Leibniz was urged by his strong supporters Johann Bernoulli and Christian Wolf to counter Newton's historical scholarship in the *Commercium Epistolicum*. They wanted Leibniz to issue an historical narrative of his own about the evolution of the "genuine calculus," and Leibniz, who admitted the wisdom of this advice, often spoke of doing so. He did manage to distribute a leaflet (or *Charta Volans* as Newton called it — a flying paper) dated July 29, 1713, in which he deprecated the *Commercium Epistolicum* and recounted the public record of what he had published on the calculus as opposed to the previously unpublished documents Newton had compiled for the *Commercium Epistolicum*. A year later, Leibniz

---

[4]G.W. Leibniz, quoted from C.I. Gerhardt, ed. *Mathematische Schriften*, vol. 5, p. 392; translated in Weil 1978, p. 227.

began to work on his own "History and Origin of the Differential Calculus," but this remained forever a fragment, nothing more than a preliminary draft.

While Bernoulli called the *Commercium Epistolicum* "twice cooked cabbage," Newton's bulldog, John Keill, took Newton's rival to task for sloppy work with the calculus. He was particularly hard on Leibniz's *Tentamen* of 1689. This, so the argument went, was proof that Leibniz did not really understand the calculus and could not have independently invented it. Instead, Leibniz must have borrowed it from Newton without having fully understood it.[5]

Joseph Raphson's posthumous *History of Fluxions* (1715) added more fuel to the Newtonians' bonfire of Leibniz's claims on the calculus. In his preface he makes it clear that the book's object was "to assert the Principal Inventions of this Method, to their First and Genuine Authors; and especially those of Sir Isaac Newton." Leibniz's priority in publication, especially his first *Acta Eruditorum* paper of 1684, was dismissed as revealing "how less apt and more laborious a method of notation, and farfetch'd symbolizing insignificant novelties (perhaps on purpose to distinguish it from the plain and easy one it was communicated to him in) he has published it to the World."[6]

In the case of the Newton-Leibniz controversy, no mathematician at the time was likely to have done justice to the debate, or to have written an objective history. Despite Weil's claim that Leibniz expected the history of science to illustrate the "art of discovery," Leibniz was being more honest when he first said that history should be written to give everyone "his due." This is often the problem when mathematicians turn to history, frequently with priority axes to grind. Their historical interest in such cases is usually limited almost entirely to questions of who did what first.

Perhaps the Newton-Leibniz case is an extreme one. It might be argued that times have changed, and that mathematicians today, more sophisticated, can be more objective in writing history to exemplify the best methods of the past rather than use it solely for their own advantage, whatever that may be.

---

[5]See John Keill, "Réponse aux auteurs des remarques, sur le différence entre M. de Leibnitz et M. Newton," *Journal Littéraire de la Haye*, **2** (July-August, 1714), pp. 445–453, and E.J. Aiton, *The Vortex Theory of Planetary Motions*, London: Macdonald, 1972, p. 138. The same sort of argument was also used by Leibniz and especially Johann Bernoulli, who did everything they could to discredit Newton's competence as a mathematician as reflected in the *Principia*. As A.R. Hall has said, "Every effort was made to convict [Newton] of error and ignorance on the ground that so feeble a mathematician could not conceivably have devised the calculus." A.R. Hall, *Philosophers at War. The Quarrel Between Newton and Leibniz*, Cambridge: Cambridge University Press, 1980, p. 193.

[6]Joseph Raphson, *The History of Fluxions, Shewing in a Compendious Manner the first Rise of, and various Improvements made in that Incomparable Method*, London: W. Pearson, 1715, p. 19. Newton also had a secret hand in Raphson's book. He was instrumental in producing a Latin version for the continent, and in a second English edition, Newton made additions of his own. As Hall has said of all this, "Newton molded Raphson's book firmly and secretly to his own ends," Hall 1980, p. 226.

Who Should Write History of Mathematics?
The Case of Abraham Robinson

Consider, in this light then, the contemporary example of a first-class mathematician – Abraham Robinson – who also had respectable interests in history of
mathematics, and was himself well-read and knowledgeable on the subject. Not
only was he commissioned by the editors of the *Dictionary of Scientific Biography*
to write several articles on important mathematicians – all of whom dealt in one
way or another with the subject of infinitesimals – but he even volunteered to
write the entry on Carnot, which had unfortunately already been assigned.[7] The
the best indicator of Robinson's interest in history of mathematics, however, is
the chapter he wrote at the end of his well-known book on nonstandard analysis
in which historical questions concerning infinitesimals are the focus of attention.[8]
Among the examples Robinson considers is the case of Cauchy's work on infinite
series, including Cauchy's famous theorem that a convergent series of continuous
functions is continuous. Robinson did not go as far as the philosopher of science,
Imré Lakatos, who later argued, based on Robinson's own analysis, that Cauchy's
proof had in fact been correct all along (Robinson only said that a nonstandard
proof of Cauchy's theorem showed it to be correct, based on an interpretation
of Cauchy's infinitesimals that would have them equivalent to Robinson's own
nonstandard infinitesimals).[9]

---

[7]Several biographical accounts of Robinson are available, including G. Seligman, "Biography
of Abraham Robinson," in H.J. Keisler *et al.*, eds., *Selected Papers of Abraham Robinson*, New
Haven: Yale University Press, 1979, vol. 1, pp. xiii–xxxii; and J. Dauben, "Abraham Robinson,"
The Dictionary of Scientific Biography, Supplement II, New York: Scribners, 1990, pp. 748–751.

[8]Abraham Robinson, *Non-Standard Analysis*, Amsterdam: North-Holland, 1966; 2nd ed.,
1974.

[9]I. Lakatos, "Cauchy and the Continuum: The Significance of Non-standard Analysis for
the History and Philosophy of Mathematics," in I. Lakatos (J. Worrall and G. Currie, eds.),
Mathematics, Science and Epistemology: Philosophical Papers, vol. 2, Cambridge: Cambridge
University Press, 1978, pp. 148–151. Reprinted in *The Mathematical Intelligencer*, **1** (1979),
pp. 151–161, with a note, "Introducing Imré Lakatos," pp. 148–151.

A more detailed discussion of Lakatos' application of nonstandard analysis in his "rational
reconstruction" of Cauchy's understanding of infinitesimals is given in J. Dauben, "Abraham
Robinson and Nonstandard Analysis: History, Philosophy and Foundations of Mathematics," in
P. Kitcher and W. Aspray, eds. *New Perspectives on the History and Philosophy of Mathematics*,
Minneapolis: University of Minnesota Press, 1987, pp. 177–200; see as well "Abraham Robinson:
Les Infinitesimaux, l'Analyse Non-Standard, et les Fondements des Mathématiques," in H. Bar-
reau, ed., *La Mathématique Non-Standard* (Fondements des Sciences), Paris: Editions du CNRS,
1989, pp. 157–184.

History from a Mathematician's Perspective

In any case, Robinson's approach to history of mathematics was not unlike that of most mathematicians. It is a perfectly natural one — namely, a mathematician is likely to be interested primarily in the history of that branch or area of mathematics in which she or he has been most active. Typical of this approach is the case Weil himself offers of how best to understand Euclid's *Elements*. According to Weil:

> It is impossible for us to analyze properly the contents of Books V and VII of Euclid without the concept of group and even that of groups with operators since the ratios of magnitudes are treated as a multiplicative group operating on the additive group of the magnitudes themselves. Once that point of view is adopted, those books of Euclid lose their mysterious character, and it becomes easy to follow the line which leads directly from them to Oresme and Chuquet, then to Neper and logarithms.[10]

But what does this sort of "mathematical history,"[11] as Weil calls it, amount to? It seems clear that whatever his "mathematical history" may be, it is not only anachronistic, but raises as well a fundamental point that further clarifies what he is doing — namely mathematics with historical examples: MHE, and not an example of the history of mathematics, EHM. Again, $MHE \neq EHM$.

Instead of being history, Weil's example is nevertheless very interesting in itself – as an example of how a mathematician naturally tends to think about any mathematical problem. Given Euclid's results, Weil surveys them with a vast repertoire of knowledge unavailable to Euclid, and sees that the whole structure of Euclid's argument works thanks to certain underlying group-theoretic principles. But the insights are mathematical ones, and do not really go beyond what Weil knows at this moment about multiplicative and additive groups. Such analysis, however, offers no new historical insights.

The same, it may be said, is equally true of Abraham Robinson's analysis of Cauchy's use of infinitesimals based upon a reconstruction using nonstandard analysis. This is very interesting mathematically, but again, it is not really history of mathematics. Similarly, if Robinson believed that nonstandard analysis made it possible to explain why Leibniz did not go wrong in using infinitesimals, the insight, if any is really to be had, is a mathematical one, not an historical one concerning what Leibniz did with — or how he conceived of — his own differential calculus.

---

[10]Weil 1978, p. 232.

[11]It should not be overlooked that throughout Weil's paper he persists in the annoying use of the term "mathematical history" when he really means "history of mathematics." The two are not equivalent. "Mathematical history," MH, is history analyzed with the tools of mathematics and statistics, generally known as "cliometrics." "History of Mathematics," HM, is any attempt to understand what mathematics was like in the past and how it came to be that way. The concepts do not commute! $MH \neq HM$. For a useful discussion of the distinction, see I. Grattan-Guinness, "Does History of Science Treat of the History of Science? The Case of Mathematics," *History of Science*, **28** (1990), pp. 149–173, esp. p. 149.

Great Ideas in Mathematics: The "Nose" Theory

If there is a problem with Weil's view of why history of mathematics should be written — "ransacking history" as I would call it for "illustrious heuristic examples" — he also seems wrong about how history of mathematics should be delimited. If history of mathematics is to consist of the "great ideas" of the subject, then it is necessary to agree upon what constitutes a "mathematical idea." On this point Weil adopts the "nose theory" of mathematics — that the mathematician "may not be able to define what is a mathematical idea, but he likes to think that when he smells one, he knows it." [12]

The infinite, for example, only smells like a mathematical idea "after Cantor defined equipotent sets and proved some theorems about them." [13] This rules out, Weil insists, anything said about the infinite as being a part of mathematics by the Greeks or Medieval philosophers, or by anyone prior to about 1880. Views of Greek philosophers about the infinite may be of interest to philosophers, he says, but Weil refuses to believe that they had any great influence on the work of Greek mathematicians.

If one is thinking of the Presocratics, for example, like Anaxamander and his rather vague ideas about the *apeiron*, then Weil may be correct. But aren't Zeno's paradoxes of motion intimately connected with the mathematical problem of the infinite, as were the Pythagoreans' troubling attempts to deal with the discovery of incommensurable magnitudes, and Eudoxus' eventual resolution of the dilemma of incommensurables by means of his theory of proportion?

Just because the infinite may not have been treated with full mathematical rigor until Georg Cantor, does this mean that earlier the infinite was not regarded as a serious mathematical problem? And what should the historian of mathematics do about the entire history of infinitesimals? Is it reasonable to argue that the subject only becomes a part of the history of mathematics after the work of Abraham Robinson and the appearance of nonstandard analysis (or with various other claimants to having developed rigorous nonarchimedean systems admitting infinitesimals in mathematically rigorous ways, like du Bois-Reymond, Veronese, or more recently, Schmieden and Laugwitz)?

When may we say that infinitesimals finally become an acceptable part of the historical record? It seems that Weil has again confused mathematics with history of mathematics — it is one thing to say that infinitesimals do not become an acceptable part of *mathematics* until the 20th century, but it is surely wrong to conclude that they are not an important part of the *history* of mathematics until their rigorous validity has been established. [14]

---

[12] Weil 1978, p. 230.

[13] Weil 1978, p. 230.

[14] On the subject of the infinite Weil adds, rather condescendingly, that it may on the other hand be of interest to philosophy, which prompts him to quip that "Some universities have established chairs for 'the history and philosophy of mathematics'; it is hard for me to imagine what those two subjects can have in common." This seems strange, indeed, since Weil, writing from his

What Makes "History"?

Weil's restricted interpretation of what constitutes a "mathematical idea" for the sake of writing history of mathematics parallels a debate that raged not long ago over the history of mechanics and who should write it. István Szabó, Max von Laue's successor in the chair for mechanics at the Technical University of Berlin from 1948-1975, wrote a book: *Geschichte der mechanischen Prinzipien*, published in 1976.

Shortly after von Laue had written his own *Geschichte der Physik* in 1946, Albert Einstein wrote to him in praise of the book which "with mastery picked out the most important." "It is really useful," Einstein continued, "that one who surveys the entire panorama with such understanding, takes the history of human thought out of the hands of the philologists and popularizers and presents the great drama cleansed from the dust of unimportant details."[15]

The same could be said of Szabó's book, *mutatis mutandis*, according to Armin Hermann, professor of History of Science and Technology at the University of Stuttgart. And yet, as Hermann charges, Szabó's qualifications as a physicist are not enough to insure that he has an appropriate understanding of the *history* of the subject.

To begin with, Szabó's history starts with Galileo, because, as he explains, he is only concerned with "what, in the development of classical mechanics really made 'history.'"[16] Nothing prior to Galileo, he claims, was sufficiently scientific to qualify as part of the true history of mechanics. This rings a familiar bell — it is virtually the same point Weil tries to make about the history of mathematics. The infinite, for example, has no place in history of mathematics until Georg Cantor.

All of this is akin to saying that the theory of phlogiston has no place in the history of chemistry, or that epicycles, deferents, and equants have no place in the history of astronomy. Similarly, can one imagine the history of astronomy without Ptolemy or Copernicus, or celestial mechanics without Descartes (whatever he may have thought about vortices!)? The problem, in both cases, for Szabó and Weil, is that each seems to assume that history should serve only the interests of the successful — as they view success. This means working backwards from what the practitioners of today deem valuable or correct, and then judging all of the past by their own present standards.

---

position at the Institute for Advanced Study in Princeton, knew among his colleagues there the commanding figure of Kurt Gödel, in whose work the history and philosophy of mathematics may be said to conmingle in an especially relevant way. But this view, that philosophy is irrelevant to mathematics, is apparently a trademark of the Bourbaki school.

[15]Unpublished letter from Albert Einstein to Max von Laue, May 15, 1947, quoted in Armin Hermann's introduction to the special edition of István Szabó, *Geschichte der mechanischen Prinzipien und ihrer wichtigsten Anwendungen*, Basel: Birkhäuser Verlag, 1976; Armin Hermann's "Begleitwort" accompanies the 2nd ed., 1979, p. xi. I am grateful to Christoph Scriba for calling my attention to the remarkable case of Hermann's introduction to this edition of Szabó's history.

[16]See the discussion of this point in Hermann's introduction to Szabó 1979, pp. xi-xiii.

History of Mathematics: The "Whiggism" View

Having objected to Weil's overly-narrow idea that history of mathematics should ransack the past for illustrious examples, and concern itself only with "real" mathematical ideas, is there any reason to doubt his claim that mathematicians should do the writing — and indeed are in the best position to do so? Reasonable though this may seem, were this to be the case, it is unlikely that much would ever get written.

Most active mathematicians have interests other than history, namely proving theorems. George Mackey, because of the "pressures of his discipline," acknowledges an interest in history, but has no time to *do* it. Moreover, for those mathematicians who do find time, history is often little more than anecdotal. Nor is accuracy all that important, especially if one is only thinking of the heuristic value of historical examples. Again, in Mackey's view, "neither the historian's nor the philosopher's detailed accuracy is beneficial pedagogically." A little history is good enough. As Mackey admits, due to the "pressures of his discipline," he is not interested in "too detailed a history."[17]

For the sake of argument, however, assume that a mathematician with "a nose for history" as Weil says, is serious about writing history with an eye open to elucidate the great results and methods of the past. What may be expected in most cases is the "going backward" approach to — or retreat from, actually — history. This method is intrinsically a-historical, and brings with it the inherent danger of writing very whiggish history — "whiggish" in the sense that the progress of mathematics leading to the present state of things was all but inevitable.

As Ivor Grattan-Guinness has said, perhaps more colorfully, mathematicians "usually view history as the record of a 'royal road to me' — that is, an account of how a particular modern theory arose out of older theories instead of an account of those older theories in their own right. In other words, they confound the question, 'How did we get here?', with the different question, 'What happened in the past?' "[18]

Doing this sort of backwards, "whiggish" history also carries with it another inherent danger. This is related to the understandably proprietary sense a mathematician may have of results leading up to her or his own work or area of special interest. In the corresponding retreat down the family tree of mathematics, from now to then, only those known or presumed predecessors to one's own work who were "significant" are likely to be included.

George Sarton[19] illustrated this problem of the mathematician's view of history with a graphic genealogical metaphor:

---

[17]G. Mackey, remarks made at the Workshop on the Evolution of Modern Mathematics, *Historia Mathematica*, **2** (1975), pp. 446–447.

[18]Grattan-Guinness 1990, p. 157.

[19]George Sarton, *The Study of the History of Mathematics*, Cambridge, Mass.: Harvard University Press, 1936; repr. New York: Dover, 1957, p. 6. Sarton's figure is on the left; the "Grattan-Guinness" version is on the right. Reproduced by permission of Harvard University Press.

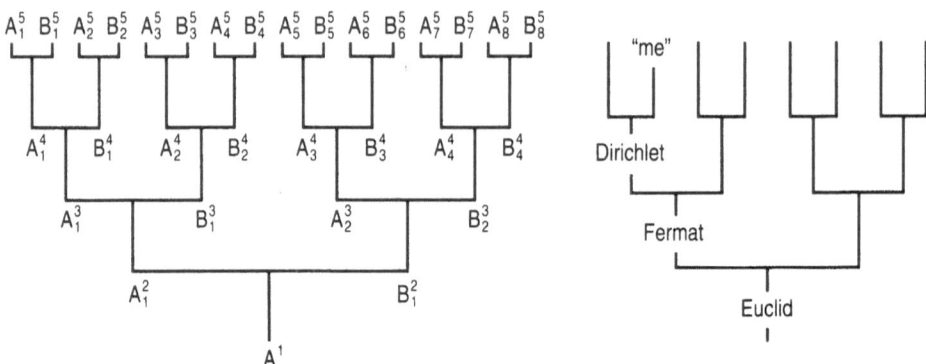

Assuming n "generations" of predecessors, of the 2^{n+1} possible paths from the first to the $(n + 1)^{\text{th}}$, the "going backwards" method will examine n predecessors and ignore $2^n - n - 1$. Wholly apart from the numbers and diagrams, to expect that the "tree" of mathematics and the history of its growth and development can be adequately covered in such an arbitrary and partial fashion is not realistically to be expected.

As a graphic counterexample to the idea that there is any straightforward thread to be followed from the historical past to the present in *either* direction, consider the elaborate "diagrams" Herbert Mehrtens devised for his *Die Entstehung der Verbandstheorie* to delineate major routes of influence in the history of lattice theory, or (as illustrated here) the deductive relations between the axiom of choice, continuum hypothesis and non-measurable sets as charted in Gregory Moore's *Zermelo's Axiom of Choice*. In each case the diagrams reflect the complex interaction of different individuals or ideas in multiple directions, sometimes simultaneously, often indirectly.[20]

---

[20]See, for example, Herbert Mehrtens, *Die Entstehung der Verbandstheorie*, Hildesheim: Gerstenberg Verlag, 1979, p. 12; and Gregory H. Moore, *Zermelo's Axiom of Choice. Its Origins, Development, and Influence*, New York: Springer-Verlag, 1982, p. 326. The diagram is reproduced by permission of Springer-Verlag, New York.

Writing History of Mathematics

Creative ability in mathematics is clearly a gift with which few are blessed. But one need not be an André Weil or a winner of the Fields Medal to understand and appreciate mathematics, an ability of a different sort altogether from the creative side of mathematics where new techniques, theorems and proofs are prized above all. Weil does not sufficiently distinguish the one from the other.

A debate with similar overtones on just this question has been going on for some time now between certain philosophers and the community interested in the frontiers of logic and computer science. The controversy was launched by the philosopher James Fetzer, who recently attacked the basic idea of program verification, saying that it was impossible to provide mathematical proofs of computer system correctness.[21] The subject itself goes back to 1969, when C.A.R. Hoare wrote a pioneering article on "An Axiomatic Basis for Computer Programming."[22]

The details of this debate about whether proofs of program correctness are possible or not can be overlooked here, but the *ad hominem* nature of one aspect, at least, of the debate is relevant. Much of the dismay among computer scientists created by Fetzer's position is related to the fact that he is a philosopher and not a mathematician. Read "historian of mathematics" for "philosopher" in the following summary of the situation by Jon Barwise, and the "ad hominem" aspect of Weil's position concerning historians of mathematics becomes clear:

> Many of the charges leveled against Fetzer's article are typical of encounters between practitioners of any field X and philosophers of X. The philosopher necessarily attempts to give an analysis of X as it presents itself to the informed outsider. The practitioner feels that the philosopher misses a (or the) main point of X. Out of frustration, he is all too often tempted to claim that one simply cannot understand X without doing X. As mathematicians (let X be mathematics), we can all surely recognize the temptation. But such reactions do not really tell against the message carried by the philosopher; they simply try to cast doubt or ridicule on the messenger.[23]

Weil, however, has not resisted the temptation. He *does* say that one cannot understand mathematics without "doing" it, and proceeds to "cast doubt or ridicule" upon those who write history of mathematics without being preferably a mathematician like himself.

Barwise, more moderately, prefers to ignore such reactions and, as he says, "get at the substance of the debate." In this case, the substance was expressed at its richest by Ken May:

---

[21] James H. Fetzer, "Program verification: The Very Idea," *Communications of the Association for Computing Machinery*, **31** (September, 1988), pp. 1048–1063.

[22] C.A.R. Hoare, "An Axiomatic Basis for Computer Programming," *Communications of the Association for Computing Machinery*, **12** (1969), pp. 576–580.

[23] Jon Barwise, "Mathematical Proofs of Computer System Correctness," *Notices of the American Mathematical Society*, **36** (September, 1989), pp. 844–851, esp. pp. 845–46.

I believe that history can and should be socially useful, to historians
of science, to policy makers, to students and users of mathematics, to
the educated layman, and above all to the mathematicians who are its
most reliable consumers and the creators of its raw material.

The history of mathematics seems to have arrived at a takeoff point
for the serious study of the recent developments, and a successful flight
requires a collaboration of historians and creative mathematicians.[24]

Basically, Weil holds what is a very old-fashioned view of mathematics —
one accepted doubtless for most of its history, but already beginning to fade at
the end of the last century. Here, his reliance (at least in the lecture he gave
in Helsinki on who should write history of mathematics for whom) on no more
recent authorities than Moritz Cantor (1829–1920), Paul Tannery (1843–1904),
and Gustav Eneström (1852–1923), may have contributed to the problem, for
what Weil seems to have in mind is the "cumulative" model of the history of
mathematics. On this view mathematics is taken to be a storehouse of "correct"
theories and theorems. The historian's task is simply to take the best of these as
great examples of results or methods, and show how they came to be. The errors,
failed experiments or faulty reasoning are all swept under the carpet.

History of Mathematics: The Historian's Perspective

But mathematics is not just *mathematics* — it is not simply a repository of correct results. If Weil were more sympathetic to philosophy of mathematics, I believe
something like Imre Lakatos' *Proofs and Refutations* would have made this clear.
Mathematics, regarded intellectually, as puzzle solving, shares something with the
experimental sciences. As mathematicians actually do mathematics, they consider
various hypotheses, possibilities, find out what "works," what "doesn't" — and
often improve upon their results thanks to social interaction with other mathematicians. The Kyoto Congress just ended is a prime example of this phenomenon in
action. In short, doing mathematics is much messier — and much more challenging
— than what gets written up as "mathematics" in books or papers.

This is all borne out in an exemplary fashion in Herbert Mehrtens' recent
analysis of the origins and development of lattice theory. As Mehrtens shows,
lattice theory arose in many different ways, as a result of diverse motives and
different approaches. The eventual acceptance of lattice theory and its emergence
as a recognized branch of mathematics in the 1930's was a social process, he says,
as well as a matter of technical mathematics.[25]

In conclusion, I prefer to take a larger view of the history of mathematics
in preference to the narrow positions of Gustav Eneström or André Weil. Here I

---

[24] Kenneth O. May, "What is Good History and Who Should Do It?" *Historia Mathematica*,
2 (1975), pp. 449–455.
[25] Mehrtens 1979.

believe George Sarton was correct when he wrote that "the history of mathematics should really be the kernel of the history of culture."[26] But this will never be the case if it is written only by mathematicians with eyes solely on the utility of past discoveries or methods for instruction or use by the current generation of mathematicians.

This alone would be all too limited a goal, especially as the methods and approaches to modern mathematics become increasingly specialized, with less and less connection to the problems and methods that occupied mathematicians of earlier generations. As certainly seems the case now for most of the subject's history prior to say 1800, and for many parts of 19th-century mathematics as well, earlier content and methods often seem foreign, even irrelevant to current work of contemporary mathematicians.

There is yet another aspect to Sarton's idea that mathematics should be central to the history of culture that also bears consideration. This is the sad fact that if history of mathematics limits itself to the interests of mathematicians, as a primarily heuristic device for the edification of its own practitioners, the history of science (and in turn the history of culture) will suffer greatly. Again, as George Sarton himself knew all too well:

> Take the mathematical developments out of the history of science, and you suppress the skeleton which supported and kept together all the rest.[27]

A Little more than fifty years ago, Sarton published his guide to *The Study of the History of Mathematics* (1936), in which he noted that the history of science was a "secret history" — and the history of mathematics a secret within the secret — for while most scholars might know something of the history of science in general, few mathematicians, scientists or even specialists in the history of science could be expected to know much about the history of mathematics.[28] If we are to make the subject less of a *secreta secretorum* and more a part of the history of science and culture generally, East and West, then we must support, not limit its development. Anyone with the means and interest to do so should be encouraged to join the growing international effort to study, teach and write the history of mathematics.

---

[26]Sarton 1936, p. 4.
[27]Sarton 1936, p. 4.
[28]Sarton 1936, p. 7.

Une Méthode de Restitution—quelques exemples dans le cas de Pascal

Hara Kokiti

Il est question ici d'une méthode qui, pour restituer le processus ignoré d'une production mathématique passée, prête particulièrement attention à la modalité qu'un document concerné montre quant à des éléments d'un ou plusieurs ensembles ordonnés.

S'il est normal, en effet, qu'un géomètre numérote les figures de son ouvrage au fur et à mesure qu'il s'en sert dans son texte, n'est-il pas vraisemblable que les écarts de cet ordre naturel soient dus à des fluctuations survenues sous la plume de l'auteur? Ne peut-on pas dès lors ériger ces normalité et vraisemblance conjointes en hypothèse de travail pour retrouver au moins un certain stade de la genèse de cet ouvrage?

La situation n'est pas si simple lorsqu'il s'agit, par exemple, de la désignation littérale des points de l'espace. Encore la même méthode méritera-t-elle d'être essayée en prenant cette fois l'ordre alphabétique pour repère, d'autant plus que la valeur d'une hypothèse de travail dépend enfin des résultats qu'elle nous permet d'atteindre.

Il va sans dire que toute autre méthode de recherche historique est aussi à utiliser pour une tentative de restitution, et il se pourra même que notre hypothèse de travail se trouve renforcée par là. Cependant, je crois avoir constaté plus d'une fois qu'elle pouvait conduire à des résultats qui m'avaient été pendant longtemps inaccessibles par d'autres voies. C'est ce qui m'est arrivé à propos des trois éminentes contributions mathématiques de Blaise Pascal: (1) découverte du théorème portant son nom en géométrie projective, (2) celle encore de l'induction mathématique, et (3) rédaction des *Lettres de A.Dettonville*, un chef-d'oeuvre de géométrie infinitésimale.

Je serais heureux si le rapport de ces expériences pouvait vous intéresser à cette méthode de restitution, dont le champ d'application est certes limité, mais qui, une fois appliquée, pourrait être fort fructueuse.

Notre hypothèse de travail s'applique très aisément au sujet 3 quant aux numéros des figures annexées.

Les *Lettres de A.Dettonville* étaient composées, les pièces liminaires mises à part, de quatre opuscules de forme épistolaire, et comportant chacun ses propres pagination et millésime. En particulier, la Lettre I, de beaucoup la plus longue, et résolvant les problèmes publiquement proposés par l'auteur lui-même en juin et octobre 1658 au sujet de la cycloïde, consiste en sept pièces. Vu d'ailleurs les matières traitées respectives, l'ordre adopté de présentation de ces lettres semble raisonnable.[1] Mais des irrégularités y abondent dans la numérotation des figures.[2]

En suivant l'ordre naturel des numéros dans la Lettre I, on obtient les Fig. 1 à 21, et les Figures restantes, de 22 à 32, s'y trouvent interposées dans les positions fort variées.

Pour ce qui est des autres Lettres, où diminue l'irrégularité, la suite des nombres de 33 à 45 fait penser qu'après la Lettre I, l'auteur les aurait rédigées dans l'ordre de IV, II, III en dépit du millésime de la Lettre II.[3]

Or, parmi tant d'irrégularités de numéros, je n'en relèverai ici que les deux suivantes, car toutes les autres sont de moindre importance, et explicables, pour la plupart, par le soin de l'auteur en vue de faire mieux comprendre son ouvrage par le commun des lecteurs.

Regardons en premier lieu l'intervention de la Fig. 28 dans la pièce 1; je l'estime la plus significative dans ces *Lettres de A.Dettonville*. Car, que prouve-t-elle sinon que l'auteur disposait en ce moment au moins des Fig. 1 à 28, et qu'à plus forte raison, la Fig. 21, la dernière en position de la Lettre I, avait déjà été dessinée, et numérotée. Cela fait penser à une rédaction assez soigneuse des solutions des problèmes du concours, puisque les numéros des figures ne devaient être nécessaires qu'à la référence explicite au cours des arguments de l'auteur. De plus, il faut qu'à la suite des Fig. 1 à 21, l'intercalation au moins des Fig. 22 à 27 ait été aussi effectuée, à quelques endroits que ce fût, et cela suppose une révision très attentive du manuscrit. Plutôt donc qu'un simple mémorandum à usage personnel, Pascal devait posséder alors une ébauche déjà soigneusement préparée en vue de la publication. Ainsi s'explique en partie l'"incroyable précipitation"[4] de notre auteur. Et cependant, il fut encore réduit à écrire: "[le] peu de loisir que j'ai de m'appliquer à ces sortes d'études."[5] Il aurait donc renoncé, lors de la rédaction définitive de la Lettre I — et aussi des trois autres — à réparer ce désordre, travail sans importance enfin, et qui risquait d'entraîner de nouvelles confusions du texte.

En deuxième lieu, que dire du décalage signalé plus haut relativement à la Lettre II? Je crois que les liens dont Pascal se liait depuis longtemps avec Roberval et Antoine Arnauld, depuis l'automne de 1657 avec Sluse,[6] et depuis l'hiver de 1658 avec Huygens,[7] et d'autre part, l'existence d'une première rédaction, en quelque sorte, de cette Lettre II depuis septembre 1658,[8] — ces diverses circonstances antérieures nous permettent de dissiper la contradiction apparente, en datant de fin 1658 la rédaction du corps principal de cette Lettre II, et de début 1659 celle d'un préambule en vue de son envoi à Huygens.

Les sujets 1 et 2 appellent l'observation de l'ordre alphabétique.

L'*Essai pour les coniques* (1640) du jeune Pascal consiste en deux définitions, trois lemmes et cinq propositions, mais sans aucune démonstration, et le "théorème de Pascal" y apparut d'abord relativement au cercle (Lemme 1), puis généralisé par projection centrale au cas de la conique (Lemme 3). Ce mémorable théorème, comment Pascal l'a-t-il démontré? Je crois possible de répondre à cette question en regardant aux données suivantes de l'*Essai*.

(1) La nomination des points géométriques, impliqués dans la Fig. 1, suggère que parmi tous ces lemmes et propositions, l'auteur aurait conçu en premier la Prop. 2 ou du moins son cas particulier relatif au cercle, que je désignerai dès maintenant par 2',[9] mais qu'il l'aurait bientôt abandonnée pour passer au Lemme 1.

(2) De plus, l'énoncé de ce Lemme 1, très différent de son acception moderne, montre plusieurs ressemblances avec celui de la Prop. 2'.[10]

(3) L'auteur déclare, immédiatement après le Lemme 3, que les cinq propositions qui vont suivre ne sont que des exemples de celles qui se déduisent "de ces 3 lemmes et de quelques conséquences d'iceux", alors que les Prop. 2 et 3 renferment explicitement une proposition qui ne semblerait pas, au premier abord, découler de ces prémisses (il s'agit de ce que j'appellerai tout à l'heure "théorème des trois sécantes de la conique").

(4) La Prop. 4 est précédée d'un aveu de l'auteur à l'égard de son maître Desargues: "je dois le peu que j'ai trouvé sur cette matière [des coniques] à ses écrits, et ... j'ai tâché d'imiter, autant qu'il m'a été possible, sa méthode sur ce sujet, ..." (Nous en avons déjà vu un exemple dans l'enchaînement du Lemme 1 au Lemme 3).[11]

Fig. 1[12]

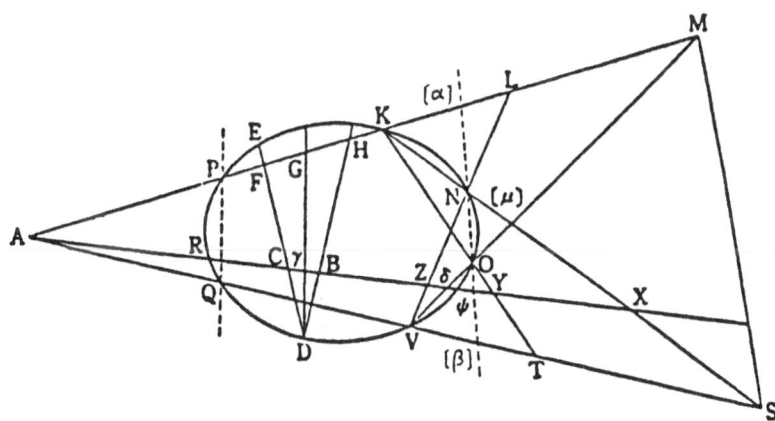

Or, de toutes les cinq propositions de l'*Essai*, la Prop. 2' est la plus rudimentaire, se démontrant aisément avec les deux théorèmes de l'Antiquité, utilisés par Desargues pour établir son propre théorème de l'involution: celui de Ménélaüs et celui des deux sécantes du cercle. Aussi les précédentes remarques 1,2,4 me portent-elles à appliquer les mêmes deux théorèmes à la prémisse du Lemme 1, et je vois alors surgir comme d'elle-même une démonstration élémentaire suivante, en supposant que la courbe de la Fig. 1 est circulaire. En coupant le triangle $A\alpha\beta$ par KS et VM,

$$\left(\frac{AS}{S\beta}\right)\left(\frac{\beta N}{N\alpha}\right)\left(\frac{\alpha K}{KA}\right) = 1 \tag{1}$$

$$\left(\frac{AV}{V\beta}\right)\left(\frac{\beta O}{O\alpha}\right)\left(\frac{\alpha M}{MA}\right) = 1.^{13} \tag{2}$$

En appliquant, d'autre part, trois fois le théorème des deux sécantes du cercle,

$$\left(\frac{AK \cdot AP}{AQ \cdot AV}\right)\left(\frac{\alpha N \cdot \alpha O}{\beta N \cdot \beta O}\right)\left(\frac{\beta Q \cdot \beta V}{\alpha K \cdot \alpha P}\right) = 1. \tag{3}$$

Ces trois égalités, multipliées entre elles, donnent par exemple

$$\left(\frac{AP}{P\alpha}\right)\left(\frac{\alpha M}{MA}\right) = \left(\frac{AQ}{Q\beta}\right)\left(\frac{\beta S}{SA}\right) \tag{4}$$

Donc, etc.

Chose remarquable d'ailleurs, cette démonstration implique la projectivité de la formule 3, puisque les trois autres sont évidemment projectives. Ainsi s'obtient, outre le Lemme 3, ce qu'on pourrait nommer théorème des trois sécantes de la conique, ou le cas le plus simple du futur "théorème de Carnot".[14] Et du même coup, la Prop. 2', qu'on peut regarder comme abandonnée à mi-chemin par l'auteur, s'élargit en la Prop. 2, et celle-ci donne lieu, immédiatement à la Prop. 3 (théorème de Carnot pour le quadrilatère), et par quelques détours à la Prop. 5 (équivalente à l'équation de la conique à centre, rapportée à une paire de diamètres conjugués).

Il est visible que les deux propositions restantes s'ensuivent du Lemme 3, pourvu qu'on rectifie la Prop. 4 (énoncé inexact du théorème de l'involution).[15] La Prop. 1 (un cas particulier de la propriété anharmonique des points de la conique) s'obtient par exemple en échangeant entre eux les points N, O de la Fig. 1, et la Prop. 4, en faisant coïncider de plus les droites AM, AS avec AX.

On dirait donc que notre hypothèse de travail, auparavant appuyée par les remarques 1,2,4, s'est enfin fortifié elle-même en dégageant la projectivité de la formule 3 pour justifier pleinement la déclaration de l'auteur signalée dans la remarque 3.

Le sujet 2 concerne la disposition des lettres.

Les écrits arithmétiques que Pascal laissa tout imprimés en deux temps, — imprimés sans doute en 1654, — et qu'on publia en 1665 sous le titre de *Traité du triangle arithmétique avec quelques petits traités sur la même matière*, utilisent trois fois l'induction mathématique.

Fig. 2[16]

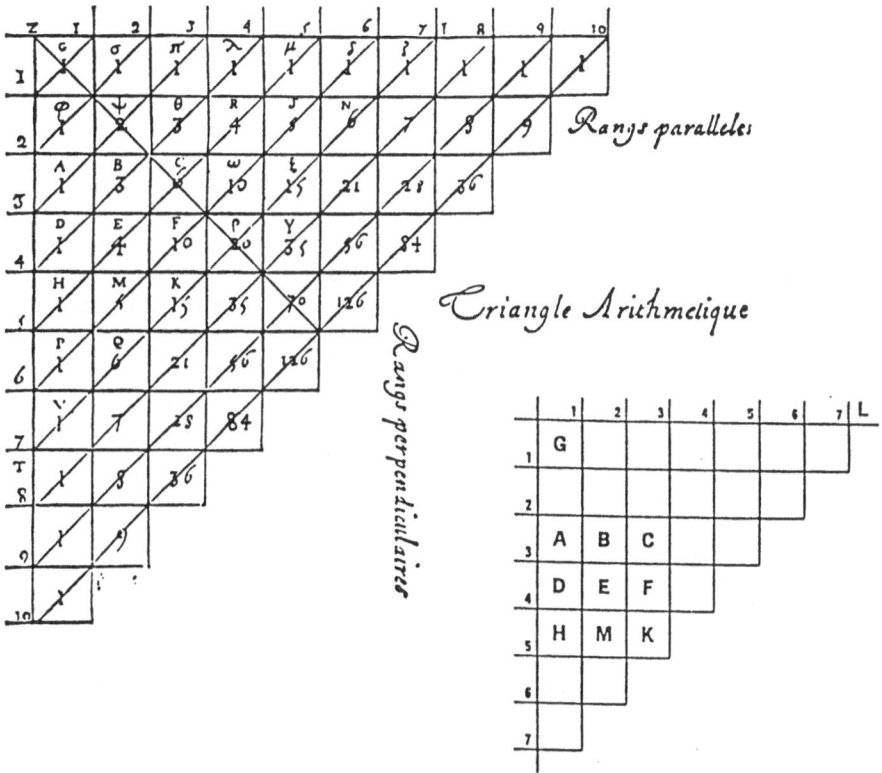

Fig. 2 bis

Désignons par $_i A_j \atop k$ le nombre de la "cellule" du triangle arithmétique, située au i^e "rang parallèle", au j^e "rang perpendiculaire", et sur la k^e "base", où toutefois un des trois indices peut manquer par suite de $i + j = k + 1$.

L'induction mathématique se trouve utilisée

(1) dans le "Triangulus arithmeticus" (1ère impression), pour démontrer

$$\underset{k}{A}_j : {}_{i-1}\underset{k}{A} = j : i - 1,{}^{[17]} \tag{5}$$

(2) dans les "Combinationes" (1ère impression), pour démontrer

$$\sum {}_i A_n = {}_k C_i \tag{6}$$

$(_1 A_1 = 1, \text{ et } n = 1, 2, \ldots, j);{}^{[18]}$

(3) dans l'"Usage du triangle arithmétique pour déterminer les partis etc." (2e impression)[19], pour démontrer

$$E_i : E_j = \sum {}_m \underset{k+1}{A} : \sum \underset{k+1}{A} {}_n \tag{7}$$

(${}_1 A_1 = 1$; E_p désigne l'espérance du joueur à qui manquent p parties;
$m = 1, 2, \ldots, j$ et $n = 1, 2, \ldots, i$).[20]

Et ces trois endroits montrent, dans l'état actuel de nos connaissances, la première application parfaite de l'induction mathématique, à ceci près qu'à son second stade d'implication $P(n) \rightarrow P(n+1)$, l'auteur prenait, faute de notation comme notre "n", un nombre particulier en exemple. Quand et comment a-t-il découvert cette importante méthode de démonstration?

L'observation attentive de la figure, annexée au *Traité* dès sa première impression, nous aide, encore une fois, à répondre à cette question.

Il aurait été tout naturel à Pascal d'inscrire d'abord aux centres des cellules des chiffres exprimant des "nombres figurés" traditionnels. Il ajouta ensuite des lettres au-dessus pour "éviter, dit-il, la confusion qui serait née l'identité des nombres placés dans plusieurs cellules".[21] Mais quelle bizarrerie de cet ajout! Et cependant, concentrons-nous sur la première partie de l'alphabet romain (Fig. 2 bis), depuis A jusqu'à L, avec le saut justifiable de I et J.[22] Il s'en dégage alors une conjecture avec facilité inespérée. En effet, la position tout isolée de la lettre G[23] n'atteste-t-elle pas déjà la naissance de l'induction mathématique, en indiquant P(1) à vérifier au départ? Et du même coup, cette lettre nous rattache à la deuxième des trois sources indiquées ci-dessus, car les deux autres exigent la vérification de P(2) marqué par φ et σ.

Or, chose encore remarquable, cette 2e proposition (Lemma 5) est la première, dans les "Combinationes", à désigner des cellules par lettres. Et la démonstration en était basée sur l'isomorphisme entre

$$_m C_n + {}_m C_{n-1} = {}_{m+1} C_n \tag{8}$$

et

$$_i \underset{k}{A} + {}_{i-1} \underset{k}{A} = {}_i \underset{k+1}{A} . \tag{9}$$

La formule récursive 9 sera bientôt adoptée par l'auteur comme définition du nombre de la cellule.[24] Mais ici, ce nombre était encore défini suivant la formule régressive traditionnelle:

$$\sum {}_{i-1} A_n = {}_i A_j \qquad (n = 1, 2, \ldots, j).[25] \tag{10}$$

En ce qui concerne toutefois l'emploi des lettres, nous importe davantage la seconde moitié du Lemma 5, où l'auteur donne, à titre d'exemple, ces deux expressions pour $_6 C_4$:

$$[1] \; D + E + F, \quad \text{et} \quad [2] \; K. \tag{11}$$

La présence de D,E,F supposant la préexistence de A,B,C, la disposition rectangulaire de ces six lettres témoigne de l'idée de nombre figuré, tandis que la lettre L, indiquant la base impliquée dans la 2e expression de 11, annonce la naissance de l'idée de triangle arithmétique.[26] Toujours est-il que Pascal a dû découvrir l'induction mathématique, j'en suis persuadé, sous la double forme de 11.

Mais cela ne veut point dire qu'il ait ainsi découvert le Lemma 5. Il se peut que des nombres figurés qu'il avait d'abord inscrits dans sa figure lui ait suggéré ce lemme, peut-être avec d'autres propositions, et qu'il l'ai même confirmé par observation combinatoire des coefficients binomiaux.[27] N'est-il pas en cherchant pour ce lemme une meilleur démonstration que Pascal a découvert l'induction mathématique? Cela me semble possible.

La disposition des lettres romaines après N, et de plusieurs lettres grecques demeure énigmatique pour moi. Mais celle des romaines avant M me porte à penser que Pascal aurait trouvé l'induction mathématique aux premiers moments de son emploi des lettres dans cette figure, et le fait que les lettres A,B,C n'y furent pas placées au premier rang horizontal serait attribuable à sa volonté de marquer la généralité du raisonnement.

Il faut d'ailleurs bien distinguer entre les deux ordres, d'"analyse" et de "synthèse" (comme disaient des anciens). Le texte des "Combinationes" présuppose celui du "Triangulus arithmeticus". Mais le fait que celui-ci porte presque au début la lettre G (indiquant cette fois seulement $_1 A_1$) me fait bien croire qu'au moment de commencer la rédaction de ce "Triangulus arithmeticus", Pascal avait déjà possédé cette nouvelle méthode de démonstration, à laquelle la postérité devait attacher de l'importance croissante.

Notes

Les abréviations suivantes seront utilisées dans ces notes.

Ed. B: Œuvres de Blaise Pascal publiées ... par MM. Léon Brunschvicg et Pierre Boutroux ..., 14 vol., Paris, Hachette, 1908-1925.

Ed. L: Pascal, Œuvres complètes ... Présentation et notes de Louis Lafuma, 1 vol., Paris, Editions du Seuil, 1963.

Ed. M, Blaise Pascal, Œuvres complètes. Texte établi, présenté et annoté par Jean Mesnard, jusqu'ici 2 vol, Paris, Desclée de Brouwer, 1964...

(1) Voici les titres et millésimes de ces Lettres:

   I *Lettre de A.Dettonville à Monsieur de Carcavy, en lui envoyant une méthode générale pour trouver les centres de gravité de toutes sortes de grandeurs...* A Paris, M.DC.LVIII.

II *Lettre de A.Dettonville à Monsieur Hugguens de Zulichem, en lui envoyant la dimension des lignes de toutes sortes de roulettes [cycloïde], lesquelles il montre égales à des lignes elliptiques.* A Paris, M.DC.LIX.

III *Lettre de A.Dettonville à Monsieur de Sluze chanoine de la Cathédrale du Liège, en lui envoyant la dimension & le centre de gravité de l'escalier,...* A Paris M.DC.LVIII.

IV *Lettre de A.Dettonville à Monsieur A.D.D.S. en lui envoyant la démonstration à la manière des anciens de l'égalité des lignes spirale et parabolique.* A Paris, M.DC.LVIII.

Après la Lettre I, la Lettre II traite encore de la cycloïde, quoique généralisée en dépassant le cadre du concours; au reste, la détermination de sa longueur avait déjà été annoncée à la fin d'une des circulaires pascaliennes relatives au concours (*éd. B*, t.VIII, p. 208, 222; *éd. L*, p. 122, 123). Cf. la note 8.

(2) Voici les numéros des figures, enregistrés selon la première apparition, et en soulignant les irréguliers:

I.1	Fig. 1, 2, 3, 4, 5, 6, <u>10</u>, 8, 9, <u>28</u>, 11.
I.2	Fig. 12, 13, <u>7</u>.
I.3	Fig. 14, 15.
I.4	Fig.<u>26</u>, <u>32</u>, 16.
I.5	Fig. 17, 18, <u>25</u>, <u>24</u>, <u>30</u>, <u>27</u>, <u>28</u>, <u>29</u>.
I.6	Fig.<u>28</u>, <u>31</u>.
I.7	Fig. 19, 20, <u>23</u>, <u>22</u>, <u>21</u>.
II	Fig.<u>41</u>, <u>42</u>, <u>43</u>.
III	Fig. 27, 44, 30, 1, 45.
IV	Fig. 33, <u>35</u>, 34, 36, 37, 38, <u>40</u>, 39.

Quant aux numéros des figures de la Lettre III, je n'y trouve rien d'irrégulier pour raison de son caractère de brèves miscellanées.

(3) Comme l'a déjà montré la note 1, seule la Lettre II porte le millésime 1659 à côté de 1658 de toutes les autres. D'autre part, l'ordre conjecturé de rédaction de I à IV est confirmé par un témoignage contemporain (voir *Revue d'histoire des sciences*, t.XV, n^os3-4, p. 330, n.1).

(4) "Il est incroyable, écrivait Gilberte Périer, avec quelle précipitation il mit cela sur le papier" (*La Vie de Monsieur Pascal, éd. B*, t.I, p. 82; *éd. L*, p. 26; *éd. M*, t.I, p. 623, §56).

(5) Vers la fin de la Lettre I (*éd. B*, t.VIII, p. 380; *éd. L*, p. 141).

(6) Voir la correspondance entre Pascal et René-François de Sluse, *éd. B*, t.VII, p. 237–238, 243–255, 331–336, et t.VIII, p. 4–13, 31–32, 115–117, 145–148, 227–229, 323–324. Voir aussi l'"engagement partiel de Sluse au concours pascalien, *ibid.*, t.VIII, p. 202, 217; *éd. L*, p. 120.

(7) Voir la correspondance entre Pascal et Christiaan Huygens, *éd. B*, t.IX, p. 162–163, 175-178; *Œuvres complètes de CH. Huygens*, publiées par la Société Hollandaise des Sciences, t.II, p. 316. Voir aussi l'engagement partiel de Huygens au concours pascalien, *éd. B*, t.VIII, p. 202, 217; *éd. L*, p. 120.

(8) Vers la fin de cette Lettre II, *éd. B*, t.IX, p. 199; *éd. L*, p. 184.

(9) La Fig. 1 se rapportant aux Lemmes 1, 3 et Prop. 1,2,4, voici les noms des points géométriques utilisés dans les Lemme 1 et Prop. 2'.

Lemme 1: A, K, M, N, O, P, Q, S, V; μ

Prop. 2': A, B, C, D, E, F, G, H, K, P, R,; γ, ψ

D'autre part, la remarque 4 et la note 11 nous permettent d'envisager d'abord la Prop. 2'.

(10) Citons le Lemme 1:

"Si dans le plan MSQ, du point M partent les deux droites MK, MV, et du point S partent les deux droites SK, SV, et que K soit le concours des droites MK, SK, et V le concours des droites MV, SV, et A le concours des droites MK, SV, et μ le concours des droites MV, SK, et que par deux des quatre points A, K, μ, V, qui ne soient point en même droite avec les points M, S, comme par les points K, V, passe la circonférence d'un cercle coupante les droites MV, MK, SV, SK, ès points O, P, Q, N, je dis que les droites MS, NO, PQ, sont de même ordre [c'est-à-dire concourantes]" (*éd. B*, t.I, p. 253; *éd L*, p. 35; *éd. M*, t.II, p. 231).

Cet énoncé montre deux aspects: (1) un assemblage de faisceaux de droites — M(K,V) et S(K,V), et (2) un cercle déterminé par rapport à des éléments du premier aspect — passant par K, V, et coupant MK, MV, SK, SV. Or, à des différences de détail près, l'aspect 1 se retrouve dans la Prop. 2' — D(F,G,H) et A(P,R), et il en est de même de l'aspect 2 — cercle passant par D, E, et coupant AP, AR.

(11) Desargues, dans le *Brouillon projet* de 1639, démontra son théorème de l'involution pour le cercle au moyen des théorèmes de Ménélaüs et des deux sécantes du cercle, (*Eléments* d'Euclide, liv.III, prop. 35,36), et le généralisa ensuite pour la conique moyennant la projectivité de l'involution qu'il avait préalablement établie. Voir R. Taton, *L'Œuvre géométrique de G. Desargues*, Paris, 1951, p. 126–129, et 143-147.

(12) Fig.I de l'*Essai*, retracée en y ajoutant les droites NO, PQ, et en nommant α, β les points de rencontre de celle-là avec AM, AS. Cf. *éd. B*, t.I, p. 252, planche.

(13) L'expression du théorème de Ménélaüs dans 1, 2 diffère de celle de Desargues (Taton, *op. cit.*, p. 126). J'ai adopté cette expression-là, et celle de 4, en tenant compte de la Prop. 1: $\left(\frac{PM}{MA}\right)\left(\frac{AS}{SQ}\right) = \left(\frac{PL}{LA}\right)\left(\frac{AT}{TQ}\right)$.

(14) Lazare Carnot, *Géométrie de position*, Paris, 1803, p. 436–438.

(15) L'auteur écrivit, sans doute par mégarde: $\frac{ZR.Z\psi}{YR.Y\psi} = \frac{\delta R.\delta\psi}{XR.X\psi}$, alors qu'il fallait écrire par exemple: $\frac{RZ.RY}{\psi Z.\psi Y} = \frac{R\delta.RX}{\psi\delta.\psi X}$ (fait signalé par R. Taton, *Revue d'histoire des sciences*, t.VIII, n°1, 1955, p. 16, n.6). On peut d'ailleurs croire qu'au plus tard en 1647, lorsque Pascal résolut le problème de Pappus *ad 4 lineas*, il avait déjà exactement conçu le théorème de l'involution, puisque visiblement, cette formule-ci satisfait à la condition de Pappus. Cf. un témoignage de Leibniz, *éd. B*, t.II, p. 221; *éd. L*, p. 37; *éd. M*, t.II, p. 1104.

(16) Notre Fig. 2 reproduit l'unique figure annexée au *Traité du triangle arithméti-que etc.* (*éd. B*, t.III, p. 446; *éd. L*, p. 50; *éd. M*, t.II, p. 1288, planche).

(17) Consect. 11 (*éd. M*, t.II, p. 1187–1188).

(18) Lemma 5 *éd. B*, t.III, p. 560–563; *éd. L*, p. 76–77; *éd. M*, t.II, p. 1238–1241).

(19) On suppose ici que deux joueurs ayant déposé chacun une somme égale d'argent, celui-là obtienne les mise totales qui aura le premier gagné un nombre déterminé de parties "à pile ou face".

(20) Problème 1 — Proposition 1 (*éd. B*, t.III, p. 488–493; *éd. L*, p. 60–61; *éd. M*, t.II, p. 1314–1317).

(21) Monitum (*éd. M*, t.II, p. 1195).

(22) Rappelons-nous que ces deux lettres faisaient défaut dans la Fig. I de l'*Essai*. Certes la Fig.II du même opuscule portait la lettre I, mais dans le présent traité, l'une et l'autre de ces lettres auraient été confondantes à cause de leur similitude avec le chiffre 1.

(23) Cette lettre ne peut pas être l'initiale de "generator", puisque la première impression ne comportait aucune désignation particulière pour $_1A_1$. Ce n'est que dans la 2e impression que Pascal le nomma "générateur" (*éd. B*, t.III, p. 448; *éd. L*, p. 51; *éd. M*, t.II, p. 1289).

(24) Au début de la 2e impression (*éd. B*, t.III, p. 448; *éd. L*, p. 51; *éd. M*, t.II, p. 1289.

(25) *éd. M*, t.II, p. 1240.

(26) La lettre H, quoique inutile dans ce lemme, témoigne de l'idée de 5e triangle, ce qui sera plus clair dans la Prop. 1 de l'"Usage du triangle arithmétique pour les combinaison" (2e impr., *éd. B*, t.III, p. 474–476; *éd. L*, p. 56–57; *éd. M*, t.II, p. 1305–1306), tandis que la lettre M, également inutile, témoigne de celle du 5e rang horizontal.

(27) Pascal écrivait à propos de la relation du triangle arithmétique et des puis-sances des binomes: "d'autres en ont déjà traité, comme Hérigone, outre que la chose est évidente d'elle-même" (*éd. B*, t.III. p. 503; *éd. L*, p. 63; *éd. M*, t.II, p. 1323).

The Birth of Maxwell's Electro-Magnetic Field Equations

W. J. Ellison

Introduction

Many modern texts on electro-magnetic field theory begin by defining the field quantities $\mathbf{E}$, $\mathbf{D}$, $\mathbf{H}$, $\mathbf{B}$ together with the equations:

$$\mathbf{D} = k \cdot \mathbf{E} \qquad \mathbf{B} = \mu \cdot \mathbf{H} \qquad \mathbf{s} = \sigma \cdot \mathbf{E}$$

and then introduce, with suitable units, as axioms, Maxwell's equations:

$$\nabla \cdot \mathbf{D} = 4\pi\rho \qquad \nabla \cdot \mathbf{B} = 0$$

$$\nabla \times \mathbf{H} = 4\pi\mathbf{s} + \frac{1}{c}\frac{\partial \mathbf{D}}{\partial t} \qquad \nabla \times \mathbf{E} = -\frac{\partial \mathbf{B}}{\partial t}$$

There is usually a quick explanation that the first set of equations are valid for homogeneous media, the first Maxwell equation represents the conservation of charge, the second that magnetic field lines are always closed loops, the third is Maxwell's modification of Ampère's law and that the fourth is Faraday's law of induction.

The explanation of the term $\frac{\partial \mathbf{D}}{\partial t}$ is often accompanied with a certain amount of hand-waving. Either there is a partial heuristic discussion of what might be happening to discrete electric charges between the plates of a condenser filled with a homogeneous dielectric and how the hypothetical displacement of these charges forms *the displacement current* or there is some appeal to 'symmetry' or the necessity to 'complete' Ampère's formula. This makes the term look like a fluke or a mathematical artifice.

After this type of introduction the mathematical machinery begins to turn at full speed: wave equations appear, boundary conditions are discussed, the equations are solved and all the principal properties of classical electromagnetic theory follow. The books usually finish with special relativity.

The average reader follows, without much trouble, the argument about slight disturbances of electric charges in non-conductors forming a displacement current, but what happens in free space where there is nothing? How on earth was Maxwell led to introduce his displacement current where there is nothing for us to displace?

25

In this lecture I would like to trace the fascinating story of how observations of physical phenomena got transformed slowly and methodically, by a combination of astute analogies into a set of abstract equations.

The fundamental philosophical question behind all of this work is: 'What do we mean when we "explain" or say that we "understand" a physical phenomenon?'. Maxwell was never naïve enough to believe that science deals in ultimate truths. For him, science describes the behaviour of things — nothing more. To carry out this description analogous thinking and theories was the best that one could do. One can make a strong argument for the thesis that 'understanding'is always based on analogies. The unknown being described by analogy with what is familiar and considered to be 'understood'. Once the process of analogical morphisms and their associated equivalence classes gets started one can quickly arrive at descriptions which one 'understands' and which apparently have little relationship with the original analogies used to start the process. (Hofstadter's book: *Gödel, Escher, Bach: An Eternal Golden Braid* is full of examples.) As we shall see, the process of going from mysterious electric and magnetic fluids and atmospheres, via Faraday's lines of force in an invisible elastic medium and Maxwell's cog wheels and ball bearings which fill the universe, to relativistic invariant field equations is a continuous sequence of analogical steps.

The three principal articles by Maxwell (1831–1879) illustrate this point quite clearly. The first article *On Faraday's Lines of Force* (1856) uses elementary and very intuitive properties of incompressible fluid flows. Almost everything, theorems and proofs, are described in words. In fact, Maxwell apologizes for the few equations which are included and he deplores the blind use of mathematical calculations. In the second paper *On Physical Lines of Force* (1861), the fluid flow gets complicated by the addition of vortices, and later the vortices get transformed into a very mechanical cog - wheel and ball bearing analogy. The electro magnetic field properties are then derived by reasoning on the purely mechanical properties of the model. Nothing is 'abstract' By now Maxwell is fully at home with the experimental facts, his models, the physical implications of his models *and* the mathematical description of these implications. In the final paper *A Dynamical Theory of the Electro-Magnetic Field* (1865) all the fluid and mechanical analogies are very much in the background, what is left is the mathematical structure, which has now become the intuitive reality. The Cheshire Cat has vanished; when newcommers see only the smile it is understandable that there is a certain air of mystery.

The Aether

A concept that was to play an important role in theoretical physics is that of the *AETHER* (αἰθήρ). The word originally meant the blue sky or upper air, as opposed to the lower air at the level of the earth. In ancient and medieval cosmologies the word had been used to denote the medium filling the celestial regions of the universe. It had no particular properties other than to fill a vacuum, which according to Aristotle, Nature abhors. The meanings given to the word during the last 400 years are legion, but from our point of view the first introduction of an aether with mechanical properties was made by Descartes (1596–1650). For Descartes a major problem was to account for the action between bodies not in contact e.g. tides and the Moon's position, the behaviour of magnets etc. He could not accept the idea of 'occult' forces between the bodies and assumed that they must be affected by the agency of pressures or impulsions. This implies that bodies can only act on each other when they are contiguous, with the consequence that the space between two bodies cannot be empty. It may be occupied by particles of matter, but the interstices between the particles of matter must be filled with particles of a more subtle nature which everywhere press upon or collide with each other. Space was thus, according to Descartes, a *plenum* being occupied by a medium which, imperceptible to our senses, is capable of transmitting force and excerting effects on material bodies immersed in it — he called this medium the aether. Before Descartes the aether just filled space, after Descartes an aether had mechanical properties and a structure. It was the essential constituent of the universe (apart from an infintesimally small fraction occupied by ordinary matter.)

Descartes assumed that the aether particles were in continuous motion. As there was no empty space for a particle to move into, he inferred that they move into the place vacated by other aether particles in motion. Thus the motion of a single aether particle involved the motion of a whole chain of particles. The motion of these chains of particles constituted vortices, which formed an important place in Descartes' cosmology.

During the next 300 years the idea of an aether with mechanical properties was to play an important role in physics. The principal preoccupation was to explain the properties of light. This gave rise to numerous highly sophisticated models of an aether necessary to explain a specific aspect of light. For a fascinating account of these theories one can do no better than consult the masterly exposition of Sir Edmund Whittaker [10]. Our object here is to indicate that by the time of Faraday and Maxwell the idea that 'space' had properties other than 'emptiness' was not at all a strange idea. It seemed natural that 'space' was a material substance which could transmit force, could be deformed by the forces of electricity and magnetism and could store energy. There was no consensus as to the nature of 'space' or as it was called the aether, just a feeling that some agency was required to explain the transmission of light and of electric, magnetic and gravitational forces between two points.

In the first decade of the twentieth century the aether concept fell into disfavour, chiefly as a result of the failure to observe the Earth's motion relative to the aether. However, with the development of quantum electrodynamics 'empty space' has become to be regarded as the source of the 'zero-point' fluctuations of electric charge and of a 'polarisation' corresponding to a dielectric constant different from unity. We are back again with a 'something' with a complex structure. A rose by any other name...

Before starting to discuss Maxwell's ideas we shall rapidly recall the historical state of affairs in electricity and magnetism up to the middle of the nineteenth century.

1 Electricity and Magnetism from 1750 to 1850

The ancient Greeks were familiar with the curious properties possessed by two minerals, amber ($\check{\eta}\lambda\epsilon\kappa\tau\rho o\nu$) and a metallic ore from the region of Magnesia ($\dot{\eta}\lambda\acute{\iota}\theta\iota o\varsigma$ $M\alpha\gamma\nu\tilde{\eta}\tau\iota\varsigma$). The former, when rubbed, attracts light bodies; the latter has the power of attracting iron.

By the latter half of the 18th century knowledge of electrostatics and magnetostatics had acquired a firm scientific basis. The distinction between conductors and insulators was well known. It was recognised that there were two kinds of electric charge: rubbing glass with silk gave vitreous electricity; rubbing amber with fur gave resinous electricity. Moreover, resinous electricity would neutralise vitreous electricity.

This gave rise to two rival theories about the origin of static electricity: the two fluid theory, associated with Du Faye (1698–1739), who was superintendent of gardens to the king of France, and the one fluid theory associated with Ben. Franklin (1706–1790). On the two fluid theory an uncharged body had equal quantities of resinous and vitreous electric fluids and electrification arose as a separation of the two fluids. One the one fluid theory, an electrostatically charged body contained either an excess or a deficit of electric fluid. A vitreous charge corresponded to an excess (positive electricity) and a resinous charge to a deficit (negative electricity).

Both theories explained all the phenomena which were known at the time equally well. Cavendish (1731–1810) favoured the one fluid theory because it was simpler mathematically and published an explanation of the known phenomena in 1771. On the other hand, Coulomb (1736–1806) favoured the two fluid theory, for which a perfectly satisfactory mathematical description was elaborated by Poisson (1781–1840) and published in 1812. The question was never really very controversial, Maxwell in the 1860's was still talking of the possibility of one or two kinds of electricity.

Franklin had shown, by his kite experiment, that lightning was electric in nature, that electric charge was conserved and that there was an action-at-a-distance between electrically charged bodies. Priestley (1733–1804) inferred from rough experimentation that the action-at-a-distance takes place with an inverse square law, just like newtonian gravitation. However, Cavendish, by means of highly ingenious experiments had shown by 1774 that if the exponent in the action-at-a-distance law was not 2,then it was contained in the interval $2 \pm 1/59$. He never published this work, it was intended to from part of a book on electricity. The publication was done by Maxwell when he edited the Cavendish MSS almost a century later.

In 1750 J. Mitchel (1724–1793) published a memoire on magnetism in which he showed that the force between two magnetic poles verified an inverse square law. The basic action-at-a-distance inverse square law for electro-statics and magnetism was generally accepted after Coulomb verified it in 1785 with his highly accurate torsion balance experiments.

At the beginning of the 19th century electrostatics and magnetism were described by a sophisticated mathematical action at a distance theory which closely paralleled newtonian gravitation.

However, charges or magnetic poles exert forces and forces produce motion. Electric charges moved quickly through conductors and experimentally it was difficult to see the effects of moving changes. At the end of the 18th century the only known method of storing electricity was with a 'Leyden jar' (a glass bottle with a conductive coating inside and out), discovered in 1745 by Musschenbrock (1692–1761). It was possible to store large quantities of electricity, but the currents were dissipated in one quick discharge.

The technological breakthrough came by accident when Galvani (1737–1798), a professor of anatomy at Bologna, stumbled upon the principle of the battery whilst disecting frogs. It was Volta (1745–1827) who followed up Galvani's investigations and by 1800 he had found that a continuous current could be produced by putting a slightly acidified damp cloth between a zinc and a copper plate. The voltaic cell provided physicists with a continuous source of electricity and marked the beginning of the quantitative study of electric currents and their relationship with magnetism.

Once the construction of batteries was possible the first investigations centred around the destructive effects of electricity on chemical bodies:— electrolysis. Davy (1778–1829) built a huge battery of 2000 voltaic cells which filled the cellar of the Royal Institution (where Faraday was later to have his laboratory).

The first unambiguous, repeatable result relating electricity to magnetism was obtained by Oersted (1771–1851) in 1820. He found that a compass needle placed parallel to a current carrying wire was deflected. This observation aroused immediate interest, notably in France, where Biot (1774–1862), Savart (1791–1841) and Ampère (1775–1836) set about experimenting, measuring, calculating and theorising about the magnetic forces generated by electric currents. By 1827 Ampère had formulated a full theory which saw magnetic effects as due to the combined effects of microscopic magnets and the microscopic magnets themselves were produced by circulating electric currents.

The next decisive step was to try and fill a missing 'symmetry'. It was known that:

1. static electricity in one body could induce an equal and opposite charge in a nearby body,

2. moving charges form an electric current,

3. an electric current produces a magnetic field.

Conversely, it was expected that a magnetic field ought to move the induced charges in a nearby conductor and so produce an induced current.

The problem was not easy to solve and it was not until 1831 that Faraday (1791–1867) showed how it was possible. Given two neighbouring copper loops, a steady current in one does *not* induce a current in the other. However, *a changing* current in one loop did induce a current in the other loop.

Oersted had found that currents i.e. moving charges, generate magnetic fields. Faraday found that moving magnetic fields generate currents. In more than one way, the circuit was closed; electric and magnetic phenomena were linked and the linkage was symmetric — they were in a mutual embrace, as Maxwell was to often say.

The complications were not yet finished. Many investigators had tried to see whether there was any connection with magnetic fields and light. Faraday started his search in 1834, but it was not until 1845 that he found his opto-magnetic effect. He placed a slab of glass between the poles of a powerful electro-magnet and found that the plane of polarisation of a light beam was rotated when the light travelled through the glass parallel to the lines of force of the magnetic field.

After the success of Newtonian gravitation and its action at a distance idea — together with the mathematical tools developed by Euler (1707–1783), Lagrange (1736–1813) and Laplace (1749–1827), a coherent and very fruitful description of physics had been forged. The programme attempted to describe physical phenomena: gravity, electricity, magnetism, heat, light, sound in terms of separate "matters" whose action was the sum of its supposed separate constituent particles. There were two modes of action. Either the constituent particles of the "matter" exerted forces at a distance on other particles, as with gravitation, or the particles of "matter" were transferred to distant points, for example light by corpuscles or heat by the movement of the caloric fluid.

In each the area image was one of direct action at a distance, whether through "forces" or by transmission of particles. This single image for each branch of physics was coherent and had a unified mathematical description, but it did not unify the different branches on the same physical basis. The "fluids" and "corpuscles" were independent. One can imagine the flux and reflux involved when considering the problem of the interaction of two hot, gravitating, luminous, electrically charged moving magnetic bodies!

In the years that followed, the action at a distance Laplacian world view, as applied to electricity and magnetism, moved from strength to strength — F. Neumann (1798–1895), B. Riemann (1826–1866), G. Kirchoff (1824–1887) and W. Weber(1804–1890) were succeeding in describing all the known electric and magnetic phenomena — they modified the simple inverse square law so as to include terms which depended on the velocity and acceleration of the charged particles which were assumed to make up electricity. The culmination was formulated in

1846 by Weber. The force between two moving charges was taken to be:

$$F = \frac{e \cdot e'}{r^2} \left(1 + r \cdot \frac{d^2 r}{dt^2} - \frac{1}{2} \left(\frac{dr}{dt} \right)^2 \right)$$

The theory was the first *Electron theory* of electro-magnetic phenomena. It was an enormous success. (So much so, that modifications of a similar kind were made to Newton's law of gravitation in order to try and account for a discrepancy in the rotation of Mercury's orbit. A problem that was not completely settled until general relativity.)

A totally different path was followed by Faraday. We cannot enter into a detailed discussion of his ideas here (see Gooding [3]). Suffice to say that Faraday viewed the seat of electro-magnetic phenomena as being in the medium surrounding the magnets and conductors. The suggestive imagery of lines of force, familiar to everyone, gave Faraday another way of describing electro-magnetic effects in terms of tensions and stresses between the lines of force.

Fortunately, Faraday who had a mathematical mind, but little or no mathematical instruction, (If he had known more formal mathematics he would probably have gone the way of Poisson and the Laplacians.) set out to analyse the structural relationships between currents and magnetism, without searching for the ultimate truth of what electricity or magnetism was. (Perhaps an important factor in Faraday's philosophy towards science as being a description of observed relationships, rather than giving a 'true' explanation of phenomena, is due to his very active membership of a fundamentalist protestant sect — the Sandemanians — for whom ultimate truths were known to God alone.) A characteristic of Faraday's work was that of careful observation, experiments designed to illustrate specific phenomena and their reproducibility. However, Faraday was never content to simply amass numerical measurements, he always formulated his observations as laws, in a form as general as possible. This aspect of his work was to be vital to Maxwell, who would take Faraday's laws as axioms and then deduce conclusions by purely formal mathematical arguments.

By the 1850's we have: electric and magnetic fluids, caloric fluid, gravitation, light corpuscles, light waves vibrating in a luminiferous aether and an electromagnetic aether. There were coherent theories for each of the subjects taken independently. The problem was that they were not independent. Oersted and Faraday's experiments showed that electricity and magnetism were linked, Davy's work on electrolysis showed that matter and electricity were linked, Joule had shown that heat and electricity were linked, Faraday's opto-magnetic work had shown that light and magnetic fields were linked. It was evident that some unified theory was (and still is) needed. The time was ripe for a 'Maxwell' to appear.

The first step in the unification came from a rather unexpected source. As a 15 year old schoolboy, W. Thompson (1824–1907), later Lord Kelvin, devoured J. Fourier's (1768–1830) masterpiece *La Théorie Analytique de la Chaleur*, published in 1822. Thompson realised that the description of heat flow in a homogeneous solid and the distribution of electrostatic force were described by the same mathematics. The isothermal surfaces of one correspond to the equipotential surfaces of the other, charges correspond to sources or sinks of heat, temperature corresponds to potential and flow of heat corresponds to the accelerating effect of attraction. What was important was not that the mathematics is the same, but that it suggests that one can look at both phenomena in two different ways.

Electrostatics had been developed in terms of an inverse square, action at a distance, law between point charges. Fourier's theory of heat was developed from the idea that heat is propagated by action between contiguous particles in a continuous medium. In his first paper, published in 1842, Thompson provided the mathematical framework for interpreting electrostatic and magnetic phenomena as a flow in a continuous medium. The idea that electrical and magnetic action takes place in a continuous medium had been used by Faraday, as a guide in his research, but it had never been accepted as 'mathematical' as it was supposed to be inconsistent with the Coulomb inverse square law and the corresponding mathematical structure due to Poisson.

2   Maxwell

One of the most fundamental aspects of Maxwell's contribution was the explicit and very conscious use of physical analogies in the elaboration of his theories. An analogy was always used to illustrate a physical phenomenon and to guide one towards a more complete description. As a young fellow of Trinity Maxwell wrote a long essay entitled *Analogies: Are there real analogies in Nature?* which was the subject of discussion by the Apostles, a Cambridge society devoted to philosophical problems.

Maxwell's first paper on electro-magnetism: *On Faraday's Lines of Force* was published in 1855, but earlier versions were discussed with P. G. Tait in 1853 and with W. Thompson in 1854. The article is in three distinct parts: 1. On Faraday's Lines of Force, 2. On Faraday's "Electro-tonic State" and 3. Examples.

In part 1 he developed an analogy for the lines of electric and magnetic force in terms of the flow of an incompressible, massless fluid through a resistive medium. He succeeded in making the flow image of forces, as suggested by W. Thompson's paper, as natural and as intuitive as the traditional action at a distance image. In this model conductors and dielectrics manifest their presence by anisotropies in the resistance to the flow of the fluid.

The second part of the paper is much less intuitive. First of all, what is the 'Electro-tonic State'? It was a notion introduced by Faraday to picture his law of induction and represented a kind of 'tension' in the space where electromagnetic phenomena were being observed. To cite Maxwell:

> When a conductor moves in the neighbourhood of a current of elec-
> tricity, or of a magnet, or when a current or magnet near a conductor
> is moved, or altered in intensity, then a force acts on the conductor

and produces electric tension, or a continuous current, according as the circuit is open or closed. This current is produced only by changes of the electric or magnetic phenomena surrounding the conductor. Still the conductor is in different states when near a current or magnet, and when away from its influence, since the removal or destruction of the current or magnet occasions a current, which would not have existed if the magnet or current had not been previously in action.

Considerations of this kind led Professor Faraday to connect with his discovery of the induction of electric currents the conception of a state into which all bodies are thrown by the presence of magnets and currents. This state does not manifest itself by any known phenomena as long as it is undisturbed, but any change in this state is indicated by a current or tendency towards a current. To this state he gave the name of the "Electro-tonic State" ...

What the physical interpretation of this state was not at all clear, as Maxwell says:

The idea of the Electro-tonic State, however, has not yet presented itself to my mind in such a form that its nature and properties may be clearly explained without reference to mere symbols, and therefore I propose in the following investigation to use symbols freely, and to take for granted the ordinary mathematical operations. By a careful study of the laws of elastic solids and of the motions of viscous fluids I hope to discover a method of forming a mechanical conception of this Electro-tonic State adapted to general reasoning.

For want of a mechanical analogy Maxwell adopts a pragmatic attitude. He took as axioms the experimental laws of Ampère, Faraday etc. and assumed that the Electro-tonic State at any point in space was described by a vector. The formulation of the laws in terms of electric and magnetic intensities, electric and magnetic quantities and the electro-tonic function was carried using the mathematical tools developed by Green (1793–1841), Stokes (1819–1903) and W. Thompson. The result was the following six laws

Law I The entire electro-tonic intensity round the boundary of an element of surface measures the quantity of magnetic induction which passes through that surface.

Law II The magnetic intensity at any point is connected with the quantity of magnetic induction by a set of linear equations.

Law III The entire magnetic intensity round the boundary of any surface measures the quantity of electric current which passes through that surface;

Law IV The quantity and intensity of electric currents are connected by a set of linear equations.

Law V The total electro-magnetic potential of a closed current is measured by
the product of the quantity of current multiplied by the entire electro-tonic
intensity estimated in the same direction round the circuit.

Law VI The electro-motive force on any element of a conductor is measured
by the instantaneous rate of change of the electro-tonic intensity on that
element, whether in magnitude or direction.

In modern terminology, the electro-tonic intensity is the vector potential function
usually denoted by **A**.

With the introduction of the electro-tonic intensity function Maxwell had
a mathematical formulation of electromagnetism for steady currents. Naturally
new questions arose immediately: What was the physical nature of the Electro-
tonic state? How were changes in the electro-tonic state transmitted through non-
conducting space? What happens with changing currents and fields? For Maxwell
it was natural to seek the answers to these questions by looking for a mechanical
model of the aether.

The answer, the famous (or infamous) molecular vortex or cog wheel and ball-
bearing model of the aether, took six years to be formulated. It was published as
a five part paper: *On Physical Lines of Force,* in 1861 and 1862. Before discussing
the model in some detail we cite two extracts from Maxwell's paper. The first
indicates his objectives and illustrates his attitude towards models, the second
indicates the heuristic reasons which incited him to consider his vortex system.

> I have in a former paper endeavoured to lay before the mind of the
> geometer a clear conception of the relation of the lines of force to the
> space in which they are traced. By making use of the conception of
> currents in a fluid, I showed how to draw lines of force, which should
> indicate by their number the amount of force, so that each line may be
> called a unit-line of force (see Faraday's Researches, 3122); and I have
> investigated the path of the lines where they pass from one medium to
> another.
>
> In the same paper I have found the geometrical significance of the
> "Electro-tonic State", and have shown how to deduce the mathematical
> relations between the electro-tonic state, magnetism, electric currents,
> and the electromotive force, using mechanical illustrations to assist the
> imagination, but not to account for the phenomenon.
>
> I propose now to examine magnetic phenomena from a mechanical
> point of view, and to determine what tensions in, of motions of, a
> medium are capable of producing the mechanical phenomena observed.
> If, by the same hypothesis, we can connect the phenomena of magnetic
> attraction with electromagnetic phenomena and with those of induced
> currents, we shall have found a theory which, if not true, can only be
> proved to be erroneous by experiments which will greatly enlarge our
> knowledge of this part of physics.

Let us now suppose that the phenomena of magnetism depend on the existence of a tension in the direction of the lines of force, combined with a hydrostatic pressure; or in other words, a pressure greater in the equatorial than in the axial direction: the next questions, what mechanical explanation can we give of this inequality of pressures in a fluid or mobile medium? The explanation which most readily occurs to the mind is that the excess of pressures in the equatorial direction arises from the centrifrugal force of vortices or eddies in the medium having their axes in directions parallel to the lines of force.

This explanation of the cause of the inequality of pressures at once suggests the means of representing the dipolar character of the line of force. Every vortex is essentially dipolar, the two extremities of its axis being distinguished by the direction of its revolution as observed from those points.

A diagram of Maxwell's model is shown in the figure. The lines of magnetic force is the axis for a vortex filament. For adjacent vortices to have the same sense of rotation, and thereby represent lines of force in the same direction, they were supposed to be separated by "idler" particles, these particles being in rolling contact with the vortex cells. The idler particles were meant to represent electricity. They were free, in all substances to rotate in place between the vortex cells. However, in conductors they could translate with more or less resistance. A stream of idler particles constituted an electric current which, because they exerted a tangential action on adjacent cells, would set them in rotation as vortices and so create a magnetic line of force around the current. The next adjacent idler particles would then be set rolling by tangential action of the first vortices (constituting electromotive force) and if the particles could not translate (being in a non conductor) they would set the next adjacent cells in rotation, and so on, thus forming lines of magnetic force around the current with larger and larger radii. Whenever the propagating effect reached a conducting region the idler particles would simply translate while rolling against the next adjacent cells and so constituting an induced current in the opposite direction from the original. Any resistance to translation would eventually set those next adjacent cells in rotation and the induced current would be stopped.

Maxwell, by using classical mechanics, showed that this heuristic machinery reproduced the required relations between currents, magnetic force and the electrotonic state, which he interpreted as the rotational momentum of the vortices. There was still another factor to introduce in the model: the mechanism of the tangential action which transmitted rotation from idler particles to the vortex cells, and from the exterior of cells through their interior. The vortices had to be fluid in order to produce the correct dynamic behaviour of the magnetic lines of force, but they could not be perfectly fluid because no tangential action could occur in their interiors. Also, the connection between idler particles and cell walls could not be perfectly rigid, because if it were, then the starting and stopping of induced currents would be instantaneous. One could not invoke a 'viscosity' as it would imply internal energy losses and heating. To avoid these problems, Maxwell proposed that the cells were elastic and that they interacted elastically with the idler cells. Such a mechanism conveniently explained the propagation of electric and magnetic effect, but it did more; it explained electrostatic effects in non conductors and in open circuits. Maxwell realised that static electricity could be interpreted as an elastic displacement or wobble of the idler particles, whereas currents involved an extended movement. While the displacement was occurring it would constitute a brief current: *the displacement current.* The effect of the displacement would propagate through space by successive induction of displacement currents. With this idea the essentials were complete for the quantitative description of all the known phenomena in electricity, magnetism and electrostatics. Maxwell supposed that his model of the aether filled the universe, the presence of matter was equivalent to local variations in conductivity — the resistance to the movement of the idler particles or to variations in the elasticity of the vortex cells — specific inductive capacity or the elasticity of the vortex cell — idler particle interaction-permeability.

By a remarkable *tour de force* Maxwell calculated the velocity of propagation of electro-magnetic effects in his model in terms of the coefficients of elasticity of the vortex cells and the idler particle — vortex cell elasticity. He found that this velocity was very close to the velocity of light (310740 km/sec for electromagnetic disturbances and 314858 km/sec for Fizeau's measurement of the velocity of light) and he inferred that *light consists in the transverse undulations of the same medium which is the cause of electric and magnetic phenomena.* That is, light is an electromagnetic phenomenon, something which had long been suspected. (apparently first seriously suggested by L. Euler in 1761).

The importance of this mechanical model was primordial. It was first invented to account for electricity and magnetism, then, for it to work as a *physical mechanism* Maxwell was forced to introduce the elasticity concepts. The elasticity idea then turned around and suggested the idea of a displacement current in the aether. It was an idea which was totally original to Maxwell and the key to any future understanding of electromagnetic effects.

What was essentially the final contribution by Maxwell to the development of 'Maxwell's equations' was the article *A Dynamical Theory of the Electromagnetic Field,* published in 1865. Gone are references to fluid flows and to complicated mechanical models, they were only aids to thought and having used them to become familiar with the properties of the electromagnetic aether they were no longer necessary for Maxwell (or for the reader who had carefully studied his papers). The paper is a masterly exposition of electromagnetic field theory from first principles.

The essential property assumed for the aether is its capacity to store, transmit and restore energy. No hypothesis is made as to *how* this is done. The energy in the aether is assumed to be of two kinds: a "kinetic" energy, due to the motions of parts of the aether and a "potential" energy consisting of the work the medium will do in recovering from displacements by virtue of its elasticity. The propagation of undulations consists in the continual transformation of one of these forms of energy into the other alternately.

This assumption led Maxwell to formulate, as the first of his general equations for the electromagnetic field, that the total motion of electricity is always the sum of the part transmitted from one body to another *plus* the rate of change of the elastic displacements. With this axiom plus those corresponding to Amp*re's and Faraday's laws Maxwell formulated, using Lagrangian mechanics, the field equations for moving bodies as they now appear (apart from the vector notation) in modern undergraduate texts. He then derived the wave equations and applied then to the standard study of condensers, mutual induction etc.

This form of the theory is virtually identical with that which was published in volume 2 of the *Treatise on Electricity and Magnetism*. For an analysis of the historical development of later electromagnetic theories see Buchwald [2] and Whittaker [10].

References

Many other references are to be found in 2 , 3 and 10.

[1]   Berry, A.J., *Henry Cavendish, his life and scientific work* Hutchison, London 1960.

[2]   Buchwald, J.Z., *From Maxwell to Microphysics- Aspects of Electromagnetic Theory in the Last Quarter of the Nineteenth Century*, Univ. of Chicargo Press, 1985.

[3]   Gooding, D et al., *Faraday Rediscovered — Essays on the life and work of Michael Faraday 1791-1867.*, MacMillan Press, London 1989.

[4]   Maxwell, J.C., *Scientific Papers 2 vols, Ed, W.D. Niven*, Cambridge Univ. Press 1890. Reprinted Dover New York 1965.

[5]   Maxwell, J.C., *Treatise on Electricity and Magnetism, 2 vols*, Clarendom Press, Oxford, 1879. Reprinted Dover New York 1954.

[6]   Maxwell, J.C., *On Faradays Lines of Force*, Trans. Camb. Philos. Soc. vol 10, 1854, p. 27, Scientific Papers vol 1, pp. 155–229.

[7]   Maxwell, J.C., *On Physical Lines of Force*, Phil. Mag. vol 21, 1861, pp. 161, 281, 338, Phil. Mag. vol 22, 1862, pp. 12, 85, Scientific Papers vol 1, pp. 451–513.

[8]   Maxwell, J.C., *A Dynamical Theory of the Electro-Magnetic Field*, Phil. Trans. Royal Soc. 1865, p. 459, Scientific Papers vol 1, pp. 526–597.

[9] Tolstoy, I., *James Clerk Maxwell — A Biography*, Cannongate Pub. Edinburgh 1981.

[10] Whittaker, E., *A History of the theories of Aether and Electricity 2 vols*, Thos. Nelson, London 1951.

[11] Wise, M.N., *The Mutual Embrace of Electricity and Magnetism*, Science, vol 203, 1979, pp. 1310–1318.

Complex Curves — Origins and Intrinsic Geometry

J. J. Gray

Introduction

One of the crucial steps in the invention of modern mathematics was the introduction of coordinate geometry by Descartes in his *La géométrie*, [1637].[1] This happy blend of algebra and geometry was to lead in the 18th century to the displacement of geometry as the central domain of mathematics and its replacement by the study of formal expressions. When in the early 19th century geometry regained some of its earlier importance, algebraic or coordinate methods were to lie at the heart of some of its most notable discoveries. So intimate has the link between algebra and geometry become, that to a modern mathematician the equation $f(x,y) = 0$ almost automatically suggests some kind of algebraic dependence of one variable, y, on another, x. A geometer will naturally think of a curve depicting this dependence in the Cartesian plane with coordinates x and y.

One reason for the success of algebra is that polynomial equations have as many solutions as their degree requires. As long as the exact solution of such an equation is not required and knowledge of the mere existence of the solutions suffices, there is a sense in which to turn a problem into algebra is to solve it. But part of the price is that the solutions are generally complex. The problems that this posed for mathematicians are well-known and are frequently described int the historical literature.[2] In the opening years of the 19th century a controversy smouldered about the nature of $\sqrt{-1}$ which was only resolved when the concept of magnitude ceased to be taken as the basis of mathematics. But if the implications of this debate for the number concept have often been written about, the implications for geometry are less often described and form one of the themes of this paper.

To understand how the implications for geometry were drawn as they were, one further aspect of 19th century mathematics must be introduced: the study of functions of a complex variable.[3] This subject began in two independent ways that were largely untouched either by each other or the problem of the nature of complex numbers. Complex function theory in its own right was the creation of the French mathematician Cauchy in the years from 1820 to his death in 1857. In the

---

[1] Well described in two papers by H.J.M. Bos, [1981] and [1984]

[2] One may start with Kline[1972]

[3] The best recent history of complex analysis is the one by Bottazzini [1986].

39

latter years other French mathematicians joined in, as did the German mathematician Riemann. The second source for complex function theory was the discovery of elliptic functions. This discovery, made independently by the Norwegian Abel and the German Jacobi, soon blossomed into a vast and exciting new domain of mathematics. But although elliptic functions were always treated as complex functions, it was only in the 1840s and 1850s that their theory was integrated with Cauchy's general theory.

The reason for the delay was that elliptic functions (as is described below) arose by the so-called inversion of elliptic integrals, and an elliptic integral involves a square root, just as the integral for the sine function does. Once variables are complex the integrand in an elliptic integral is therefore necessarily two-valued, and it was widely felt that this made them hopelessly ambiguous and beyond the reach of Cauchy's theory.[4] Until the ideas on the one hand of Liouville and on the other of Riemann, more algebraic methods were therefore preferred. The work of Abel in particular focussed on the locus defined by the equation $f(z, w) = 0$, where z and w are complex variables, but never considered this locus geometrically, only algebraically.

Yet despite these firmly algebraic origins, today the situation is that to a modern mathematician the equation $f(z, w) = 0$ suggests more or less all that the equation $f(x, y) = 0$ does, except that the choice of z and w evokes the idea that the variables are complex, and the geometrical realization of the curve is not quite so intuitive or elementary. There is an approach via Riemann surfaces, or one may just regard the locus as a subset of $\mathbf{C} \times \mathbf{C}$. My concern in this talk is: just when and how did this consensus about the complex case begin to emerge? Who gave us complex curves and complex space? I shall show that the answer to this question involves us in looking again at the debate about the meaning of complex numbers, and that the answer is not von Staudt, as it usually said. When I have answered it, I shall look at just one of the implications for research in geometry, the study of plane curves of degree four.

Although the consensus whose emergence I shall describe is the modern one, it did not by any means eliminate its rivals at once. Other interpretations of complex numbers in geometry were offered throughout the second half of the 19th century. I need not describe them, their successes, and the reasons for their ultimate collapse, because the task was done definitively by Coolidge in his [1924]. For once in the history of mathematics it is the history of the survivors that still needs to be told, while that of the defeated has long since been written.

Complex Points on Curves (1) — The 18th Century?

It might seem that the answer to the question 'Who invented complex points on a curve?' should have been given during the 18th century. To Euler, for example, the very definition of a curve was via formulae. Euler in his influential *Introductio in*

---

[4]How this perception was developed and ultimately abandoned is described in Gray[1992].

analysin infinitorum [1748] defined conic sections as the points of the (Cartesian) plane satisfying a second degree equation in x and y. To an author like Euler or Cramer, what we call Bezout's theorem was something that required proof, whereas to earlier authors (such as Newton) this was something that enabled you to define the degree of a curve. Bezout's theorem, to Euler, was the natural claim that a curve defined by an equation of degree m met another, defined by an equation of degree n, in exactly mn points, counted with suitable multiplicity. In 1748 he published an attractive paper in which he confronted must of the elementary problems that such a view poses. He showed that one must add points at infinity, and that one must define the multiplicity of an intersection. Finally, and for our present purposes must interestingly, the necessary intersections may not exist; a line may fail to meet a conic at all. The points of intersection must are then complex; Euler talked of imaginary intersections.

However, this clear formulation of the case did nothing to stimulate investigation of the complex points on a curve. The view seems to have been that a curve is a real thing. It comes with an equation, it may even be defined as the points of the plane satisfying this equation, but when you study its intersections with another curve, don't be misled. Algebra may give you intersection points that are not there, they are fictitious imaginary ones.

The Foundational Debate about Complex Numbers at the Start of the 19th Century

Then, at the start of the 19th century the debate around complex numbers associated with such names as Argand, Franais, Mourey, and Warren began. An interesting and neglected aspect of this story is that these writers are in a certain sense minor. Their names tend not to crop up anywhere else in mathematics, and by the same token the 'big' names of the period are silent on the subject until rather later. This division of labour is striking and interesting and should not have been allowed to pass without comment for so long. What we have here is an issue in the philosophy of mathematics, about the nature of certain objects.

It is also striking that the debate lasted so long. It suggests that while most mathematicians found it easy to *use* complex numbers, explaining what they are challenged the two philosophies of mathematics then current. One, the more classical, regarded all mathematical quantities as types of quantity or magnitude; but is a complex number a *number* (in the strict sense of the term)? Is it even a magnitude if it cannot be compared in point of size with another? The other, then enjoying a vogue, regarded mathematics as a language of particular clarity which manipulated empty symbols. On this perspective, complex numbers offered no difficulties to the mind, but this philosophy of mathematics has a particularly thin, even disagreeable ontology.

What the explanations put forward did was to give a realist interpretation to complex numbers, either as directed line segments or as points in the plane. A minority view was that a complex number was an ordered pair of numbers and the sign $\sqrt{-1}$ was a sign of separation. In 1837 Hamilton published the most austere presentation of them all, defining a complex number to be an ordered pair of real numbers and defining the basic operations upon them. From this position it was easy to pass to the graphical representation, but not necessary to say that a complex number *was* a point of the plane. The result of all this debate, and I think of mathematical practice as much as all this philosophy, was the gradual abandonment of the idea that mathematics, and algebra in particular, worked because it was about magnitudes in the Greek sense. What it was about was not so clear, but, as is the way with foundational debates, that ceased to seem so important.

The case of Gauss is interesting. His first proof of the fundamental theorem of algebra [1799] treats complex quantities as pairs of real ones throughout. But in the *Disquisitiones Arithmeticae* of 1801 he had no problem (not even a pedagogic one) in presenting the 17th and 19th roots of unity as complex numbers in the plane, with their coordinates neatly calculated to many decimal places. In 1832 he published his theory of (what we call) the Gaussian integers, giving them a firmly geometrical base and handing out a stern, no-nonsense paragraph or two about the validity of so doing. But he made clear in letters at the time that he knew the difference between representing a complex number by a point in the plane and defining a complex number.[5]

Geometry Applied to Complex Function Theory; The Case of Elliptic Functions

The acclaim that greeted the work of Abel and Jacobi on elliptic functions heralded the start of an unbroken tradition of work on these important functions of a complex variable. The thoroughly complex nature of their work was expressed in purely algebraical terms.[6] However, and this seems not to have been said in this context before, exactly what enabled Gauss to say of Abel with justice that Abel had only come a third of the way in his theory of elliptic functions,[7] was that Abel's theory began by inverting elliptic integrals with a real modulus only. The generalisation to the case where the modulus may be complex and so the periods arbitrary, in short the identification of doubly periodic functions with elliptic functions, was only taken by Gauss. He managed it by generalising his theory of the arithmetico-geometric mean to the complex setting. When this happens the agm becomes many-valued, and its inverse automorphic. This theory, which amounts almost to the discovery of the fundamental domain for the modular function, is

---

[5]See Schlesinger [1912], p. 176.

[6]For such an endorsement of Gauss's impact see Kline [1972], 631, and Bottazzini [1986] 166. For such an assessment of the impact of elliptic function theory, see Stillwell [1989] and the writings of many historians at the start of the 20th century. Since then rapture has been tempered by the alleged inaccessibility of the theory to undergraduate students of mathematics.

[7]Gauss to Schumacher, 30 May 1828, quoted in gauss *Werke* III, p. 495.

necessarily complex; Gauss could come to it because he was able to think of complex numbers geometrically.[8]

Complex Points on Curves (2) — The First Half of the 19th Century

Geometers were also ambiguous about the meaning of complex quantities in geometry. Plücker was a leading exponent in the revival of analytical or algebraic geometry that got underway at the start of the 19th century. In 1834 he announced a result that is still attractive, but is also illustrative of our problem today: a plane algebraic curve of degree 4 has 28 bitangents. In his book of 1839 he gave a detailed account of these bitangents, showing that all 28 could be real, but that they need not be. He enumerated all the cases that can arise of N real bitangents and $28 - N$ imaginary ones. His methods were entirely algebraic. When imaginary tangents and their imaginary points of contact were to be investigated, Plücker wrote expressions of this form

$$p' + p''\sqrt{-1} \tag{1}$$

All this suggests that he considered as imaginary points (and lines) those whose coordinates (or coefficients) are imaginary. And indeed he spoke of not merely the contact points of the bitangents but whole branches of a curve being imaginary.[9] But this impression does not exactly fit the case. We must ask: what impression did he suppose he conveyed by writing about imaginary branches of a curve, imaginary points on it, and imaginary asymptotes? For Plücker the original curve was always a real curve. He therefore felt he had to confront the question: what is the real (i.e. true) interpretation of an imaginary tangent? The answer he gave in 1828, but not (so far as I can see) in 1839 was in terms of involutions.

An involution on the real projective line is a map of period 2. As such it has fixed points; two real distinct ones, or a real repeated one, or two conjugate imaginary ones. In this last case the involution is said to be elliptic. Plücker interpreted pairs of conjugate imaginary points (say, the imaginary contact points of a bitangent) as meaning that there was an elliptic involution with those points as its fixed points. In 1847 he had shifted his position.

> Wolten wir aber unmittelbar das imaginär in die geometrische Discussion aufnehmen, so hiesse das algebraisch verfahren. [But even the imaginary points that arise lie on real lines, so] Es ist dies die geometrische Einkleidung eines algebraisches Factums.

These are two rather different positions, but even the later one falls short of putting complex points on a par with real ones; complex points are still to be

---

[8] For Gauss's work on elliptic functions, see Schlesinger, [1912].

[9] 'Die beiden Curven-Zweige und zugleich mit ihnen also auch die beiden Berürungspuncte sind imaginär.' [1839], p. 231. 'Wir finden also auch hier die Bestätigung davon, dass einer Curve vierter Ordnung, wenn sie vier Doppelpuncte hat, in zwei, reelle oder imaginäre Kegelschnitte zerfällt.' [1839], p. 233

explained geometrically in terms of something real. I do not want to discuss this theory at all except to point out how vital it was in its day. It was taken up by von Staudt in the 1850s, and it is usually he who is credited with the creation of complex geometry as a result. In his *Beiträge zur geometrie der Lage*, [1856–60] he developed an intricate theory of complex lines and planes in what has been called complex projective three-space,[10] all in terms of involutions of various pencils (of points, lines, or planes).

Von Staudt was not an easy writer. Such was Klein's opinion, as it was of others, and through the 1870s the newly-founded *Mathematische Annalen* carried a number of articles explaining and extending it. There was paper by Stolz, two long ones by Lüroth, and a number by Klein in which he tried to draw what he called new Riemann surfaces associated to real algebraic curves. Klein's approach, applied to an ellipse, went as follows.[11] Count the number of imaginary tangents to the curve from each point of the plane; it is 2 for points inside the curve, 0 for points outside. Associate to each point as many points of the new surface as there are imaginary tangents to the curve. This produces a real surface which you can think of as an ellipsoid, and Klein claimed that in this way you will always get a real surface equivalent to the Riemann surface of a curve in Riemann's sense. Klein seems to have thought this enterprise was worthwhile, and recapitulated it in his *Entwicklung* towards the end of his life. But it is plainly not the modern point of view which we all slip fairly comfortably into, and that tells us that this vigorous and (in its day) attractive theory, so closely tied to one's sense of geometry as being about real things, was a solution to the question 'what are the complex points of a curve?'. The problem was solved, which is why the modern solution did not get a look-in.

The Introduction of Complex Points

To see where the modern solution comes from, we must return to elliptic function theory, and a piece of that story which is, as it happens, is not so well-known today. It concerns a search conducted in the 1830s and 40s by Jacobi, Cayley, Cauchy, and Eisenstein to find the best foundations for the new theory. What provoked them to such a search was the feeling that a complex integral involving a square root was a hopelessly obscure thing. They all protested this point quite vigorously. Of them all, it was Cauchy who came closest to confronting this difficulty head-on. In a remarkable paper of 1848 he made what may be the first statement of the matter that resembles what we have been expecting to hear all along.

What one calls the inverse function of the integral

$$t = \int_0^x \frac{dx}{\sqrt{(1 - x^2)(1 - k^2 x^2)}}$$

---

[10]In, for example, 'Staudt', W. Burau, *Dictionary of Scientific biography*, XIII, 1976, 4–6, p. 5.

[11]See Klein [1874].

k being real and < 1, is not the value of x obtained from the complete integral of the differential equation

$$dx = \sqrt{(1 - x^2)(1 - k^2 x^2)}dt$$

but the value of x which provides the complete integral of the differential equation

$$dx = ydt$$

if one requires that the variable y:

1. satisfies the finite equation

$$y^2 = (1 - x^2)(1 - k^2 x^2)$$

2. varies with x by insensible degrees [i.e. is continuous as a function of x], and if one requires moreover that x, y, t simultaneously take the initial values

$$x = 0, \quad y = 1, \quad t = 0.$$

Although Cauchy was at the time engaged in thinking through and systematizing the theory of complex functions he had created in a rather disordered way over the previous 20-odd years, he failed to appreciate the merit of this insight, and shortly retreated to his theory of *lignes d'arrêt*, which deals with many-valued expressions by cutting up the z-plane. However, in 1850 Puiseux, largely inspired by Cauchy's work, wrote a long paper of his own in which it is quite clear that a function $f(u, z) = 0$, as he wrote it, of degree m in u, can be thought of as a m-sheeted covering of the z-plane branched over a number of points.

But analysis is not geometry. It is still profitable to speculate on how a complex theory of geometrical curves came about. The main protagonist in this development seems to have been Riemann. Let me consider this claim from the three angles I have been using: foundational questions, elliptic function theory, and the geometry of algebraic curves. Riemann's idea that an algebraic curve with equation $f(s, z) = 0$ of degree n in s and m in z, where s and z are complex variables, can be thought of as an n-fold covering of the complex z-plane (or, better, z-sphere) is, of course, a great idea. It makes available for algebraic functions all the resources of complex function theory. First, note that Riemann surfaces are a good example of Riemann's philosophy of magnitudes and quantities at work. He had secure starting point in his ontology of n-fold extended quantities. Second, Riemann was of course extremely interested in the theory of elliptic functions, upon which he lectured, and their generalisation to Abelian functions, upon which he wrote one of his most important papers, the one that is our source of the idea of a Riemann surface. Third, from Riemann's standpoint there was no need for a real interpretation of complex points on a curve. For the first time, a curve may simply have complex points on it.

Reception of Complex Points

Historians rightly point out that Riemann's ideas took time to get across. The main reason for this, we are assured, is the problematic role played by Dirichlet's principle in Riemann's presentation of his ideas. All this is true and is usually documented by reference to the analytical school around Weierstrass and Kronecker. But geometers too found them difficult. I think that this was because of the difficulty in accepting that the real and complex points satisfying a given equation be accepted equally as points of a surface.[12]

Such an idea is literally implicit in the work of Abel and Jacobi on elliptic functions, but it was never interpreted geometrically. As we have seen, there was a hint of it in some remarks of Cauchy's, but that hint was later covered over by a much feebler interpretation of the matter. There was a geometrical theory of complex quantities, but that tradition (in Germany from Plücker to von Staudt and beyond) sought to give a real interpretation to complex points as involutions. It was entirely acceptable by mid-century to regard a complex quantity as a point in a real plane (often called the Gaussian plane). But the acceptance of complex points on a curve, and their presentation as a surface, now seems to me to have been entirely novel and in many ways to have been difficult to accept. Difficult to a trained geometer, because there was a theory in place that had the virtue of starting with *real* points. This was a theory people had invested heavily in, as we saw, and would not lightly be abandoned. Difficult because it broke entirely with the ontology that started from real objects. And difficult because, indeed, it is hard, given an equation for a complex curve, to see and get to grips with the associated Riemann surface.

If it is correct that the first person to see an equation of the form $f(z, w) = 0$ as defining a surface of complex points in $\mathbf{C} \times \mathbf{C}$ was Riemann then this novelty would have been an extra barrier to the ready reception of his ideas.[13] It gives a further point to the question of how did his ideas reach the mainstream. The answer, so far as algebraic geometry is concerned, seems to have been Clebsch, who took them up quite deliberately in 1862. From then until his death 10 years later at the age of 39 Clebsch was the leader of growing group of German geometers, and he and his colleague founded the *Mathematische Annalen* in 1869 as a forum for their views. This is the milieu within which several of the authors I have already quoted (notably Lüroth and Klein) wrote.

To conclude, let me sketch briefly how the Riemannian theory of complex curves went, and how it was taken up by Clebsch, Weber, and others. I have chosen Riemann's treatment of the 28 bitangents to a plane quartic, a topic he covered in lectures in February 1862 and which was published in the first edition

---

[12]For the response of the analysts see Bottazzini [1986]. For the geometers' response, see Gray [1989].

[13]I note that Cayley, when explaining in his [1878, 32] how a curve can be thought of as a set of points in $\mathbf{C} \times \mathbf{C}$, explicitly remarked that 'I was under the impression that the theory was a known one; but I have not found it anywhere set out in detail'.

of his *Werke* of 1876 in a version based on notes taken by Roch.[14] His paper on Abelian functions of 1857 underlay his account of algebraic curves. Let X be a non-singular plane curve given by an equation $f(x, y) = 0$ of degree n. A holomorphic 1-form on it will be of the form $\frac{a\,dx}{(\partial f/\partial y)}$ where a is of degree $n - 3$. If $n = 4$ then the genus is 3 and $n - 3 = 1$, so a basis for these things will be found on letting $a = x$, $a = y$, and $a = 1$. The general form for a will therefore be $mx + ny + k$, which equated to zero is the equation of a line. Such a line meets a quartic in 4 points, illustrating the theorem that a holomorphic 1-form on a curve of genus g has $2g - 2$ zeros.

Riemann's general theory gave him $2^{2g} = 64$ cases where something strange occurs. Riemann showed that what happened could be described by his theory of θ-functions, more precisely, by his theory of θ-functions with characteristic. This is a generalisation of Jacobi's theory in the elliptic case to the general setting. Riemann's thinking was somewhat along these lines. The quotient of two 1-forms will be a/a', say, a function on X whose zeros are along the line with equation $a = 0$. If a and a' are chosen carefully, these zeros can be made to occur in pairs, and then the line is a bitangent. The careful choice is made (in a way not to be described here) by the theory of characteristics. But there was more. In homogeneous coordinates x, y, z the original curve may have an awful equation. By letting a, a', and $a'' = 0$ be coordinate axes it is possible to simplify the equation, indeed to Plücker's form — but in a more reliable way. Moreover, the labeling on the bitangents now enables one to deduce theorems about the families[15] they occur in, while the transition to Plücker's form enabled Riemann to write down their equations explicitly. All told, a rich geometrical harvest.

As I have indicated, the decisive influence on spreading this way of thinking was that of Clebsch. His most important paper, though not the first of his on complex geometry, came out in 1863. There he generalised and extended what Riemann had done in lectures, and dealt with a wide variety of topics in the theory of curves by means of Riemann's theory of theta functions. In particular, he pushed the theory of the bitangents to a plane quartic further, and re-derived results of Hesse's. He was followed almost immediately by Roch [1864] and in 1876 by Weber, who however, confined their attention to the quartic and its bitangents.

Perhaps of greater significance was Clebsch's working-through of Abel's theorem, and the way he and Gordan cast it into geometrical language. Where Abel had talked of one equation (fixed) and another whose coefficients depend (linearly) on some parameters, Clebsch and Gordan spoke of a linear family of curves sweeping out sets of points on the fixed curve. They interpreted the functions $a(x, y, z)$ that enter expressions like $\frac{a\,dx}{(\partial f/\partial y)}$ as curves of certain degrees, lying (when the degree is $n-3$) in an adjoint system to the curve and being constrained for analytic

---

[14]To be found in Riemann, *Werke*, 3rd edition, nr. XXXI, 487–504

[15]They occur in families or groupings of 6 pairs, any two of which reduce the equation to Plücker's form; they are recognised as the sets of characteristics all pairs of which have the same sum. Sets of 4 bitangents determine 8 points on both the curve and a conic, sets of 6 are connected with cubics, and so forth, according to a rich geometric theory developed by Hesse [1855].

reasons to pass $k - 1$ times through each k-fold singular point of the fixed curve. In this way analysis and geometry were firmly brought together, and the concept of a complex curve was created in something like its modern form.

Riemann and Roch, his most able pupil, died in 1866; Clebsch in 1872, but Brill, Noether and Klein lived long productive lives. These accidents of history gave a particular twist to the question of whether one had a complex curve or a Riemann surface — but that, as one says, is another story.[16]

Bibliography

Bottazzini, U., 1986 *The Higher Calculus: a history of real and complex analysis from Euler to Weierstrass*, Springer-Verlag, New York.

Cauchy, A.L., 1846 Considerations nouvelles sur les intégrales définies qui s'étendent à tous les points d'une courbe fermée, et sur celles qui sont prises entre des limites imaginaires, *Comptes Rendus*, **23**, 689 = *Oeuvres*, (1), 10, nr. 345, 153–168.

Cayley, A., 1878 On the geometrical representation of imaginary variables by a real correspondence of two planes, *Proceedings of the London Mathematical Society*, **9**, 31–39 = *Collected Mathematical Papers*, X, nr. 689, 316–323.

Clebsch, R.F.A., 1863 Ueber die Anwendung der Abelschen Functionen in der geometrie, *Journal für Mathematik*, **63**, 189–243.

Clebsch, R.F.A. and Gordan, P., 1866 *Theorie der Abelschen Functionen*, Teubner, Leibzig.

Coolidge, J.L., 1924 *The geometry of the complex domain*, Clarendon Press, Oxford.

Cramer, G., 1750 *Introduction á l'analyse des lignes courbes algébriques*, Geneva.

Descartes, R., 1637 *La géométrie*, reprinted as *The geometry of René Descartes*, Dover, New York, 1954.

Euler, L., 1748 *Introductio in analysin infinitorum*, 2 vols = *Opera Omnia*, (1) **8** and **9**, English translation by J.D. Blanton, *Introduction to the analysis of the infinite*, Springer-Verlag, 1988, 1990.

Gauss, C.F., 1799 Demonstratio nova theorematis, etc, *Werke*, **3**, 3–30.

_____, 1801 *Disquisitiones Arithmeticae*, *Werke*, **1**, English translation by A.A. Clarke and W.C. Waterhouse, *Disquisitiones Arithmeticae*, Springer-Verlag, 1986.

_____, 1832 Theorie der biquadratischen Reste, 2, Comm. soc. reg, sci. Göttingen, **7**, = *Werke*, **2**.

---

[16]For which one may consult Gray [1989].

Gray, J.J., 1989 Algebraic geometry in the late nineteenth century in *The history of modern mathematics*, D.A. Rowe and J. McCleary, eds. Academic Press.

_____, 1992 Cauchy — elliptic and algebraic integrals, *Cahiers d'histoire et de philosophie des sciences*.

Hamilton W.R., 1837 Theory of conjugate function of algebraic couples, etc., *Trans Roy Irish Academy*, **17**, = *Mathematical Papers*, **3**, 1967, 3–96.

Hesse, L.O., 1855 Ueber die Doppeltangenten der Curven vierter Ordnung, *Journal für Mathematik*, **49**, 243–264 = *Gesammelte Werke*, nr. 24, 319–344, 2nd ed. Chelsea, New York, 1972.

Klein, C.F. 1874 Über eine neue Art der Riemannschen Flächen, *Mathematische Annalen*, **7**, = *Gesammelte Mathematische Abhandlungen*, **2**, nr. 38, 89–98, Springer Berlin 1922.

Kline, M., 1972 *Mathematical Thought from Ancient to Modern Times*, Oxford.

Plücker, J., 1828 *Analytischen-geometrische Entwicklungen*, **1**, Essen.

_____, 1834 Solution d'une question fondamentale concernant la théorie générale des coubes, *Journal für Mathematik*, **12**, 105–108 = *Gesammelte Werke*, **1**, nr. 21, 298–301, Teubner Leipzig, 1895.

_____, 1839 *Theorie der algebraischen Curven*, Bonn

_____, 1849 Ueber Curven dritter Ordnung und analytische Beweisführung, *Journal für Mathematik*, **34**, 329–336 = *Gesammelte Werke*, **1**, nr 27, 404–412, Teubner Leipzig, 1895.

Puiseux, V., 1850 Recherches sur les fonctions algébriques, *Journal des mathématiques*, **15**.

_____, 1851 Suite, *ibid*, **16**.

Riemann, B., 1851 Grundlagen für eine allgemeine Theorie der Functionen einer veränderlichen complexen Grösse, Inauguraldissertation, Göttingen = *Gesammelte Mathematische Werke*, 3–45.

_____, 1857 Theorie der abel'schen Functionen, *Journal für Mathematik*, **54** = *Gesammelte Mathematische Werke*, 88–144 1990 *Gesammelte Mathematische Werke*, 3rd ed., ed. R. Narasimhan, Springer-Verlag, New York.

Roch, G., 1864 Ueber die Anzahl der Willkürlichen Constanten in algebraischen Functionen, *Journal für Mathematik*, **64**, 372–376.

Schlesinger, L., 1912 Über Gauss' Arbeiten zur Funktionentheorie, in Gauss, *Werke*, X.2.

Stillwell, J., 1989 *Mathematics and its history*, Springer-Verlag, New York.

von Staudt, K.G.C., 1856–1860 *Beiträge zur Geometrie der Lage*, 3 vols, Nuremberg.

Weber, H., 1876 *Theorie der Abel'schen Functionen vom Geschlecht 3*, Berlin.

From Gauß to Weierstraß:
Determinant Theory and Its Historical Evaluations

Eberhard Knobloch

Introduction

The mathematicians of the 19th century were especially interested in linear problems. This applies to matrices, algebraic forms, invariants, quaternions, hypercomplex numbers, new algebras, and is shown in more than 2000 publications dealing with determinants. The most important authors of this century were among the contributors to determinant theory like Carl Friedrich Gauß, Augustin Louis Cauchy, Carl Gustav Jacob Jacobi, Arthur Cayley, Charles Hermite, James Joseph Sylvester, Karl Weierstraß, Leopold Kronecker. This theory, in contrast to the invariant theory which came out of it, has up to this day kept its great value for mathematical-physical problems even though a shift in its fields has in the meantime taken place. If one seeks its sources one comes up against not only contradictory reports of the origin of the name, but also and especially, of which person or persons are to be recognized as its founders and which problems led to its development.

A diligent historical inquiry shows

1. that we have to distinguish between the emergence of a notion, the applications of a method, and the development of a theory if we are to avoid contradictory evaluations of the contributions,

2. that there are at least five different roots of this "branch of analysis" (number theory, linear equations, elimination theory, permutation theory, geometry),

3. that there is a slow development from the solution of special problems to generalizations and axiomatizations which lead to a further generalization (invariant theory).

We have to accept the conclusion that we must relate the use of a mathematical concept historically; that we can make historical statements about mathematical theories only when at the same time reference is made to the developmental state of the theory concerned, that by an unhistorical approach is mostly without further ado in the present time of the speaker. In spite of having the same name, under one concept can later be understood something different from what the original author meant by it. Strictly speaking the Jacobian theory of 1841 is not the theory dealt with a hundred years earlier, or the other way around: the theory that was there at the beginning, that of the 17th or 18th century, is not already the theory of 1841. As a consequence it seems to be worthwhile for historians of mathematics to study the history of determinant theory. It seems to be worthwhile for mathematicians, too, because of the revival of algorithmic thinking in contemporary mathematics. James Joseph Sylvester rated it particularly highly. He said in 1851: It is wonderful that a theory so purely analytical should originate in a geometrical speculation. My friend M. Hermite has pointed out to me, that some faint indications of the same theory may be found in the Recherches Arithmétiques of Gauß. The notation which I have employed for determinants is very similar to that of Vandermonde, with which I have become acquainted since writing the above, in Mr. Spottiswoode's valuable treatise On the elementary theorems of determinants. Vandermonde was evidently on the right road. I do not hesitate to affirm, that the superiority of his and my notation over that in use in the ordinary methods is as great and almost as important to the progress of analysis, as the superiority of the notation of the differential calculus over that of the fluxional system.

For what is the theory of determinants? It is an algebra upon algebra; a calculus which enables us to combine and foretell the results of algebraical operations, in the same way as algebra itself enables us to dispense with the performance of the special operations of arithmetic. All analysis must ultimately clothe itself under this form. [Sylvester 1851b, 246–247]

It is obvious from this quotation that Sylvester had only a very weak indirect knowledge of his many predecessors who had dealt with determinants. While he alludes to Alexandre Théophile Vandermonde (1776) and William Spottiswoode (1851) he does not mention Gabriel Cramer (1750), Etienne Bézout (1767; 1779), Joseph Louis Lagrange (1775a, b, c), Pierre Simon Marquis de Laplace (1776) or Augustin Louis Cauchy (1815b; 1821; 1841). Otherwise he would not have been so astonished that the origins of determinant theory have to do with geometry, as his own paper was partly stimulated by his earlier article on 'Contacts' [Sylvester 1851a]. Lagrange's contributions to that theory are mainly to be found in an article on triangular pyramids, i.e. they were motivated by 'geometrical speculations' [Lagrange 1775a]. A geometrical problem also stimulated the Swiss mathematician Gabriel Cramer to develop the notion of a determinant because he had to solve a system of linear equations:

he tried to determine a curve of nth degree which goes through $(\frac{1}{2}n^2 + \frac{3}{2}n)$ arbitrarily given points. [Cramer 1750, 656–676]

It is true that Cramer neither created a designation for the homogeneous polynomial of n^2 variables nor gave a general definition of determinants, let alone any definition which was identical with the later or modern determinant notion. It took shape empirically as Bourbaki put it. [Bourbaki 1971, 76]

Sylvester's words make clear, too, that he was not acquainted with Gauß's 'Arithmetical Investigations' which appeared in 1801 [Gauß 1801]. In order to get a better idea of how determinant theory developed during the 19th century we have to have a look at these 'faint indications'. This seems all the more to be necessary because in 1930 the American group theorist George Abram Miller saw Gauß as the founder of determinant theory because Gauß had used — as he said — for the first time the modern dual concept of a determinant, i.e. a quadratic matrix with the polynomial belonging to it [Miller 1930]. As we shall see, later authors like Cauchy (1815b), Binet (1813) and Jacobi (1841a) refer indeed to Gauß (1801). As a consequence I would like to discuss the following four aspects:

1. Carl Friedrich Gauß: the number-theoretical approach

2. Augustin Louis Cauchy: the function-theoretical approach

3. Carl Gustav Jacob Jacobi: the algorithmic approach

4. Karl Weierstraß and Leopold Kronecker: the axiomatic approach

1 Carl Friedrich Gauß: the Number-Theoretical Approach

According to his own statement Gauß had dealt with investigations on this subject, which he published in 1801 in his classical number-theoretical book since 1795. He referred explicitly to the earlier studies of Pierre de Fermat, Leonhard Euler, Lagrange, and Adrien Marie Legendre. In the fifth section 'On forms and indefinite equations of second degree' he considers binary and ternary forms:

(1) $$ax^2 + 2bx + cy^2$$

$$ax^2 + a'x'^2 + a''x''^2 + 2bx'x'' + 2b'xx'' + 2b''xx'$$

He named their discriminants

(2) $$b^2 - ac$$

$$ab^2 + a'b'^2 + a''b''^2 - aa'a'' - 2bb'b''$$

as we now say following Sylvester's suggestion the determinants of such and such forms. To this extent it is correct to say that Gauß coined the name "determinant" [Struik 1976, 162], but not in the modern sense of the word. Gauß spoke only of determinants of forms, as George Salmon had already pointed out [Salmon 1885, 338–339]. His considerations were based on Lagrange's "Recherches d'arithmétiques" [Lagrange 1775/77]. Gauß used the same title translating it into Latin. Hence we have to have a look at Lagrange's inquiries. Lagrange studied those numbers which can be represented by

(3) $$Bt^2 + Ctu + Du^2$$

that is, by binary forms. He investigated the relationship between the discriminants of two forms when the second form arises from the first by two linear transformations. His result which is nowadays often formulated by means of Hessean determinants reads as follows:

The discriminants differ only by the square of the module of the linear transformations. Gauß extended this theorem to ternary forms. He chose a quadratic array (see (6)) of the nine coefficients of the three linear transformations

$$
\begin{aligned}
(4) \qquad x &= \alpha y + \beta y' + \gamma y'' \\
x' &= \alpha' y + \beta' y' + \gamma' y'' \\
x'' &= \alpha'' y + \beta'' y' + \gamma'' y''
\end{aligned}
$$

Their module $\alpha'\beta'\gamma'' + \beta\gamma'\alpha'' + \gamma\alpha'\beta'' - \gamma\beta'\alpha'' - \alpha\gamma'\beta'' - \beta\alpha'\gamma''$ that is to say the value of the determinant in the modern sense of the word of this coefficient matrix is called k. Gauß deduced the theorem [1801, 302]:

$$
(5) \qquad\qquad\qquad E = k^2 D
$$

D is the discriminant of the ternary form $ax^2 + a'x'^2 + a''x''^2 + 2bx'x'' + 2b'xx'' + 2b''xx'$ (see (1)) that is $D = ab^2 + a'b'^2 + a''b''^2 - aa'a'' - 2bb'b''$ (see (2)). E is the discriminant of the new ternary form, or D and E are the determinants in the Gauß's sense of the two forms. This is apparently not the multipication theorem of determinants in the modern sense of the word. But we can speak of an implicit use of matrix multiplication (the notion matrix does not exist in 1801) — Gauß of course does not use this expression -- when Gauß shows that we can replace two linear transformations which are carried out one after the other by only one linear transformation. Its coefficient scheme can be obtained by the modern matrix multiplication. Let f, f', f'' be ternary forms, let

f be transformed in f' by $\qquad$ f' be transformed in f'' by

$$
(6) \qquad
\begin{matrix}
\alpha, & \beta, & \gamma \\
\alpha', & \beta', & \gamma' \\
\alpha'', & \beta'', & \gamma''
\end{matrix}
\qquad\qquad
\begin{matrix}
\delta, & \epsilon, & \zeta \\
\delta', & \epsilon', & \zeta' \\
\delta'', & \epsilon'', & \zeta''
\end{matrix}
$$

then f is transformed in f'' by

$$
\begin{matrix}
\alpha\delta + \beta\delta' + \gamma\delta'', & \alpha\epsilon + \beta\epsilon' + \gamma\epsilon'', & \alpha\zeta + \beta\zeta' + \gamma\zeta'' \\
\alpha'\delta + \beta'\delta' + \gamma'\delta'', & \alpha'\epsilon + \beta'\epsilon' + \gamma'\epsilon'', & \alpha'\zeta + \beta'\zeta' + \gamma'\zeta'' \\
\alpha''\delta + \beta''\delta' + \gamma''\delta'', & \alpha''\epsilon + \beta''\epsilon' + \gamma''\epsilon'', & \alpha''\zeta + \beta''\zeta' + \gamma''\zeta''
\end{matrix}
$$

We apply the same procedure when we multiply two determinants. Insofar Cauchy could base his generalization on Gauß's formulae [Cauchy 1815b, 168]. These are indeed 'faint indications' of determinant theory as Sylvester had heard from Hermite but neither the notion, the name, nor the methods were made explicit or used in the modern sense of the word. Instead they remained hidden behind calculating devices.

I would like to add that there are no connections between his number-theoretical determinants or matrix-like transformations and his elimination theory which he indicated in his "Theory of the Motion of Celestial Bodies" in 1809 [Gauß 1809] and which he fully developed in his "Inquiry into the Elliptic Elements of Pallas" in 1811 [Gauß 1811].

2   Augustin Louis Cauchy: the Function-Theoretic Approach

During the 19th century the second phase of determinant theory began in 1812. On November 30 Jacques Philipp Marie Binet [1813] and the three years younger A. L. Cauchy read two papers at the Institut de France [1815b]. Binet called his treatise "On a System of Analytic Formulae and their Applications to Geometric Considerations". He wanted to establish a series of analytical relations which hold between quantities deduced from each other according to a certain rule. He said that these relations are based on a theorem. He is going to state namely the multiplication theorem for determinants [1813, 281]:

> The product of two resultants of the same order is again a resultant of this order.

Similar to Cauchy he hinted at the earlier preparatory works by Vandermonde [1776], Laplace [1776], and Gauß [1801]. But his true model was Lagrange. His main aim was to give a generalization of certain geometrical relations given by Lagrange. His paper was overshadowed by Cauchy's paper [1815b] entitled "On Functions Which Can Assume but Two Equal Values of Contrary Signs by Means of Transpositions Carried Out between the Enclosed Variables". There are several reasons for this:

1. Binet's proof of the multiplication theorem was cumbersome and unsatisfying.

2. In spite of the title the paper had not the analytical tendency as that of his friend and rival Cauchy:

   a Two thirds of Binet's paper dealt with geometrical investigations, rhomboids, parallelepipeds and surfaces of second degree which do not play any role in Cauchy's article. On the other hand Binet did not discuss the solution of equation systems which are of fundamental importance for Cauchy.

   b In contrast to Binet, Cauchy did not even cite Lagrange in his article of 1812, and altogether only once in his later 17 treatises dealing with determinant theory. But this did not apply to Lagrange's mentioned articles on triangular pyramids [1775a], rotating bodies [1775b] or spheroids [1775c] where formulae are to be found which coincide in substance with statements on determinants of three rows.

3. The main reason for Cauchy's superiority was certainly his method and aim when working on his article. It consists of two parts which deal systematically firstly with alternating symmetric functions (1st part), and secondly with those amongst these functions which he called determinants (2nd part). Symmetric functions are called 'alternating' in contrast to 'permanent symmetric functions' if they change their sign after a permutation of variables. That means he had at that time the following logical scheme in mind:

$$\text{functions}\begin{cases}\text{symmetric}\begin{cases}\text{alternating symmetric}\begin{cases}\text{determinants}\\ \ldots\end{cases}\\ \text{permanent symmetric}\end{cases}\\ \ldots\end{cases}$$

He admitted that he had been stimulated by Gauß to write this article: He wanted to generalize Gauß's formulae [Cauchy 1815b, 168]. He succeeded in doing this by proving the multiplication theorem and by applying certain systems of symmetric equations. He defined determinants in a way that was criticized by Jacobi in 1841 [Jacobi 1841b]. We shall come back to this criticism later on. Cauchy's construction of determinants consisted of three steps:

Definition:

a. Take the product of $2^{n(n-1)/2}$ differences of n different quantitites $a_1, \ldots, a_n$

(7) $(a_2 - a_1)(a_3 - a_1) \cdots (a_3 - a_1)(a_3 - a_2) \cdots (a_n - a_2) \cdots (a_n - a_{n-1})$

b. Multiply this product by $a_1 \cdots a_n$ and designate the resulting alternating symmetric function by

(8) $S(\pm a_1 a_2^2 a_3^3 \cdots a_n^n)$

c. Replace the exponents by identical second indices. The new expression

(9) $S(\pm a_{1,1} a_{2,2} a_{3,3} \cdots a_{n,n})$

is called a determinant. The operation symbol S refers to the left indices. The sign rule is based on the number of cycles into which the permutation of the left indices of a special term can be decomposed: Consider say $\begin{pmatrix} 1 & 2 & 3 & \ldots & n \\ \alpha & \beta & \gamma & & \zeta \end{pmatrix}$ Let g be the number of these cycles. The term is positive if $n - g$ is even, otherwise negative.

Cauchy discussed such cyles in the first part of his paper thus giving a graph-theoretical interpretation of permutations [Cauchy 1815b, 99]. He knew that Cramer had used another sign-rule based on the numbers in inversions of a permutation [Cramer 1750, 658]. He could not know that Leibniz had already used such graphical representations of permutations in 1683/84 in order to formulate a sign-rule for his combinatorial aggregates which we now call determinants [Knobloch 1980, 44].

Cauchy gave the expression its modern meaning. Apparently following Gauß's model he arranged the m^2 summands in the form of a quadratic matrix and developed systematically the relevant theory. His explanations of systems of derivative quantities or systems of derivative equations and their determinants broke completely new ground.

Thus Cauchy laid the foundation for the systematic development of this new theory of determinants as Kronecker said in his lectures on this subject published in 1903 [Kronecker 1903 I, 4]. Cauchy obtained to a certain extent the same result as in the case of permutation theory. Immediately beforehand he had written a treatise "On the Number of Values which Might be Assumed by a Function if the Quantitites Enclosed by it are Permuted in Every Way". The combinatorial aspect of his initial considerations did not arise by chance. At the beginning of his article on determinants he refers to this immediately preceding article on permutations [1815a]. By the time of this article he had consolidated permutation theory as a special mathematical discipline in three respects, though Ruffini had the priority over many relevant writings by Cauchy:

1. His writings were far more influential on the mathematical community than those of Ruffini (*effectiveness*).

2. He systematized the results found by his predecessors and stated a great number of general theorms (*systematization, generalization*).

3. He elaborated a *general, universal terminology.*

The question is whether the same is true of Cauchy's publication on determinants. The answer is: no. Only the second statement applies to this publication.

As to the first point his article certainly overshadowed Binet's article (Jacobi, for example, does not cite Binet in his famous articles of 1841 [1841a, b, c]), but itself remained nearly unnoticed for the next 30 years.

As to the third point he returned to the same subject several times later on but dealt with it in other ways and used other designations. Though he gave currency to the designation 'determinant' by the time of his article of 1812, this name was no longer mentioned nine years later in his "Analyse algébrique" [Cauchy 1821]. He used — just as Binet in 1812 — the expression 'resultant'. Already Laplace had used this expression in [1776] and even G. W. Leibniz about 1712 [Knobloch 1980, 271]. The "Essay on Alternating Sums, Known under the Name of Resultants" [1841 b] allows us to recognize the following logical scheme:

$$(10) \quad \text{functions} \begin{cases} \text{alternating} \begin{cases} \text{alternating sums} \begin{cases} \text{resultants} \\ \dots \end{cases} \\ \dots \end{cases} \\ \text{symmetric} \\ \dots \end{cases}$$

The expression 'permanent' has disappeared. The determinants are now called 'resultants' and form a subgroup of the new quantities 'alternating sums'. However Cauchy doesn't mention this change with a single word.

As Kronecker supposed [1903, 5] this practice — being quite typical for Cauchy — was probably one of the possible reasons for the limited effectiveness of Cauchy's paper because he made it difficult for the readers to find lasting results, not to mention that his new kind of definition of determinants involved some complications, for example in the case of the formulation of Cramer's rule. But other reasons might have played a role, too, namely

2. the degree of difficulty of the article,

3. the then still limited possibility of propagating research results by scientific journals.

3 Carl Gustav Jacob Jacobi: the Algorithmic Approach

There can be no doubt that Cauchy had

1. developed a new mathematical discipline,

2. set out a system of theorems whose details were deduced from some propositions stated at the beginning.

As a consequence he was named the true founder of determinant theory by Josef Frantisek Studnicka in [1876] and later authors like Ernesto Pascal [1900, p. XV]. Other authors disagreed. Thomas Muir favoured Vandermonde [1906-1923 I, 131]. He asserted that Vandermonde was the first to give a coherent presentation of the theory and to have defined the functions independently of their connection with other topics. Domenico Fontebasso said in [1873, p. III] that it was Jacobi who in 1841 raised the few scattered bits of knowledge so to speak to the dignity of a theory, "pervenne ad elevare per cosi dire, alla dignita di teoria, poche e sparse nozioni". Cayley published a compromise opinion: "The memoirs of Cauchy and Jacobi contain the greatest part of their (sc. of the determinants) known properties, and may be considered as constituting the general theory of the subject" [1843, 63]. A similar view was taken by Carl Boyer [1956, 242]. I cannot discuss here this interesting methodological question but I want to mention this evaluation problem. According to my research, Jacobi used the concept 'determinant theory' for the first time and did this in 1841 when the critical phase of determinant theory during the 19th century began [1841, 319 footnote): What he presumes "e theoria aequationum linearium et Determinantium algebraicorum" is found in the preceding work, namely [1841 a]. In 1841 he published three decisive treatises: "On the Formation and the Properties of Determinants", "On Functional Determinants", "On Alternating Functions and their Division by the Product of the Differences of the Elements" [1841 a, b, c]. His aim was to improve the roundabout and unclear presentations of his predecessors of those algorithms which, as he said, serve in the solution of systems of linear equations [1841a, 285]. To this extent he pursued the same idea as many of his predecessors. While Cramer believed in 1850 that he had found a "rather comfortable and general rule" [1750, 60] Laplace having in mind Bézout's article [1767] declared in 1772 that the general rules of Cramer and Bézout were unpractical and only proven by induction [Laplace 1776, 395 f.].

Vandermonde emphasized that one was still far away from a general, unambiguous elimination formula, and that it was therefore his goal to show the possibility of obtaining such a formula [1776, 516]. He read his paper already on 12 January 1771 before the Académie des Sciences, in the same year in which he became a member. In the published version, however, he left out several things, since they had in the meantime been published by other mathematicians. With this he probably alluded to the work of Laplace, who, strangely enough, did not mention him, although his essay appeared in the same year of the journal.

Bézout found fault in 1779 with the fact that up to then only successive elimination methods were known [1779, p. V].

It was Jacobi's achievement that determinant theory became well known among the mathematical community. The algorithmization of determinant theory is due to him, as Heinrich Weber and Julius Wellstein wrote in their "Encyklopädie der Elementarmathematik" [1903-1907 I, 126]. He laid the foundation for a treatment of the determinant theory, as Francesco Brioschi wrote in his relevant textbook [1854, p. IV]. Jacobi was obviously stimulated by Cauchy's paper of 1812, and he himself stimulated the aging Cauchy to write again on this subject.

Like Cauchy he started by discussing difference products, Cauchy's sign rule for permutations included, but he did not use these products for the definition of a determinant. Instead he

a) took quite generally $(n + 1)^2$ quantitites $a_k^{(i)}$, $i, k = 0, \ldots, n$

b) constructed the $(n + 1)!$ products

$$(11) \qquad aa_1' a_2'' ... a_n^{(n)}$$

by permuting the lower or upper indices in every possible way.

c) prescribed + or − according to Cauchy's aforementioned sign rule for even or odd permutations.

Def: The aggregate

$$(12) \qquad R = \sum \pm aa_1' a_2'' ... a_n^{(n)}$$

Eberhard Knobloch

is called a determinant. In this he followed, as he said, Gauß and others, whereby Cauchy is meant. But we know that there is no connection between Gauß and this definition or denotation.

In [1841 c] he critized Cauchy's method of introducing determinants through difference products by letting indices take the place of exponents, but he did this without mentioning Cauchy's name. He held the reverse procedure to be correct: If we develop a difference product consisting of $2^{\frac{n(n+1)}{2}}$ terms only $(n+1)!$ will remain while all the others will cancel themselves out. As a consequence it is better to explain the development of such a product by saying that it behaves like a determinant rather than the other way round. The determinant R is a linear expression with regard to the single quantities $a_k^{(i)}$. As a consequence, partial differentiation with respect to these quantities leads to

$$(13) \qquad R = a^{(i)}\frac{\partial R}{\partial a^{(i)}} + a_1^{(i)}\frac{\partial R}{\partial a_1^{(i)}} + \cdots + a_n^{(i)}\frac{\partial R}{\partial a_n^{(i)}}$$

The analyst Jacobi did indeed extensively use partial differentiations in order to deduce his theorems. His approach implied a decisive level of abstraction. His set-theoretical definition left entirely open what kind of thing the determinant elements were, for example, numerical coefficients or "the partial differentials of a system of as many functions as these functions have variables with respect to these variables" [1841 c, 319].

The second case led to what Jacobi called 'functional determinants' which since Sylvester [1853, 506] have been called Jacobians in honour of Jacobi:

$$(14) \qquad \sum \pm \frac{\partial f \partial f_1}{\partial x \partial x_1} \cdots \frac{\partial f_n}{\partial x_n}$$

Jacobi's third article [1841 c] was devoted to them. He underlined the correlation between algebraic and functional determinants, because the properties of the former could be deduced from the properties of the latter and vice versa. At the end of his [1841 c] he applied his results to the transformation of multiple integrals and deduced the relation

$$(15) \qquad \int U \partial f \partial f_1 \ldots \partial f_n = \int U \cdot (\sum \pm \frac{\partial f}{\partial x}\frac{\partial f_1}{\partial x_1} \cdots \frac{\partial f_n}{\partial x_n} \partial x \partial x_1 \ldots \partial x_n$$

U is a given function of the variables $f, f_1, \ldots f_n$. The theorem enables us to avoid successive integrations. Nowadays we would write

$$\int \int \cdots \int_G g(x_1 \ldots x_{n+1}) dx_1 dx_2 \ldots dx_{n+1} =$$

$$\int \int \cdots \int_{G^*} g(\psi_1(y_1 \ldots y_{n+1}) \ldots \psi_{n+1}(y_1 \ldots y_{n+1})) \frac{\partial(x_1 \ldots x_{n+1})}{\partial(y_1 \ldots y_{n+1})} dy_1 \ldots dy_{n+1}$$

An important change of emphasis had taken place: No longer were special problems like elimination, linear equations or integral transformations in the foreground but the development of a theory which could be applied to special problems.

4 Karl Weierstraß and L. Kronecker: the Axiomatic Approach

A whole mathematical school developed on the basis of Jacobi's writings. F. J. Richelot, L. O. Hesse, R. F. A. Clebsch belonged to it. But a thorough knowledge of determinant theory was essentially limited to Jacobi's pupils around the middle of the 19th century.

All the more important was the fact that in 1841, too, Arthur Cayley published the first English contribution to the new theory [Cayley 1841]. His article was entitled "On a Theorem in the Geometry of Position". He was the first investigator to frame the array of constituents by two upright lines and to launch it into mathematical works as a new entity. He created as MacMahon said in [1927, 273 f.] a new algebra and showed that the new entity was something to conjure with. According to MacMahon, Cayley gave the determinant an enduring place in mathematical analysis. MacMahon was right in stressing the importance of a suitable notation, but his statement seems to be exaggerated.

Cayley neither knew of nor mentioned his predecessors. When stating the multiplication theorem he did at least however doubt whether it was new. Only a few years later he really founded two new theories: matrix theory and invariant theory. In 1845, he called invariants hyperdeterminants [1845, 82] because he looked for properties of determinants when studying other algebraic entities. The invariance turned out to be an extension of the multiplication theorem of determinants, the determinants turned out to be the first and simplest invariants in the history of mathematics. This historical process of generalization is reflected in Paul Gordan's two volumes "Lectures on Invariant Theory" [Gordan 1885-1887]: The first volume deals with determinants, the second with binary forms.

Only in 1903 did a completely new approach to determinants become known. Kurt Hensel published posthumously Leopold Kronecker's 21 "Lectures on Determinant Theory" [Kronecker 1903]. As a consequence, it is not astonishing that Kronecker was deeply interested in determinants. His lectures on them were one of his three standard lectures. Harold Edwards recently underlined Kronecker's emphasis on the exploration of algorithms and internally consisting operations [Edwards 1989, 72].

What is interesting here is that he drew three characteristic properties as a main conclusion from his theory of linear equations. These characteristics enabled him to make the transition from quadratic systems to rectangular matrices, that is, to seek those functions of the mn matrix elements which have the same characteristic properties with regard to these elements.

Eugen Netto said in his article 'Combinatorics' for the "Encyclopaedia of Mathematical Sciences" [1898-1904, 36] that Kronecker had based his lectures on a function-theoretical explanation of determinants. But this is not true. His conclusions must not be confused with a definition.

This achievement is due to his famous colleague Karl Weierstraß. It is really a remarkable coincidence that in the same year (1903) Johannes Knoblauch published the third volume of the mathematical works of his teacher Weierstraß which contained a communication entitled "On Determinant Theory" [Weierstraß 1903]. This note was an excerpt from a Weierstraßian lecture "On the Theory and Application of Bilinear and Quadratic Forms", given during the winter term 1886/87. Weierstraß begins by saying that the aim of his communication is to give a definition of those peculiar arithmetical entities called determinants which is useful in many cases and to apply this definition to the demonstration of some theorems of determinant theory.

He took as a basis a system of n linear equations and n unknowns. It did not matter for his considerations whether this system was inhomogeneous and solvable which was presupposed only tacitly. He posed the problem to find a function A of the quantities $a_{\alpha\beta}$ with the following characteristics:

1. A is an entire linear homogeneous function with regard to the elements of every row.

2. A changes nothing except the sign whenever two rows are exchanged with each other.

3. If $a_{11} = a_{22} = ... = a_{nn} = 1$ and all other elements are equated with 0, A has the value 1.

First Weierstraß demonstrates that 1), 2) are equivalent to 1) 2')

2'. $A = 0$ if two rows are equal.

Then he demonstrates that A is uniquely determined by these three characteristics and that it is identical with the determinant .

We owe the knowledge to Ferdinand Georg Frobenius who underlined Weierstraß's priority over Kronecker that Weierstraß used his defintion since 1864 at the latest and that precisely Kronecker objected to it somewhat even ten years later [Frobenius 1905, 353 f.]. It is the modern definition of a determinant as a normed linear mapping which led to the embedding of determinant theory into linear algebra.

Bibliography

Bézout, E. 1767. *Recherches sur le degré des équations résultantes de l' évanouisse-ment des inconnues, et sur les moyens qu'il convient d'employer pour trouver ces équations. Histoire de l'académie royale des sciences Année 1764*, 288–338.

Bézout, E. 1779. *Théorie générale des équations algébriques.* Paris: Ph.-D. Pierres.

Binet, J.-Ph. M. 1813. Mémoire sur un système de formules analytiques, et leur application à des considérations géométriques. *Journal de l'Ecole polytechnique 16e cahier* 1813, 280–354 (Lu à l'Institut le 30 Novembre 1812).

Boyer, C. B. 1956. *History of analytic geometry.* New York: Scripta mathematica.

Bourbaki, N. 1971. *Elemente der Mathematikgeschichte*, translated by A. Ober-schelp. Göttingen: Vandenhoeck & Ruprecht.

Brioschi, F. 1854. *La teorica dei determinanti e le sue principali applicazioni.* Padua: Eredi Bizzoni.

Cauchy, A. L. 1815a. Mémoire sur le nombre des valeurs qu'une fonction peut acquérir lorqu'on y permute de toutes les manières possibles les quantités qu'elle renferme. *Journal de l'Ecole polytechnique 17e cahier 10*, 1–28. Reprinted in *Œuvres complètes*, Académie des Sciences, Ed., vol. II, 1, pp. 64–90. Paris: Gauthier-Villars, 1905.

Cauchy, A. L. 1815b. Mémoire sur les fonctions qui ne peuvent obtenir que deux valeurs égales et de signes contraires par suite des transpositions opérées entre les variables qu'elles renferment. *Journal de l'Ecole polytechnique 17e cahier 10*, 29–112 (Lu à l'Institut le 30 Novembre 1812). All page references are to *Œuvres complètes*, Académie des sciences, Ed., vol. II, 1, pp. 91–169. Paris: Gauthier-Villars, 1905.

Cauchy, A. L. 1821. *Cours d'analyse de l'Ecole Royale Polytechnique, 1ère partie Analyse algébrique.* Paris: Debure frères. Reprinted as *Œuvres complètes*, Académie des Sciences, Ed., vol. II, 3. Paris: Gauthier-Villars, 1897.

Cauchy, A. L. 1841. Mémoire sur les sommes alternées, connues sous le nom de résultantes. *Exercises d'analyse et de physique mathématique 2*, 160–176. All page references are to *Œuvres complètes*, Académie des Sciences, Ed., vol. II, 12, 183–201. Paris: Gauthier-Villars, 1916.

Cayley, A. 1841. On a therorem in the geometry of position. *Cambridge Mathematical Journal 2*, 267–271. All page references are to *The collected mathematical papers*, vol. I, 1–4. Cambridge: University Press, 1889. Reprint New York: Johnson, 1963.

Cayley, A. 1843. On the theory of determinants. *Transactions of the Cambridge Philosophical Society 8*, 1–16. All page references are to *The collected mathematical papers*, vol. I, 63–79. Cambridge: University Press, 1889.

Cayley, A. 1845. On the theory of linear transformations. *Cambridge Mathematical Journal 4*, 193–209. All page references are to *The collected mathematical papers*, vol. I, 80–112. Cambridge: University Press, 1889.

Cramer, G. 1750. *Introduction à l'analyse des lignes courbes algébriques*. Geneva: Cramer and Cl. Philibert.

Edwards, H. 1989. *Kronecker's views on the foundations of mathematics*. In *The history of modern mathematics*, D. E. Rowe and J. McCleary, Eds., vol. I, 66–77. Boston-San Diego etc.: Academic Press.

Fontebasso, D. 1873. *I primi elementi della teoria dei determinanti e loro applicazioni all' algebra ed alla geometria proposti agli alunni degli istituti tecnici*. Treviso: L. Zoppelli.

Frobenius, F. G. 1905. Zur Theorie der linearen Gleichungen. *Journal für die reine und angewandte Mathematik 129*, 175–180. All page references are to Gesammelte Abhandlungen, J.-P. Serre, Ed., vol. III, 349–354. Berlin-Heidelberg-New York: J. Springer, 1968.

Gauß, C. F. 1801. *Disquisitiones arithmeticae*. Leipzig: G. Fleischer. All page references are to Gauß, C. F. 1863. *Werke*, vol. I, Königliche Gesellschaft der Wissenschaften zu Göttingen, Ed. Göttingen: Dieterich (Second edition 1870).

Gauß, C. F. 1809. *Theoria motus corporum coelestium in sectionibus conicis solem ambientium*. Hamburg: F. Perthes u. I. H. Besser. Reprinted in (Gauß 1863–1929 VII, 1–288).

Gauß, C. F. 1811. Disquisitio de elementis ellipticis Palladis ex oppositionibus annorum 1803, 1804, 1805, 1808, 1809. *Commentationes societatis regiae scientiarum Gottingensis recentiores 1* (1808–1811), Tomus I classis mathematicae, pp. 1–26. Reprinted in (Gauß 1863–1929 VI, 1–24).

Gauß, C. F. 1863–1929. *Werke*, Königliche Gesellschaft der Wissenschaften zu Göttingen, Ed. 12 vols. Göttingen: Dieterich and Teubner.

Gordan, P. 1885–1887. *Vorlesungen über Invariantentheorie*, G. Kerschensteiner, Ed. 2 vols. Leipzig: Teubner.

Horiuchi, A. 1989. Sur un point de rupture entre les traditions chinoise et japonaise des mathématiques. *Revue d'Histoire des Sciences 42*, 375–390.

Jacobi, C. G. J. 1841a. De formatione et proprietatibus determinantium. *Journal für die reine und angewandte Mathematik 22*, 285–318. Reprinted in (Jacobi 1881-1891 III, 355–392). German translation in Jacobi, C. G. J. 1896. Über die Bildung und die Eigenschaften der Determinanten, translated by P. Stäckel, pp. 1–49. Leipzig: W. Engelmann (Ostwald's Klassiker N. 77).

Jacobi, C. G. J. 1841b. De functionibus alternantibus earumque divisione per productum e differentiis elementorum conflatum. *Journal für die reine und angewandte Mathematik 22*, 360–371. Reprinted in (Jacobi 1881–1891 III, 439–452). German translation in Jacobi C. G. J. 1896. Über die Bildung und die Eigenschaften der Determinanten, translated by P. Stäckel, pp. 50–65. Leipzig: W. Engelmann (Ostwald's Klassiker N. 77).

Jacobi, C. G. J. 1841c. De determinantibus functionalibus. *Journal für die reine und angewandte Mathematik 22*, 319–352. Reprinted in (Jacobi 1881–1891 III,

393–438). German translation Jacobi, C. G. J. 1896. Über die Functionaldeterminanten, translated by P. Stäckel. Leipzig: W. Engelmann (Ostwald's Klassiker N. 78).

Jacobi, C. G. J. 1881–1891. *Gesammelte Werke*, C. W. Borchardt, A. Clebsch, E. Lottner, K. Weierstraß, Eds. 8 vols. Berlin: G. Reimer. Reprint New York: Chelsea, 1969.

Knobloch, E. (Hrsg.) 1980. *Der Beginn der Determinantentheorie, Leibnizens nachgelassene Studien zum Determinantenkalkül.* Textband. Hildesheim: Gerstenberg.

Kronecker, L. 1903. *Vorlesungen über die Theorie der Determinanten*, prepared and continued by K. Hensel. 2 vols. Leipzig: Teubner.

Lagrange, J. L. 1775a. Solutions analytiques de quelques problèmes sur les pyramides triangulaires. *Nouveaux mémoires de l'académie royale des sciences et belles-lettres (de Berlin).* Année 1773, 149–176. All page references are to the reprint (Lagrange 1867–1892 III, 659–692). Reprint Hildesheim-New York: Olms, 1973.

Lagrange, J. L. 1775b. Nouvelle solution du problème du mouvement de rotation d'un corps de figure quelconque qui n'est animé par aucune force accélératrice. *Nouveaux mémoires de l'académie royale des sciences et belles-lettres (de Berlin).* Année 1773, 85–120. Reprinted in (Lagrange 1867–1892 III, 577–616).

Lagrange, J. L. 1775c. Sur l'attraction des sphéroides elliptiques. *Nouveaux mémoires de l'académie royale des sciences et belles-lettres (de Berlin).* Année 1773, 121–148. Reprinted in (Lagrange 1867–1892 III, 617–658).

Lagrange, J. L. 1775/1777. Recherches d'arithmétique. *Nouveaux mémoires de l'académie royale des sciences et belles-lettres (de Berlin).* Années 1773, 265–312 and Année 1775, 323–356. Reprinted in (Lagrange 1867–1892 III, 693–795).

Lagrange, J. L. 1867–1892. *Œuvres*, J.-A. Serret and G. Darboux, Eds., 14 vols. Paris: Gauthier-Villars. Reprint Hildesheim-New York: Olms, 1973.

Laplace, P. S. Marquis de. 1776. Recherches sur le calcul intégral et sur le système du monde. *Histoire de l'académie royale des sciences.* Année 1772, Seconde partie pp. 267–376. All page references are to *Œuvres complètes*, Académie des Sciences, Ed., vol. VIII, 367–477. Paris: Gauthier-Villars, 1891.

MacMahon, P. A. 1927. The structure of a determinant. *The Journal of the London Mathematical Society 2*, 273–286.

Miller, G. A. 1930. On the history of determinants. *The American Mathematical Monthly 37*, 216–219.

Muir, Th. 1906–1923. *The theory of determinants in the historical order of development.* 4 vols. London: Macmillan & Co. Reprint New York: Dover, 1960.

Netto, E. 1898–1904. Kombinatorik. In: *Encyklopädie der mathematischen Wissenschaften mit Einschluss ihrer Anwendungen.* Vol. I, part 1, chapter A, Arithmetik, part 2, 29–46. Leipzig: Teubner.

Pascal, E. 1900. *Die Determinanten. Eine Darstellung ihrer Theorie und Anwendungen mit Rücksicht auf die neueren Forschungen.* Deutsche Ausgabe von H. Lietzmann. Leipzig: Teubner.

Salmon, G. 1885. *Lessons introductory to the modern higher algebra.* 4th edition. Dublin: Hodges & Figgis. Reprint New York: Stechert, 1964.

Spottiswoode, W. 1851. *Elementary theorems relating to determinants.* London: Longman, Brown, Green, etc. A revised and enlarged version was published in *Journal für die reine und angewandte Mathematik* 51 (1856), 209–271, 328–381.

Struik, D. J. 1976. *Abriß der Geschichte der Mathematik.* 6. Auflage. Berlin: Deutscher Verlag der Wissenschaften.

Studnicka, J. F. 1876. *A. L. Cauchy als formaler Begründer der Determinanten-Theorie. Eine literarisch-historische Studie.* Prag: Verlag der königl. böhmischen Gesellschaft der Wissenschaften. (*Abhandlungen der königl. böhmischen Gesellschaft der Wissenschaften vom Jahre 1875 und 1876.* 6. Folge, 8. Bd. B. Abhandlungen der mathematisch-naturwissenschaftlichen Classe. Prag: Verlag der königl. böhmischen Gesellschaft für Wissenschaften, 1877).

Sylvester, J. J. 1851a. An enumeration of the contacts of lines and surfaces of the second order. *Philosophical Magazine 1*, 119–140. All page references are to *The collected mathematical papers*, vol. I, 219–240. Cambridge: University Press, 1904. Reprint New York: Chelsea, 1973.

Sylvester, J. J. 1851b. On the relation between the minor determinants of linearly equivalent quadratic functions. *Philosophical Magazine 1*, 295–305. All page references are to *The collected mathematical papers*, vol. I, 241–250. Cambridge: University Press, 1904.

Sylvester, J. J. 1853. On a theory of the syzygetic relations of two rational integral functions, comprising an application to the theory of Sturm's functions, and that of the greatest algebraical common measure. *Philosophical Transactions of the Royal Society of London 143*, 407–548. All page references are to *The collected mathematical papers*, vol. I, 429–586. Cambridge: University Press, 1904.

Vandermonde, A. Th. 1776. Mémoire sur l'élimination. *Histoire de l'académie royale des sciences.* Année 1772, Seconde partie, pp. 516–539.

Weber, H. & Wellstein, J. 1903–1907. *Encyklopädie der Elementar-Mathematik.* 3 vols. Leipzig: Teubner.

Weierstraß, K. 1903. Zur Determinantentheorie. Posthumously published in *Mathematische Werke*, vol. III, 271–287, J. Knoblauch, Ed. Berlin: Mayer & Müller. Reprint Hildesheim-New York: Olms & Johnson, 1967.

The Reciprocity Law from Euler to Eisenstein

Günther Frei

0 Introduction

The Reciprocity Law plays a very central rôle in number theory. It grew out of the theory of quadratic forms. The Quadratic Reciprocity Law was first formulated by *Euler* and *Legendre* and proved by *Gauss* and partly by *Legendre*. The search for higher reciprocity laws gave rise to the introduction and study of the Gaussian integers and more generally of algebraic numbers. Analytic methods introduced by *Euler* and *Dirichlet* in connection with the study of sets of prime numbers and primes in arithmetic progressions and their generalization by *Dedekind* and *Weber* to algebraic number fields led to a general form of the reciprocity law found and proved by *Artin*. This Reciprocity Law of Artin which can be considered as being an abelian reciprocity law plays a central rôle in class field theory. It is the starting point for a search of a more general non-abelian reciprocity law, a hint to whose existence is given by *Langlands' program*.

In this paper we will indicate in the First Part how the Quadratic Reciprocity Law developed out from the theory of quadratic forms and how it was generalized to higher reciprocity laws by *Euler, Gauss, Jacobi* and *Eisenstein*.

In a Second Part we will take up the development of the analytic methods leading from *Euler* and *Dirichlet* to *Dedekind, Weber* and *Artin*.

1 Quadratic Reciprocity Law

1.1 Diophantos of Alexandria (ca. 250 A.D.)

In his *Arithmetica* (comprising 13 books), *Diophantos* implicitly applies the following theorem (see e.g. [Dio-1893], Problem XIV in Book 6), which contains a part of the First Complementary Law of the Quadratic Reciprocity.

Theorem 1.1 *(1) If a is an integer of the form $a = 4n + 3$, then $a = x^2 + y^2$ has no solutions with integers $x, y \in \mathbf{Z}$.*

In particular:

(2) *If p is a prime number of the form $p = 4n + 3$, then $p = x^2 + y^2$ has no integral solutions $x, y \in \mathbf{Z}$.*

This is equivalent to say:

(3) *If $p \neq 2$ is a prime number and $p = x^2 + y^2$ has integral solutions $x, y \in \mathbf{Z}$, then p must be of the form $p = 4n + 1$.*

Remarks: (a) In Problem XIV of Book 6 *Diophantos* applies the Theorem 1.1(1) to $a = 15$.

(b) Theorem 1.1 was explicitly stated for the first time by *Fermat* in a letter to *Mersenne* in 1638.

(c) Diophantos does not give a proof for this theorem. However, a proof can easily be given by the theory of even and odd numbers developed by the *Pythagoreans* and which was known to Diophantos.

If we use Euler's theorem that $\mathbf{Z}/p\mathbf{Z}$ is a field if p is a prime number (1775), and *Legendre's symbol*, introduced by Legendre in 1798 (see Definition 1.16), we can deduce immediately the following form of this theorem.

Theorem 1.1 (4) *If $p \neq 2$ is a prime number and $p = x^2 + y^2$ has a solution with integers $x, y \in \mathbf{Z}$, then $\left(\frac{-1}{p}\right) = 1$.*

Hence we have the following diagram of logical implications.
Let $p \neq 2$ be a prime number.

$$p = 4n + 1$$

$$\overset{(3)}{\nearrow}$$

$$x^2 + y^2 = p \text{ is solvable with } x, y \in \mathbf{Z} \qquad \updownarrow (5)$$

$$\underset{(4)}{\searrow}$$

$$\left(\frac{-1}{p}\right) = 1$$

The equivalence (5) is the *First Complementary Law* of the Quadratic Reciprocity. It was first stated by *Fermat* in 1640 and first proved by *Euler* in 1750.

1.2   Fermat (1601–1665)

From 1636 on *Fermat* began studying carefully the new edition of Diophantos' Arithmetica, written in Greek and edited by *Bachet* (1581–1638) in 1621 together with a Latin translation and a commentary. Motivated by this book Fermat found the following essential part of the *First Complementary Law* of the Quadratic Reciprocity as stated in a letter to his friend *Mersenne* (1588-1648) on 25 December 1640.

Theorem 1.2 *If $p = 4n + 1$ is a prime number, then $p = x^2 + y^2$ has a unique solution with natural numbers $x, y \in \mathbf{N} = \{1, 2, 3, \ldots\}$.*

Similarly, he stated in the same letter:

Theorem 1.3 *(1) If $p = 12n + 1$ or $p = 12n + 7$ is a prime number, then $p = x^2 + 3y^2$ has a unique solution with $x, y \in \mathbf{N}$.*

(2) If $p = 8n + 1$ or $p = 8n + 3$ is a prime number, then $p = x^2 + 2y^2$ has a unique solution with $x, y \in \mathbf{N}$.

Remarks: (a) *Fermat* does not give any proof for these statements, but he indicates that they can be proved by infinite descent.

(b) Proofs for Theorem 1.2 were given by *Euler* in 1750, 1760 and 1758. Euler also proved a converse of Theorem 1.2 in 1758 which gives a primality criterion for odd numbers m of the form $m = 4n + 1$.

(c) Similarly *Euler* was able to prove Theorem 1.3(1) in 1763. He also found a converse, stated by Euler in 1759 and 1772, but proved only by *Lagrange* in 1775. Theorem 1.3(2) was proved by Euler in 1761 and 1774 together with a proof of a converse statement, except for the part that if $p = 8n + 3$ is a prime, then $p = x^2 + 2y^2$ is uniquely solvable. This part was again first proved by Lagrange in 1775.

(d) Let's remark that in modern terminology, a necessary condition for a prime p to be representable as $p = x^2 + Ny^2$ with $N \in \mathbf{Z}$ and $x, y \in \mathbf{N}$ is that $\left(\frac{-N}{p}\right) = 1$, where $\left(\frac{\cdot}{p}\right)$ is the Legendre symbol (see Definition 1.16). This follows from Euler's theorem that $\mathbf{Z}/p\mathbf{Z}$ is a field if p is a prime number.

(e) We also notice that $p = x^2 + Ny^2$ is solvable with $x, y \in \mathbf{N}$ for a prime p and a given integer $N \in \mathbf{Z}$ if and only if

$$p = x^2 + Ny^2 = (x + \sqrt{-N}y)(x - \sqrt{-N}y)$$

i.e. if and only if p is the norm of an integer

$$\alpha = x + \sqrt{-N}y \in \mathbf{Z}(\sqrt{-N}) = \{r + s\sqrt{-N} : r, s \in \mathbf{Z}\}$$

1.3 Euler (1707–1783)

It was *Kronecker* (1823-1891) who, in 1875, hinted at the fact that the Quadratic Reciprocity Law was already stated by *Euler* and even stated in essentially the same form as was done later by Legendre (see [Kro-1895], Vol.2, I, 1–10). Euler was led to this law by his study of quadratic forms inspired by Fermat's investigations on primes p represented as $p = x^2 + Ny^2$ for $N = 1, \pm 2, 3$ with integers $x, y \in \mathbf{Z}$.

Around 1741, Euler begin his study of the following two problems.

Question 1.4 *Given $N \in \mathbf{Z}$, describe the primes $p \neq 2$ for which $p = x^2 + Ny^2$ is solvable with $x, y \in \mathbf{Z}$.*

A weaker form of this problem is:

Question 1.5 *Given $N \in \mathbf{Z}$, describe the primes $p \neq 2$ for which p divides m, $p|m$, where m is any integer of the form $m = x^2 + Ny^2$, with $x, y \in \mathbf{Z}$.*

The question 1.4 is more difficult to answer, since it involves the class number of the quadratic field $K = \mathbf{Q}(\sqrt{-N})$.

As to the question 1.5, which leads to the reciprocity law, Euler introduced the following notations.

Definition 1.6 *Let p be a prime, $p \neq 2$, and $N \in \mathbf{Z}$. $p|x^2 + Ny^2$, i.e. p is a divisor of the form $x^2 + Ny^2$, if there exist integers $a, b \in \mathbf{Z}$ with $(a, b) = 1$, that is whose greatest common divisor is 1, such that $p|a^2 + Nb^2$.*

In order to make it easier for the modern reader, we already introduce the *Legendre symbol* as defined by Legendre in 1798.

Definition 1.7 *Let $p \neq 2$ be a prime number and $n \in \mathbf{Z}$ an integer with $(n, p) = 1$.*

$$\left(\frac{n}{p}\right) := \pm 1 \text{ such that } \left(\frac{n}{p}\right) \equiv n^{(p-1)/2} (\text{modulo } p).$$

Then we can describe the primes in Question 1.5 in the following way.

Proposition 1.8 *Let $N \in \mathbf{Z}$ be given and let p be a prime with $(N, p) = 1$. Then*

$$p|x^2 + Ny^2 \Leftrightarrow \left(\frac{-N}{p}\right) = 1$$

Proof : This follows from Euler's theorem (1775) that $\mathbf{Z}/p\mathbf{Z}$ is a field, namely:
$p|x^2 + Ny^2 \Leftrightarrow p|a^2 + Nb^2$ for some $a, b \in \mathbf{Z}$ with $(a, b) = 1 \Leftrightarrow a^2 + Nb^2 \equiv 0$ modulo $p \Leftrightarrow \frac{a^2}{b^2} \equiv -N$ modulo $p \Leftrightarrow \left(\frac{-N}{p}\right) = 1$.

Let's introduce for a given integer $N \in \mathbf{Z}$ the following set of primes.

Definition 1.9

$$P_N := \{primes \ p \neq 2 : p|x^2 + Ny^2\} = \{primes \ p \neq 2 : \left(\frac{-N}{p}\right) = 1\}$$

Question 1.5 then takes on the following form.

Question 1.10 *Give a description of the set of primes P_N for a given $N \in \mathbf{Z}$.*

Notice that by the Definition 1.7 the symbol $\lambda_p(n) := \left(\frac{n}{p}\right)$ is a function of $n \in \mathbf{Z}$ for a given prime p, whereas in the description of P_N one has to determine the primes p for which $\left(\frac{n}{p}\right) = 1$ for a given $n \in \mathbf{Z}$, i.e. $\chi_n(p) := \left(\frac{n}{p}\right)$ appears as a function of p for a given $n \in \mathbf{Z}$. The relation between $\lambda_p(n)$ and $\chi_n(p)$ is the quadratic reciprocity discovered by Euler. Fermat's and Euler's fundamental observation was that the primes $p \in P_N$ can be described by congruence conditions modulo $4|N|$. This was first discovered for $N = 1, 2, 3$. It already follows from the theorems 1.2 and 1.3, namely:

Theorem 1.11 *Let p be a prime number, $p \neq 2$.*

(1) $p|x^2 + y^2 \Leftrightarrow p = 4n + 1$

(2) $p|x^2 + 2y^2 \Leftrightarrow p = 8n + 1$ or $p = 8n + 3$

(3) $p|x^2 + 3y^2$ (for $p \neq 3$) $\Leftrightarrow p = 3n + 1 \Leftrightarrow p = 12n + 1$ or $p = 12n + 7$

Remarks: If we already suppose the quadratic reciprocity law in the form given by *Legendre* or by *Gauss*, we can easily deduce from the properties of the Legendre symbol that P_N is described by congruence classes modulo $4|N|$. Namely, if $N = (\pm 1)2^r q_1 \cdots q_s$ is the prime decomposition of N with odd primes $q_1, \ldots, q_s$, then the primes p in P_N, i.e. for which one has

$$1 = \left(\frac{-N}{p}\right) = \left(\frac{\pm 1}{p}\right)\left(\frac{2}{p}\right)^r \left(\frac{q_1}{p}\right) \cdots \left(\frac{q_s}{p}\right),$$

are described modulo $4|N|$, since each symbol on the right hand side is described modulo $4|a|$ for $a = \pm 1, 2, q_1, \ldots, q_s$.

Euler's findings between 1741 and August 1742 lead to the following *First* or *Implicit Form* of the *Quadratic Reciprocity Law* (see [Eu-1911], I-2, 194–222 (1744/1746; 1751): Theoremata circa divisores numerorum in hac forma $pa^2 \pm qb^2$ contentorum).

Theorem 1.12 *Let $N \in \mathbf{Z}$ be a given integer and p be a prime number with $(p, 2N) = 1$. Then the following properties hold.*

(1) $p \in P_N \Leftrightarrow p \in 4|N|\mathbf{Z} + r = \{4|N|t + r : t \in \mathbf{Z}\} =: C(r)$ for certain remainders $r \in \mathbf{N}$ modulo $4|N|$ with $0 < r < 4|N|$ and $(r, 4N) = 1$.

More precisely, let $\bar{p}$ be the congruence class of p modulo $4|N|$ and let's introduce the set R_N of congruence classes modulo $4|N|$ containing prime numbers p with $\left(\frac{-N}{p}\right) = 1$, i.e.

$$R_N \quad := \quad \{\bar{p} : p \in P_N\} = \{\bar{p} : p \text{ prime}, (p, 2N) = 1, \left(\frac{-N}{p}\right) = 1\}$$

$$= \{C(r) : C(r) \cap P_N \neq \emptyset, 0 < r < 4|N|\}$$

and

$$E_N = (\mathbf{Z}/4|N|\mathbf{Z})^\times \text{ the unit group modulo } 4|N|.$$

Then

(2) R_N *is a subgroup of* E_N *of index* $[E_N : R_N] = 2$.

(3) *If* $\bar{p} = \bar{r}^2$ *for some* $\bar{r} \in E_N$, *then* $\bar{p} \in R_N$.

(4) $\overline{-1} \in R_N \Leftrightarrow N < 0$.

(5) *If* $N = 4n - 1$, *then* P_N *is already characterized by congruence conditions modulo* $|N|$. *If* $N \neq 4n-1$, *then* P_N *is characterized by congruence conditions modulo* $4|N|$ *but not yet modulo* $|N|$.

(6) *Except for* $4|N|$ *and* $|N|$ *(if* $N = 4n - 1$*) there are no other moduli* $m \in \mathbf{N}$ *which characterize* P_N.

This first form of the reciprocity law gives a full description of the structure of R_N, but not an explicit description of R_N or of P_N, e.g. if $N = 2$, then $E_N = \{\bar{1}, \bar{3}, \bar{5}, \bar{7}\}$. Since $\bar{1} = \bar{1}^2 \in R_N$, we must have $R_N = \{\bar{1}, \bar{3}\}$ or $\{\bar{1}, \bar{5}\}$ or $\{\bar{1}, \bar{7}\}$. With the help of (4) we can exclude $R_N = \{\bar{1}, \bar{7}\} = \{\bar{1}, \overline{-1}\}$, but we cannot decide whether $R_N = \{\bar{1}, \bar{3}\}$ or $R_N = \{\bar{1}, \bar{5}\}$. As a matter of fact, Euler (and before him Fermat) determined experimentally that $R_2 = \{\bar{1}, \bar{3}\}$ and $R_{-2} = \{\bar{1}, \bar{7}\}$.

In the cases $N = 1, 2, -2$, Euler got an explicit description of R_N for which he also gave proofs. For $N = 1$ this is the *First* and for $N = -2$ this is the *Second Complementary Law* of the *Quadratic Reciprocity*, namely:

Theorem 1.13 (1) $R_1 = \{\bar{1}\}$, *i.e.* $p | x^2 + y^2 \Leftrightarrow p = 4n + 1$.

(2) $R_{-2} = \{\bar{1}, \bar{7}\}$, *i.e.* $p | x^2 - 2y^2 \Leftrightarrow p = 8n + 1$ *or* $p = 8n - 1$.

Proof: *Euler* first proved this theorem with the help of Fermat's little theorem, proved by Euler in 1742, together with the observation that $x^m + 1 \equiv 0$ modulo p for a prime p cannot be satisfied for all integers $x \in \mathbf{Z}$ with $(x, p) = 1$ if $0 < m < p - 1$, an observation proved by Euler in 1749.

Later in 1772, Euler got the *Second* or *Explicit Form* of the *Quadratic Reciprocity Law*, published in his *Opuscula Analytica* in 1783 after his death (see Eu-[1911], I-3, 512 (1783): Observationes circa divisionem quadratorum per numeros primos).

Theorem 1.14 *Let* p *and* q *be two distinct odd primes.*

(1) *If* $p = 4n + 1$ *then* $p | x^2 - qy^2 \Leftrightarrow q | x^2 - py^2$.

(2) If $p = 4n + 3$ then $p|x^2 - qy^2 \Leftrightarrow q|x^2 + py^2$.

This means in terms of the Legendre symbol :

(3) If $p = 4n + 1$ then $\left(\frac{q}{p}\right) = \left(\frac{p}{q}\right)$.

(4) If $p = 4n + 3$ then $\left(\frac{q}{p}\right) = \left(\frac{-p}{q}\right)$.

1.4 Legendre (1752–1833)

The first who gave at least a partial proof of the Quadratic Reciprocity Law was *Legendre*. Legendre studied Euler's work from 1785 on. His work on the *Quadratic Reciprocity Law*, a term coined by Legendre, can be found in

(1) Recherches d'Analyse Indéterminée (1785/88),

(2) Essai sur la Théorie des Nombres (1798).

The second reference is the first book ever entirely dedicated to number theory. It contains four parts.

Legendre cites Euler's *Opuscula Analytica* in 1808, in the second edition of this book, but it is not clear whether he knew about Euler's Reciprocity Law before that date (see also [Kro-1895], Vol. 2, I, 1–10).

Legendre's proof is based on the following lemma which also played a crucial rôle in the discovery of the *Local-Global Principle* by *Hasse* (1923).

Theorem 1.15 *Let $a, b, c \in \mathbf{Z}$ be integers, not all of the same sign and such that abc is square free.*

Then $ax^2 + by^2 + cz^2 = 0$ has a non-trivial integral solution $x, y, z \in \mathbf{Z}$ if and only if the following three conditions are simultaneously satisfied:

(A) $-bc$ is a quadratic residue modulo $|a|$,

(B) $-ca$ is a quadratic residue modulo $|b|$,

(C) $-ab$ is a quadratic residue modulo $|c|$.

In 1798 *Legendre* also introduced the *Legendre symbol* in the way we have already given, namely:

Definition 1.16 *Let $p \neq 2$ be a prime number and $N \in \mathbf{Z}$ an integer not divisible by p. Then*

$$\left(\frac{N}{p}\right) := \pm 1 \ such \ that \ \left(\frac{N}{p}\right) \equiv N^{(p-1)/2} (\ modulo\ p)$$

Legendre showed that this symbol is well defined and that it is multiplicative. Then he went on to prove the Quadratic Reciprocity Law by using Theorem 1.15. He distinguished 8 cases, depending on whether $p, q \equiv \pm 1 \pmod{4}$ and $\left(\frac{p}{q}\right) = \pm 1$. With the help of Theorem 1.15 he then deduced the sign of $\left(\frac{q}{p}\right)$ given that $\left(\frac{p}{q}\right)$. However, his proof is complete only in the following two cases:

Theorem 1.17 *Let p and q be two distinct odd primes and let $q = 4m + 3$.*

(1) If $p = 4n + 1$ and $\left(\frac{p}{q}\right) = -1$ then $\left(\frac{q}{p}\right) = -1$,

(2) If $p = 4n + 3$ and $\left(\frac{p}{q}\right) = 1$ then $\left(\frac{q}{p}\right) = -1$.

Proof: (1) Consider the equation $x^2 + py^2 - qz^2 = 0$. Since $x^2 + py^2 - qz^2 \equiv x^2 + y^2 + z^2 \pmod{4}$ and $x^2 + y^2 + z^2 \equiv 0 \pmod{4}$ has no integral solution $x, y, z \in \mathbf{Z}$ with $(x, y, z) = 1$, one gets that $x^2 + py^2 - qz^2 = 0$ has no non-trivial solution with integers $x, y, z \in \mathbf{Z}$. Hence not all conditions (A), (B) and (C) can be satisfied at the same time for the equation $x^2 + py^2 - qz^2 = 0$. Since pq is a quadratic residue modulo 1, (A) is satisfied. Also (C) is satisfied, because $-p$ is a quadratic residue modulo q, since

$$\left(\frac{-p}{q}\right) = \left(\frac{-1}{q}\right)\left(\frac{p}{q}\right) = (-1)(-1) = 1$$

Therefore (B) cannot be satisfied, that is $\left(\frac{q}{p}\right) = -1$.

(2) In this case Legendre considers the equation $x^2 - py^2 - qz^2 = 0$ and deduces similarly that $\left(\frac{q}{p}\right) = -1$.

In the remaining six cases Legendre considered in a similar way an equation of the form $ax^2 + by^2 + cz^2 = 0$ with $a \equiv b \equiv c \equiv 1 \pmod{4}$. This equation does not allow a non-trivial solution since $ax^2 + by^2 + cz^2 \equiv x^2 + y^2 + z^2 \pmod{4}$. But in these cases Legendre had to make use of an auxiliary prime t such that $t \equiv r \pmod{4}$, $\left(\frac{t}{p}\right) = \epsilon$, $\left(\frac{t}{q}\right) = \delta$ for appropreate values $r, \epsilon, \delta \in \{+1, -1\}$, i.e. Legendre had to choose t in a certain congruence class modulo $4pq$. This presupposes, as Gauss pointed out in his *Disquisitiones Arithmeticae* (1801) (see [Ga-1801], art.151 and art.297; see also the German edition by H. Maser of 1889, p.449 and [Kro-1895], Vol.2,I,1–10) a theorem, proved only in 1837 by *Dirichlet*, namely that any arithmetic progression modulo m, $m \in \mathbf{N}$, and prime to m contains at least one prime. Legendre was aware of this fact, but since he could not prove it and since it seemed quite evident to him, he took this theorem as a kind of axiom in his book of 1798.

1.5   Gauss (1777–1855)

It was *Gauss* who gave the first complete proof of the *Quadratic Reciprocity Law* in 1801 in his treatise *Disquisitiones Arithmeticae*, where he, in fact, furnished two entirely different proofs. Altogether Gauss found 8 different proofs, 6 of which were published by him and two others were found after his death in his papers.

   We start with giving a summary of his discoveries concerning the Quadratic Reciprocity Law in the form of the following table. D.A. stands for *Disquisitiones Arithmeticae* (see [Ga-1801]) and M.D. for his *Mathematische Tagebuch (Mathematical Diary)* (see [Gr-1984]).

1792 Construction of tables of remainders $r = r(p, q)$ for primes p and q such that $p \equiv r(\text{modulo } q)$, $1 \leq r < q$, for $p \leq 997$, $q \leq 503$.

1795 Beginning: Discovery of $\left(\frac{-1}{p}\right) = (-1)^{(p-1)/2}$.

1795 March: Discovery of the *Theorema Fundamentale* (Quadratic Reciprocity Law).

1796 8 April: *Proof I*, by induction, 8 cases

(D.A., art.131–144; M.D., no.2).

1796 22 June: *Proof II*, with the genus of quadratic forms

(D.A. art.262; M.D., no.16).

1796 2 September (before): *Proofs VII and VIII*, with cyclotomy and higher power residues

(M.D., no.30).

1801 Mid May: *Proof IV*, with Gaussian sums; a sign remains indetermined

(Summatio quarundam serierum singularium (see [Ga-1889], 463–495); M.D., no.118).

1805 30 August: Determination of the sign

(M.D., no. 123).

1807 6 March ? : *Proof III*, simple proof by means of Gauss' lemma

(Theorematis arithmetici demonstratio nova, 1808 (see [Ga-1889], 457–462); M.D., no. 134)

1808 27 August (after 6 May 1807): *Proof V*, with the Gaussian lemma

(Theorematis fundamentalis in doctrina de residuis quadraticis demonstrationes et amplificationes novae, 1818 (see [Ga-1889], 496–510))

1808 or before (according to Kronecker before 6 May 1807): *Proof VI*, with cyclotomy in the field $\mathbf{Q}[x]/(1 + x + \cdots + x^{p-1})$

(Theorematis fundamentalis in doctrina de residuis quadraticis demonstrationes et amplificationes novae, 1818 (see [Ga-1889], 496–510))

We will look at the second and the sixth proof which are particularly important for the further development of the Reciprocity Law.

1.5.1   Second Proof: Genus of Quadratic Forms.

In his *first proof*, Gauss proceeds by induction whereby he distinguishes 8 cases as
did Legendre before him. The proof is elementary but quite complicated.

The *second proof* uses the theory of quadratic forms. *Euler* had already studied quadratic forms of the type $f(x,y) = ax^2 + cy^2$ with integers $a, c \in \mathbf{Z}$ in
connection with the theory of convenient numbers (1772) (see [Fr-1985]) and *Lagrange* studied more generally quadratic forms of the type $f(x,y) = ax^2 + bxy + cy^2$
with integers $a, b, c \in \mathbf{Z}$, whereby Lagrange introduced the notions of equivalence,
of equivalence classes, of reduction and of the determinant. He showed that for a
given determinant d there are only finitely many classes (1773 and 1775). *Gauss*
took up the subject in more generality, whereby he introduced the fundamental
notions of proper equivalence, of orders and of genera.

His theory of genera is intimately connected with the Reciprocity Law. It
goes as follows (see [Ga-1801], section 5, art. 153–334):

Definition 1.18 *Let* $f = f(x,y) = ax^2 + 2bxy + cy^2 = [a, b, c]$ *with* $a, b, c \in \mathbf{Z}$ *be
a binary integral quadratic form.*

(1) f is called properly primitive *if* $(a, 2b, c) = 1$.

(2) $d = \det(f) := b^2 - ac$ is called the determinant *of f.*

(3) f and f' are called properly equivalent, $f \equiv f'$, *if there exists a linear integral
transformation* $T = \begin{pmatrix} \alpha & \beta \\ \gamma & \delta \end{pmatrix} \in Mat_{2\times 2}(\mathbf{Z})$ *with determinant* $\det(T) = 1$,

such that $f(x,y) = f(T(x',y')) = f'(x',y')$; that is if $M_f = \begin{pmatrix} a & b \\ b & c \end{pmatrix}$ is the

symmetric matrix associated with f and $M_{f'} = \begin{pmatrix} a' & b' \\ b' & c' \end{pmatrix}$ is the symmetric

*matrix associated with the form f', where $f = [a, b, c]$ and $f' = [a', b', c']$ then
$M_{f'} = T^t M_f T$, where T^t denotes the transpose of T.*

(The two forms f and f' are called equivalent *if $\det T = \pm 1$).*

(4) f is called ambiguous *if* $a | 2b$.

(5) $f_p(\mathbf{Z}^2) := \{m = ax^2 + 2bxy + cy^2 : x, y \in \mathbf{Z} \text{ and } (m, p) = 1\}$ is the set of integers m prime to a given integer p and represented by the form $f = [a, b, c]$.

The key observation made by Gauss for the notion of a genus is the following property (see [Ga-1801], art.229).

Theorem 1.19 Let $f = [a, b, c]$ be a primitive form, i.e. $(a, b, c) = 1$, with the non-square determinant $d = \det(f) = b^2 - ac \neq e^2$ and $p \neq 2$ a prime with $p|d$.
Then one has

$$\text{either} \quad \left(\frac{m}{p}\right) = 1 \quad \text{for all } m \in f_p(\mathbf{Z}^2)$$

$$\text{or else} \quad \left(\frac{m}{p}\right) = -1 \quad \text{for all } m \in f_p(\mathbf{Z}^2).$$

Proof: This follows from the composition law for quadratic forms.

If $m, m' \in f_p(\mathbf{Z}^2)$, then

$$m = ax^2 + 2bxy + cy^2 \quad \text{with } x, y \in \mathbf{Z}$$
$$m' = ax'^2 + 2bx'y' + cy'^2 \quad \text{with } x', y' \in \mathbf{Z}.$$

Hence $mm' = [axx' + b(xy' + x'y) + cyy']^2 - d[xy' - x'y]^2$.

Since $p|d$ we get $\left(\frac{mm'}{p}\right) = 1$ and therefore $\left(\frac{m}{p}\right) = \left(\frac{m'}{p}\right)$.

Remarks: (a) The use of the *Legendre symbol* in this context is due to *Dirichlet* (1839). Gauss uses the letters R for residue, and N for non-residue.

(b) There is a similar but more complicated theorem for the prime $p = 2$ which gives a characterization of the integers $m \in f_2(\mathbf{Z}^2)$ modulo 8.

This property gives rise to the notion of a *character* and to the notion of the *genus*, both introduced by Gauss in 1801 (see [Ga-1801], art. 231).

Definition 1.20 Let $f = [a, b, c]$ be a primitive form with determinant $d = \det(f) = b^2 - ac$ and $p \neq 2$ a prime dividing d, $p|d$. Then

$$\epsilon_p(f) := \left(\frac{m}{p}\right) \qquad \text{with } m \in f_p(\mathbf{Z}^2)$$

is called the character of f with respect to the prime p with $p|d$.

Remark: There is also a character $\epsilon_2(f)$ for $p = 2$ depending on the congruence class of d modulo 8 in the case where $d \not\equiv 1$ modulo 4. Following *Hilbert* and *Hasse* there is also an *infinite character* in the case $d < 0$, namely

$$\epsilon_\infty(f) = \begin{cases} +1 & \text{if } m \geq 0 \quad \text{for any } m \in f(\mathbf{Z}^2) \\ -1 & \text{if } m \leq 0 \quad \text{for any } m \in f(\mathbf{Z}^2) \end{cases}$$

where $f(\mathbf{Z}^2) = \{m = ax^2 + 2bxy + cy^2 : x, y \in \mathbf{Z}\}$.

The *genus* is then defined as follows:

Definition 1.21 *Let f and f' be two quadratic forms with the same determinant d.*

(1) *f and f' are in the same genus, $f \sim f'$, if $\epsilon_p(f) = \epsilon_p(f')$ for all p dividing d, including $p = 2$ if $d \not\equiv 1(modulo\ 4)$ and $p = \infty$ if $d < 0$.*

(2) *$f_0 = [1, 0, -d]$ is called the* principal form *of determinant d.*

The proper class $[f_0] = P$ of f_0 is called the proper principal class *of determinant d.*

The genus $[[f_0]] = E$ of f_0 is called the principal genus *of determinant d.*

If two forms f and f' are properly equivalent , $f \equiv f'$, or simply equivalent, then $\det(f) = \det(f')$ and $f_p(\mathbf{Z}^2) = f'_p(\mathbf{Z}^2)$ for any $p \in \mathbf{Z}$. Hence f and f' are then in the same genus.

The Quadratic Reciprocity Law is contained in the following fundamental fact (see [Ga-1801], art.264 and art.287).

Theorem 1.22 *(1) There are precisely 2^{t-1} different non-empty genera of determinant d, where t is the number of different prime divisors of d (including $p = 2$ if $d \not\equiv 1$ modulo 4 and $p = \infty$ if $d < 0$), or else*

(2) *There is exactly one linear relation among the t characters $\epsilon_p(f)$, $p|d$, namely*

$$\prod_{p|d} \epsilon_p(f) = \prod_p \epsilon_p(f) = 1$$

where p ranges over all primes (including $p = 2$ and $p = \infty$) in the second product.

Remarks: (a) Let g be the number of genera for a given determinant d, $g = g(d)$. In a first step, Gauss proved that $g \leq 2^{t-1}$. This is based on the theory of ambiguous forms and ambiguous classes and is equivalent to the Reciprocity Law. Indeed, in art.262 of the D.A. (see [Ga-1801]), Gauss derives the reciprocity law from this inequality. This inequality is essentially the *first inequality* of class field theory for quadratic number fields. It is equivalent to Dirichlet's density theorem, or else to Dirichlet's theorem on primes in arithmetic progressions. We have already seen in Legendre's proof that Dirichlet's theorem on primes in arithmetic progressions implies the reciprocity law.

(b) The *second inequality* $g \geq 2^{t-1}$ lies deeper and is obtained by Gauss from the representation theory of binary quadratic forms by ternary quadratic forms (see [Ga-1801], art.266–287).

In class field theory, the second inequality was obtained by *Takagi*, or by *Herbrand*, in the case of cyclic extensions by a cohomological argument.

(c) The formula (2) in Theorem 1.22 is called the *Hilbert Reciprocity Law*, whereby the symbols $\epsilon_p(f)$ should be viewed as *Hilbert symbols* (see [Fr–1979]).

The Theorem 1.22 together with the following theorem (see [Ga-1801], art.286), to which it is related, must be considered as the *Fundamental Theorem of Genera*. It also has to be viewed as the *Fundamental Theorem of Class Field Theory* of quadratic number fields.

Theorem 1.23   *(1) Every square of a quadratic form of determinant d lies in the principal genus. Conversely, any form in the principal genus E is the square of a quadratic form of determinant d, or else, in terms of (proper) classes,*

(2) $E = C^2 = \{[f]^2 : f = \text{form of determinant } d\}$, *where C are the proper classes of determinant d and [f] denotes the proper class of f.*

A translation of the theory of quadratic forms of determinant d into the language of ideals in the quadratic number field $K = \mathbf{Q}(\sqrt{d})$ with discriminant d was given by Dedekind in 1894 (see [De-1894], Art.182, 186, 187).

Let $\mathcal{Q} = \{x\alpha_1 + y\alpha_2 : x, y \in \mathbf{Z}\} =: [\alpha_1, \alpha_2]$ be an ordered integral ideal in $K = \mathbf{Q}(\sqrt{d})$ generated by algebraic integers $\alpha_1, \alpha_2 \in K$, where ordered means that $\alpha_1 \alpha_2' - \alpha_2 \alpha_1' = N(\mathcal{Q})\sqrt{d}$ is positive or positive imaginary. Hereby $N(\mathcal{Q}) := [\mathcal{O}(K) : \mathcal{Q}]$ denotes the norm of the ideal $\mathcal{Q}$, $\mathcal{O}(K)$ denotes the integers in K and α' stands for the conjugate of α in $\mathbf{Q}(\sqrt{d})$, that is, if $\alpha = a + b\sqrt{d}$, then $\alpha' = a - b\sqrt{d}$. Then Dedekind associates to the integral ideal $\mathcal{Q}$ in $\mathbf{Q}(\sqrt{d})$ the binary quadratic form

$$f_\mathcal{Q} := \frac{(\alpha_1 x + \alpha_2 y)(\alpha_1' x + \alpha_2' y)}{N(\mathcal{Q})} = ax^2 + bxy + cy^2.$$

He then showed that the proper class $[f_\mathcal{Q}]$ of $f_\mathcal{Q}$ does not depend on the ordered basis chosen for $\mathcal{Q}$ and that two ordered integral ideals $\mathcal{Q}$ and $\mathcal{Q}'$ belong to the same proper, i.e. narrow ideal class in $Q(\sqrt{d})$ if and only if $f_\mathcal{Q}$ and $f_{\mathcal{Q}'}$ are properly equivalent. The theory of the genus of binary quadratic forms can therefore be translated into a theory of the genus of ideals in a quadratic number field.

From this bijection between proper classes of binary quadratic forms of determinant d and narrow ideal classes in $\mathbf{Q}(\sqrt{d})$ follows that the set of proper classes of forms can be equiped with a group structure. The composition law is a special case of Gauss' composition of two forms or of two classes. For more details about the theory of the genus, see [Fr-1979].

1.5.2   Sixth Proof: Gauss Sums in $\mathbf{Q}[x]/1 + x + \cdots + x^{p-1}$

Proof VI introduces, as *Gauss* says in his introduction (see [Ga-1889], 496–510), a completely new idea, which allows to give an analogous proof for the cubic and the biquadratic reciprocity law. In this paper of 1818 (see [Ga-1889], 497), Gauss promised to publish soon these analogous proofs, but he never fulfilled his promise. Gauss works in the ring

$$\mathbf{Q}[x]/1 + x + \cdots + x^{p-1},$$

where p is an odd prime, but Eisenstein pointed out (see also [Ei-1989],I,478) in 1845 that this amounts to the same as working in the field of the p-th roots of unity $\mathbf{Q}(\zeta_p)$, $\zeta_p = e^{2\pi i/p}$, since

$$\mathbf{Q}[x]/1 + x + \cdots + x^{p-1} \cong \mathbf{Q}(\zeta_p).$$

We therefore translate Gauss' proof into terms of $\mathbf{Q}(\zeta_p)$. The new idea consists in introducing the so called *Quadratic Gauss Sum* γ in $\mathbf{Q}(\zeta_p)$ defined by Gauss as follows (see [Ga-1889], 502).

Definition 1.24 *Let $p \neq 2$ be a prime, $\zeta = \zeta_p := e^{2\pi i/p}$ and $< g >= (\mathbf{Z}/p\mathbf{Z})^{\times}$, i.e. g is a primitive root modulo p. Then*

$$\gamma := \zeta - \zeta^g + \zeta^{g^2} - \zeta^{g^3} + \cdots - \zeta^{g^{p-2}} = \sum_{t=1}^{p-1} \left(\frac{t}{p}\right) \zeta^t \in \mathbf{Z}[\zeta]$$

is called a quadratic Gauss Sum.

The fundamental properties of the Gauss Sum γ are the following.

Theorem 1.25   *(1)* $\gamma^2 = (-1)^{(p-1)/2} p =: p^*.$

(2) $\gamma = \pm\sqrt{p^*} \in \mathbf{Z}[\zeta]$, *hence* $\mathbf{Q}(\sqrt{p^*}) \subseteq \mathbf{Q}(\zeta)$.

(3) $\gamma^q \equiv \left(\frac{q}{p}\right) \gamma(modulo\ q)$ *for any odd prime* $q \neq p$.

From these properties Gauss deduced the quadratic reciprocity law for two distinct odd primes p and q as follows.

Proof VI: $\gamma^{q-1} = (\gamma^2)^{(q-1)/2} = p^{*(q-1)/2} \equiv \left(\frac{p^*}{q}\right) (modulo\ q)$

by property (1) and Definition 1.16.

$\gamma^q \equiv \left(\frac{p^*}{q}\right) \gamma \equiv \left(\frac{q}{p}\right) \gamma(modulo\ q)$

by property (3). After multiplication with γ one obtains

$\left(\frac{p^*}{q}\right) p^* \equiv \left(\frac{q}{p}\right) p^*(modulo\ q)$

and after division by p^*

$\left(\frac{p^*}{q}\right) \equiv \left(\frac{q}{p}\right) (modulo\ q).$

Hence $\left(\frac{p^*}{q}\right) = \left(\frac{q}{p}\right)$, since $q \neq 2$.

Therefore

$$\left(\frac{q}{p}\right) = \left(\frac{p^*}{q}\right) = \left(\frac{-1}{q}\right)^{(p-1)/2} \left(\frac{p}{q}\right) = (-1)^{\frac{p-1}{2}\frac{q-1}{2}} \left(\frac{p}{q}\right).$$

Remarks: (a) The Reciprocity Law is essentially obtained from the fact that the quadratic field $\mathbf{Q}(\sqrt{p^*})$ is contained in the cyclotomic field $\mathbf{Q}(\zeta_p)$ and by comparing the decomposition law in $\mathbf{Q}(\zeta_p)$ with that of $\mathbf{Q}(\sqrt{p^*})$.

(b) This proof was generalized to the cubic and biquadratic reciprocity law by *Jacobi* and *Eisenstein* following the indications given by Gauss.

2 Cubic Reciprocity Law

2.1 Euler (1707–1783)

In Euler's *Tractatus* (1749; published in 1849; Tractatus de Numerorum Doctrina Capita Sedecim quae supersunt (see [Eu-1911], I–5, 182)), there are two tentative chapters on cubic and biquadratic residues. Euler gives conditions for 2,3,5,6,7,10 to be residues modulo a prime p of the form $p = 3n+1$ (see [Eu-1911], I–5, 250–251, nos. 407–410).

One of his findings is the following.

Theorem 2.1 *Let $p = 3n+1$ be a prime. Then 2 is a cubic power residue modulo p, i.e. $p|x^3-2$ if and only if there exist natural numbers $a, b \in \mathbf{N}$ with $p = a^2 + 27b^2$.*

2.2 Gauss (1777-1855)

Gauss mentions the *Cubic Reciprocity Law* only in a footnote in his second paper on the Biquadratic Reciprocity Law (1832) (see [Ga-1889], 534–586, in particular 541). He says that the Cubic Reciprocity Law can be deduced similarly as the Biquadratic Reciprocity Law by considering the ring $A = \mathbf{Z}[\rho]$, where $\rho = e^{2\pi i/3} = \frac{-1+\sqrt{-3}}{2}$ and that the arithmetic in A has to be established in order to formulate and prove the Cubic Reciprocity Law. More generally, he says, the arithmetic in $A = \mathbf{Z}[\zeta], \zeta = e^{2\pi i/m}$, has to be developed in order to formulate the reciprocity law for m-th power residues. His investigations on the Cubic Reciprocity Law were only found after his death in his papers (see Gauss' Mathematical Diary (see [Gr-1984]), no. 138, 137, 132, 131, 130 and Gauss' notebook "Uraniae sacrae" (see also [Ga-1870], Vol. VIII, 5–14 and [Ba-1911], 46–48)). They are summarized in the following table.

1805 Experimental tables on cubic residues.

1807 15 February: Systematic study (in the notebook "Uraniae sacrae")

1818 Publication of the sixth proof of the Quadratic Reciprocity Law with the announcement of the Cubic Reciprocity Law.

1832 Second paper on the Biquadratic Reciprocity Law with a footnote saying that the Cubic Reciprocity Law has to make use of the arithmetic in $\mathbf{Q}(\sqrt{-3})$.

In his notebook, Gauss gives two approaches to the Cubic Reciprocity Law, namely by

(i) Representation of a prime p by certain binary quadratic forms,

(ii) Cyclotomy by means of Cubic Gauss Sums.

In order to prove the Cubic Reciprocity in $\mathbf{Z}[\rho]$, $\rho = \zeta_3 = e^{2\pi i/3}$ for primes $p \equiv 1 \pmod 3$ and $q \equiv \pm 1 \pmod 3$, Gauss first established the following fact.

Theorem 2.2 *$A = \mathbf{Z}[\rho]$ admits a Euclidean algorithm, hence the Fundamental Theorem of Arithmetic is true in A.*

Gauss' proof of the Cubic Reciprocity is based on his sixth proof of the Quadratic Reciprocity, namely on properties of the *Cubic Gauss Sum*

$$\alpha = (\rho, \zeta) = \zeta + \rho\zeta^g + \rho^2\zeta^{g^2} + \cdots + \rho^{p-2}\zeta^{g^{p-2}}$$

where $< g >= (\mathbf{Z}/p\mathbf{Z})^\times, \rho = e^{2\pi i/3}, \zeta = e^{2\pi i/p}$ for a prime $p \neq 2$.

Remarks: (a) From Theorem 2.2 Gauss deduced Fermat's conjecture that $x^3 + y^3 + z^3 = 0$ has no integral solutions $x, y, z \in \mathbf{Z}$ with $xyz \neq 0$.

(b) Gauss also studied the arithmetic in the cubic field $K = \mathbf{Q}(\sqrt[3]{n})$ for $n \in \mathbf{Z}, n \neq m^3$, and the units in K.

2.3 Jacobi (1804–1851)

Motivated by the remark made by Gauss in his sixth proof of the Quadratic Reciprocity, *Jacobi* began working on the Cubic Reciprocity. Jacobi's evolution in that matter can be summarized as follows:

1828 8 February: Letter to Gauss on Rational Cubic Reciprocity for primes $p = 3n + 1$ and $q = 2, 3, 5, 7, 11, \ldots, 37$.

Introduction of Jacobi sums.

1828 22 June: Publication of the Rational Cubic Reciprocity (in Latin) in Crelle's Journal, vol.2.

1836 October — February 1837: Lectures on Number Theory given at the University of Königsberg.

1837 16 October: Introduction (in print) of Jacobi sums and formulation of the Cubic Reciprocity Law (Reprinted in Crelle's Journal, vol.30 (1846) with additions dating of October 1845).

Jacobi was the first to give a statement of the Cubic Reciprocity Law in print, namely as follows (see Über die Kreistheilung und ihre Anwendung auf die Zahlentheorie, [Ja-1881], Vol.6, 254–274 or Crelle 30 (1846), 166–182).

Theorem 2.3 *Let* π, π' *be two primes in* $\mathbf{Z}[\rho], \rho = e^{2\pi i/3}$, *of the form* $\pi = \frac{a+b\sqrt{-3}}{2}, 3|b$ *and* $\pi' = \frac{a'+b'\sqrt{-3}}{2}, 3|b'$. *Let for an algebraic integer* $\alpha = x + y\sqrt{-3}$, $\left(\frac{\alpha}{\pi}\right)_3$ *be the cubic residue character, i.e.*

$$\left(\frac{\alpha}{\pi}\right)_3 := 1, \rho, \rho^2 \text{ such that } \left(\frac{\alpha}{\pi}\right)_3 \equiv \alpha^{\frac{N(\pi)-1}{3}} \text{(modulo } \pi)$$

where $N(\pi)$ *denotes the norm of* π. *Then*

$$\left(\frac{\pi'}{\pi}\right)_3 = \left(\frac{\pi}{\pi'}\right)_3 .$$

In a footnote in his paper in Crelle's Journal, Vol.30 (1846) (see [Ja-1881], Vol.6, 261), Jacobi adds that he proved this theorem in his course on Number Theory given in the winter term 1836/37 at the University of Königsberg. Only two copies of the notes taken by Rosenhain of these lectures seem to have survived. One is in Berlin and the other is in the library of the ETH in Zürich. The first published proof , however, is due to Eisenstein, who published it in March 1844 in Crelle's Journal, Vol.27, 289–310 (see [Ei-1989], I, 59–80). For the priority dispute between Jacobi and Eisenstein about the Cubic Reciprocity Law, see [Ja-1881], 261–262, [Ei-1989], I, 477–478 and [Col-1977].

2.4 Eisenstein (1823–1852)

We confine ourself to give a short summary of Eisenstein's contribution to the Cubic Reciprocity Law. *Eisenstein* was motivated by Gauss' remark made in 1818 and by Jacobi's statement of 1837 as well as by Gauss' two papers on the Biquadratic Reciprocity Law of 1828 and 1832 and by a remark made by Gauss in his introduction to the last chapter in his *Disquisitiones Arithmeticae*.

1844 March: *Proof I* based on Gauss' sixth proof (1818) and on Gauss' indications about the arithmetic in $\mathbf{Z}[\rho], \rho = e^{2\pi i/3}$ (Beweis des Reciprocitätssatzes für die cubischen Reste in der Theorie der aus dritten Wurzeln der Einheit zusammengesetzten complexen Zahlen, Crelle's Journal 27 (1844), 289–310 (see [Ei-1989], I, 59–80; see also [Ei-1989], I, 81–88).

1844 September: Sketch of *Proof II* by means of a theory of genera and by a study of the arithmetic of cyclic cubic fields (theory of units, regulator, class number, L-series and genera in: allgemeine Untersuchungen über die Formen dritten Grades mit drei Variablen, welche der Kreistheilung ihre Entstehung verdanken, Crelle's Journal 28 (1844), 289–374 and 29 (1845), 19–53 (see [Ei-1989], I, 167–286)).

1845 February: Sketch of *Proof III* by means of elliptic functions related to the differential equation $dx = dy\,\sqrt{1 - x^3}$ (in French: Applications de l'Algèbre à l'Arithmétique transcendante, Crelle's Journal 29 (1845), 177–184 (see [Ei-1989], I, 291–298)).

3   Biquadratic Reciprocity Law

3.1   Euler (1707–1783)

In his *Tractatus* (1749; published 1849) (see [Eu-1911], I-5, 182) *Euler* has two tentative chapters on cubic and biquadratic residues. He gives conditions for 2, 3 and 5 to be biquadratic residues modulo p for a prime p of the form $p = 4n + 1$ (see [Eu-1911], I-5, 258–259, nos 456–457).

Among his discoveries are the following.

Theorem 3.1   *(1) Let $p = 4n + 1$ be a prime. Then 2 is a 4th power residue modulo p, i.e. $p\,|\,x^4 - 2$ if and only if there exist natural numbers $a, b \in \mathbf{N}$ with $p = a^2 + 64b^2$.*

(2) Let $p = 4n + 1$ be a prime. Then there exist $a, b \in \mathbf{N}$ unique with $p = a^2 + 4b^2$.

Furthermore

(a) If $p\,|\,x^4 - 2$, then a and b are characterized modulo 4.

(b) If $p\,|\,x^4 - 3$, then a and b are characterized modulo 6.

(c) If $p\,|\,x^4 - 5$, then a and b are characterized modulo 5.

A proof of the property (1) was published by Gauss in 1828 (see [Ga-1889], 511–533).

3.2   Gauss (1777–1855)

Gauss studied the Biquadratic Reciprocity Law together with the Cubic Reciprocity Law, but he published results concerning only the former, the proof of the latter being similar. We summarize Gauss' results on the Biquadratic Reciprocity Law as follows.

1805 Experimental tables on the cubic and biquadratic reciprocity laws.

1818 Publication of the fifth and sixth proof of the quadratic reciprocity law (see [Ga-1889, 496–510). These were the fruits of Gauss' tentatives to prove the cubic and biquadratic reciprocity law. In that article, Gauss says that he finally achieved a proof for these higher reciprocity laws and that he hopes to publish it soon. Publications with proofs on the subject appeared however only in 1828 and in 1832, but they concern only the two complementary laws of the biquadratic reciprocity, namely the determination of the biquadratic character of −1 and of 2.

1828 First paper on the Biquadratic Reciprocity Law, already submitted in 1825 (see [Ga-1889], 511-533). It contains the determination of the biquadratic character of 2 and −1.

1832 Second paper on the Biquadratic Reciprocity Law (see [Ga-1889], 534–586). It contains proofs of properties for the arithmetic of the Gaussian integers $A = \mathbf{Z}[i]$ and a statement of the Biquadratic Reciprocity Law.

A third paper was planned by Gauss that would have to contain all the proofs. These were however only found in his papers after his death.

The fundamental new idea introduced by Gauss is the realization that the Biquadratic Reciprocity Law can only be fully stated in the field $\mathbf{Q}(i)$ of the 4th roots of unity. For that purpose, Gauss had at first to develop the arithmetic in $\mathbf{Q}(i)$, i.e. he had to determine the primes, the units and the decomposition law in the ring of Gaussian integers $\mathbf{Z}[i] = A$. A further fundamental property proved by Gauss in his paper of 1832 is the following (see [Ga-1832], art.37).

Theorem 3.2 $\mathbf{Z}[i]$ *admits a Euclidean algorithm. Hence the Fundamental Theorem of Arithmetic holds true in* $\mathbf{Z}[i]$.

Next Gauss introduced the 4th power residue symbol (see [Ga-1832], art.61) following the definition of Legendre for the quadratic residue symbol.

Definition 3.3 *Let* $\pi = a + bi$ *be a prime in* $\mathbf{Z}[i]$, $\pi \neq 1 \pm i$. *Let* $\alpha \in \mathbf{Z}[i]$ *be a Gaussian integer with* $(\alpha, \pi) = 1$. *Then*

$$\left(\frac{\alpha}{\pi}\right)_4 := i^r \ with \ 0 \le r \le 3,$$

such that

$$\left(\frac{\alpha}{\pi}\right)_4 \equiv \alpha^{\frac{N(\pi)-1}{4}} (\text{modulo } \pi),$$

where $N(\pi)$ *denotes the norm of* π.

Then Gauss states the Biquadratic Reciprocity Law (see [Ga-1832], art.62 and art.67).

Theorem 3.4 *(1) $x^4 \equiv \alpha(\text{modulo } \pi)$ is solvable with $x \in \mathbf{Z}[i]$ for a given $\alpha \in$ $\mathbf{Z}[i]$ and a prime $\pi \in \mathbf{Z}[i]$ with $(\alpha, \pi) = 1$ if and only if $\left(\frac{\alpha}{\pi}\right)_4 = 1$.*

(2) If α and β are two primes in $\mathbf{Z}[i]$ with $\alpha, \beta \neq (1 \pm i)$ and $\alpha, \beta \equiv 1(\text{modulo} $ $(1 + i)^3), \alpha \neq \beta$, then

$$\left(\frac{\alpha}{\beta}\right)_4 = (-1)^{\frac{N(\alpha)-1}{4} \cdot \frac{N(\beta)-1}{4}} \left(\frac{\beta}{\alpha}\right)_4.$$

Remarks: (a) Gauss did not publish any proof of this theorem. He planned to do so in a third memoir. It was found later in his papers (see [Ba-1911], 39–45).

(b) Jacobi gave a proof of the Biquadratic Reciprocity Law in his lectures of 1836/37 on number theory at the University of Königsberg. Two copies of this lecture written by Rosenhain still exist. One is in Berlin and the other in Zürich in the Library of the ETH.

3.3   Eisenstein (1823-1852)

Eisenstein gave five different proofs for the Biquadratic Reciprocity Law. Three of them are based on the theory of *elliptic functions*. Eisenstein was influenced by remarks made by Gauss. Eisenstein's proofs can be summarized as follows.

Proof I: with Jacobi and Gauss sums and with the theory of cyclotomic fields (Crelle's Journal, vol.28, 1844; see [Ei-1989], I, 126–140: Lois de réciprocité).

Proof II: with Jacobi and Gauss sums (Crelle's Journal, Vol.28, 1844; see [Ei-1989], I, 141-163: Einfacher Beweis und Verallgemeinerung des Fundamentaltheorems für die biquadratischen Reste).

Proof III: by means of the elliptic (lemniscatic) function or more generally by means of curves symmetric to the axes with four congruent arcs that can be divided into p and q equal parts (Crelle's Journal, Vol.29, 1845; see [Ei-1989], I, 291–298: Applications de l'Algèbre à l'Arithmétique transcendante).

Proof IV: by means of the division of the lemniscate and by means of a generalization of Gauss' lemma (Crelle's Journal, Vol.30, 1846; see [Ei-1989], I, 299–324: Ableitung des biquadratischen Fundamentaltheorems aus der Theorie der Lemniscatenfunctionen, nebst Bemerkungen zu den Multiplications- und Transformationsformeln).

Proof V: by means of Eisenstein series (infinite products) (Crelle's Journal, Vol.35, 1847; see [Ei-1989], I, 357–478: Genaue Untersuchungen der unendlichen Doppelproducte, aus welchen die elliptischen Functionen als Quotienten zusammengesetzt sind).

4 Higher Reciprocity Laws

4.1 Survey

1837 Jacobi: On m-th power residues for $m = 5, 8$.

1845 Eisenstein: On m-th power residues for $m = 7, 16$.

1845 Kummer: Introduction of ideal numbers.

1845 Kummer: Definition of the l-th power residue symbol for an odd prime l and a prime ideal $\mathcal{P}$ and a rational number m: $\left(\frac{m}{\mathcal{P}}\right)_l$.

1850 Eisenstein: Rational reciprocity law for p-th power residues for a prime $p \neq 2$, a rational number a and an integer α in the p-th cyclotomic field.

1859 Kummer: General reciprocity in the ring $\mathbf{Z}[\zeta_p]$ of the p-th cyclotomic field $\mathbf{Q}(\zeta_p), \zeta_p = e^{2\pi i/p}$ for a regular prime p.

1902–13 Furtwängler: Reciprocity in fields K containing the p-th cyclotomic field: $\mathbf{Q}(\zeta_p) \subseteq K$.

1927 Fueter: Reciprocity for imaginary quadratic fields $K = \mathbf{Q}(\sqrt{-d})$.

1922 Takagi: Reciprocity in cyclotomic fields.

1927 Artin: General reciprocity in abelian extension fields.

1924–28 Hasse: Complementary reciprocity laws in abelian extension fields and generalization of Hilbert's Reciprocity Law.

4.2 Eisenstein's Rational Reciprocity Law

In 1850 *Eisenstein* found a quite general higher p-th reciprocity law for rational numbers in the p-th cyclotomic field for a prime $p \neq 2$ (see [Ei-1989], II, 712–721 : Beweis der allgemeinsten Reciprocitätsgesetze zwischen reellen und complexen Zahlen. Bericht der Preuss. Akad. der Wiss. zu Berlin (1850)).

Definition 4.1 *Let $K = \mathbf{Q}(\zeta)$ be the m-th cyclotomic field, where $\zeta = \zeta_m = e^{2\pi i/m}, m \in \mathbf{N}$. Let $\mathcal{Q}$ be an integral prime ideal in K with $(\mathcal{Q}, m) = 1$ and let q be the norm of $\mathcal{Q}$, i.e. $N(\mathcal{Q}) = [A : \mathcal{Q}] = q$, where $A = \mathbf{Z}[\zeta_m]$ stands for the integers in K.*

Then $q \equiv 1 (\text{modulo } m)$ and $q = p^s$ for a prime p.
Let $\alpha \in A = \mathbf{Z}[\zeta_m]$ with $(\alpha, \mathcal{Q}) = 1$. Then we define

$$\left(\frac{\alpha}{\mathcal{Q}}\right)_m := \zeta_m^i$$

such that

$$\left(\frac{\alpha}{\mathcal{Q}}\right)_m \equiv \alpha^{\frac{q-1}{m}} \,(\text{modulo } \mathcal{Q})$$

with $0 \le i \le m - 1$.

Definition 4.2 *Let* $\mathcal{A} = \mathcal{Q}_1 \cdots \mathcal{Q}_t$ *be the decomposition of an integral ideal* $\mathcal{A} \subseteq A = \mathbf{Z}[\zeta_m]$ *into prime ideals* $\mathcal{Q}_i$ *and let* $\alpha \in A$ *with* $(\alpha, \mathcal{A}) = 1$. *Then we define*

$$\left(\frac{\alpha}{\mathcal{A}}\right)_m := \left(\frac{\alpha}{\mathcal{Q}_1}\right)_m \cdots \left(\frac{\alpha}{\mathcal{Q}_t}\right)_m$$

Definition 4.3 *Let* $m = p$ *be a prime number and* $\alpha \in A = \mathbf{Z}[\zeta_p]$. α *is called* p-primary *if*

(i) $(\alpha, p) = 1$.

(ii) $\alpha \equiv r(\text{modulo } (1 - \zeta_p)^2)$ *for some* $r \in \mathbf{Z}$.

Eisenstein's Rational Reciprocity Law now reads as follows.

Theorem 4.4 *Let* p *be a prime,* $p \ne 2$. *Let further* $a \in \mathbf{Z}$ *be a rational integer with* $(a, p) = 1$ *and let* $\alpha \in A = \mathbf{Z}[\zeta_p]$ *be a* p-primary *algebraic integer. Then*

$$\left(\frac{\alpha}{a}\right)_p = \left(\frac{a}{\alpha}\right)_p$$

For a modern presentation, see [IR-1990].

References

[Ba-1911]   Paul *Bachmann*: Über Gauss' zahlentheoretische Arbeiten. In 'Materi-
            alien für eine wissenschaftliche Biographie von Gauss. Gesammelt von
            F. Klein und M. Brendel'. Leipzig 1911.

[Col-1977] Mary Joan *Collison*: The Origins of the Cubic and Biquadratic Reci-
            procity Laws. Archive Hist. Exact Sciences 17 (1977), 63–69.

[De-1894]   Richard *Dedekind*: Über die Theorie der ganzen algebraischen Zahlen.
            Supplement XI of Dirichlets Vorlesungen über Zahlentheorie. Braun-
            schweig (Vieweg) 1894.

[Dio-1893] *Diophanti* Alexandrini Opera cum omnia Graecis commentariis, ed.
            Paulus Tannery. Leipzig (Teubner) 1893–1895, 2 vol.

[Ei-1989]   Gotthold *Eisenstein*: Mathematische Werke. New York (Chelsea) 1989,
            2nd edition (first edition 1975).

[Eu-1911]   Leonardi *Euleri*: Opera Omnia, ..., 1911– , (Teubner, Birkhäuser)

[Fr-1979]   Günther *Frei*: On the Development of the Genus of Quadratic Forms.
            Ann. Sci. Math. Québec 3 (1979), 5–62.

[Fr-1985]   Günther *Frei*: Leonhard Euler's convenient numbers. Math. Intelli-
            gencer 3 (1985), 55–58 and 64.

[Ga-1801]   Carl Friedrich *Gauss* : Disquisitiones Arithmeticae. Leipzig 1801. Also
            in Vol. 1 of Gauss, Werke, Göttingen 1870 and in 'Untersuchungen
            über höhere Arithmetik', ed. H. Maser, Berlin 1889 (see [Ga-1889]) or
            the new English edition by Waterhouse, Springer 1989.

[Ga-1832]   Carl Friedrich *Gauss* : Theorie der biquadratischen Reste. Zweite Ab-
            handlung. In 'Untersuchungen über höhere Arithmetik', ed. H. Maser,
            Berlin 1889, 534–586 (see also [Ga-1889]).

[Ga-1870]   Carl Friedrich *Gauss* : Werke. 12 vol. Göttingen 1870–1929.

[Ga-1889]   Carl Friedrich *Gauss* : Untersuchungen über höhere Arithmetik, ed. H.
            Maser, Berlin 1889. Reprints New York (Chelsea) 1965, 1981.

[Gr-1984]   J.J. *Gray*: A Commentary on Gauss's Mathematical Diary, 1796–1814
            with an English translation. Expositiones Mathematicae 2 (1984), 97–
            130.

[IR-1990]   Kenneth *Ireland*, Michael *Rosen*: A Classical Introduction to Modern
            Number Theory. Springer 1990. 2nd edition. First edition 1981.

[Ja-1881]   C.G.J. *Jacobi*'s Gesammelte Werke. ed. K. Weierstrass. 8 vol. Berlin
            1881–1891. Reprint New York (Chelsea) 1969.

[Kro-1895] Leopold *Kronecker*'s Werke. ed. K. Hensel. Leipzig 1895–1931. 5 vol.

[Le-1798]   A.M. *Legendre*: Essai sur la Théorie des Nombres. Paris (Duprat) 1798.

[Wei-1983] André *Weil*: Number Theory. An approach through history
            ... Birkhäuser 1983.

This paper was supported by a grant from the Canadian Research Council.
I would also like to thank François Grondin for putting my manuscript into TEX .

Three Aspects of the Theory of Complex Multiplication

Takase Masahito

Introduction

In the letter addressed to Dedekind dated March 15, 1880, Kronecker wrote:

> — the Abelian equations with square roots of rational numbers are exhausted by the transformation equations of elliptic functions with singular modules, just as the Abelian equations with integral coefficients are exhausted by the cyclotomic equations. ([Kronecker 6] p. 455)

This is a quite impressive conjecture as to the construction of Abelian equations over an imaginary quadratic number fields, which Kronecker called "my favorite youthful dream" (ibid.). The aim of the theory of complex multiplication is to solve Kronecker's youthful dream; however, it is not the history of formation of the theory of complex multiplication but Kronecker's youthful dream *itself* that I would like to consider in the present paper. I will consider Kronecker's youthful dream from three aspects: its history, its theoretical character, and its essential meaning in mathematics.

First of all, I will refer to the history and theoretical character of the youthful dream (Part I). Abel modeled the division theory of elliptic functions after the division theory of a circle by Gauss and discovered the concept of an Abelian equation through the investigation of the algebraically solvable conditions of the division equations of elliptic functions. This discovery is the starting point of the path to Kronecker's youthful dream. If we follow the path from Abel to Kronecker, then it will soon become clear that Kronecker's youthful dream has the theoretical character to be called the *inverse problem of Abel*.

Next, I will consider what the solution of Kronecker's youthful dream really means, that is, its essential meaning in mathematics (Part III). From the purely theoretical viewpoint, Abel's theory of Abelian equations is only one origin of the youthful dream; however, the essential meaning of the youthful dream can be found not in Abel's theory but in Eisenstein's theory, the latter of which gives us a proof of the biquadratic reciprocity law in the Gaussian number field, based on the division theory of the lemniscatic function. Today's theory of class fields teaches us that, if we can know all the relative Abelian number fields over the Gaussian number field, then we will be able to obtain a proof of the biquadratic reciprocity law; so we must recall here another statement of Kronecker's, made before his youthful dream, by which we can know that every relative Abelian number field over the Gaussian number field is generated by a special value of the lemniscatic function. This statement is very important, for it clearly shows that Kronecker was deeply interested in the reason why Eisenstein's theory succeeded. Kronecker wanted further to know why the lemniscatic function was enough to generate all the relative Abelian number fields over the Gaussian number field and he proposed the youthful dream with the aim of answering this question; therefore we will be able to say that the youthful dream shows why there can be a transcendental proof of the biquadratic reciprocity law. This is the essential meaning of Kronecker's youthful dream, or we can say, if we like, this is the essential meaning of the theory of complex multiplication.

Thus there is an intimate internal relation between the youthful dream and the reciprocity law; so, before the consideration of the essential meaning of the youthful dream, I would like to review the history of the reciprocity law from Gauss to Artin and observe how the difficulties that Kummer had faced were overcome by Hilbert's theory of class fields(Part II). The class field theory is the ultimate key to the investigation of the theory of reciprocity law and its prototype was given by Kronecker's youthful dream. This situation is never accidental. I firmly believe it will back up my opinion that the essential meaning of the youthful dream can be found in Eisenstein's theory.

I   The Way to Kronecker's Youthful Dream —From Abel to Kronecker—

1   Discovery of Abelian Equations (1) Gauss's Message

In 1801, Gauss developed the division theory of a circle in his great work *Disquisitiones Arithmeticae*, Chap. VII. At the beginning of this chapter he left a mysterious message to his successors. This message is the birthplace of the concept of an Abelian equation and so I would like to begin by reviewing it. Gauss wrote:

> The principles of the theory which we are going to explain actually extend much farther than we will indicate. For they can be applied not only to circular functions but just as well to other transcendental functions, e.g. to those which depend on the integral $\int [1/\sqrt{(1-x^4)}]dx$ and also to various types of congruences. ([Gauss 1] p. 407)

The inverse function $x = \phi\alpha$ of the lemniscatic integral

$$\alpha = \int_0^x \frac{dx}{\sqrt{1-x^4}}$$

will be called the lemniscatic function; so, according to Gauss, there will be the same theory for the lemniscatic function as for circular functions on the basis of common principles. As far as I know, there are only two young mathematicians, Abel and Eisenstein, who furthered the essence of Gauss's message. Let's first survey Abel's investigation.

When we observe Gauss's division theory of a circle from the viewpoint of the theory of algebraic equations, we can find there two important things; one is the establishment of the fact that every division equation of a circle is algebraically solvable, and the other is the complete knowledge of all cases in which a division equation of a circle is solvable using just square roots. This is the algebraic aspect of Gauss's theory that Abel succeeded. In a letter addressed to Holmboe written in December, 1826, Abel vividly wrote:

> I found that with a ruler and a compass we can divide the lemniscate into $2n + 1$ equal parts, when this number is prime. The division depends on an equation of degree $(2^n + 1)^2 - 1$. But I found the complete solution by the help of square roots. This made me penetrate at the same time the mystery which enveloped Gauss's theory on the division of a circle. ([Abel 5] p. 261)

What Abel stated here is nothing but the division theory of a lemniscate; but I think such a statement will be possible only when it is supported with a general theory. Abel lifted the veil of mystery which covered Gauss's theory and saw the essence of algebraic solvability. *The ultimate principle which controls the algebraic solvability of an algebraic equation is some specific relation among the roots.* On this basic recognition, Abel first discovered the concept of a cyclic equation and then that of an Abelian equation (See [Abel 4]); thus he succeeded in throwing light on one of the principles of Gauss's division theory of a circle.

2 Discovery of Abelian Equations (2) Division Theory of a Lemniscate

The division theory of a lemniscate by Abel was developed in his monumental paper *Recherches sur les fonctions elliptiques* (See [Abel 1]). Let $\phi\alpha$ again be the lemniscatic function and $4\nu + 1$ a prime number. Following Abel, I would like to review the method of dividing the periods of $\phi\alpha$ into $4\nu + 1$ equal parts in order to see how Abel discovered the concept of an Abelian equation (See [Abel 1] pp. 352–362). By a theorem of Fermat's we can express $4\nu + 1$ as a sum of two squared integers in this way

$$4\nu + 1 = \alpha^2 + \beta^2;$$

and so we can further express it as a product of two odd prime Gaussian integers in the form

$$\alpha^2 + \beta^2 = 4\nu + 1 = (\alpha + \beta i)(\alpha - \beta i).$$

In general, let $\alpha + \beta i$ be an arbitrary Gaussian integer; then we can make the division equation of the periods of the lemniscatic function concerning $\alpha + \beta i$ by using two fundamental properties of the lemniscatic function. One is that the lemniscatic function satisfies the addition theorem; the other is that it has complex multiplication in the Gaussian number field. The latter means the complex multipliers of the lemniscatic function are all in the Gaussian number field. Realistically speaking, this means we have the functional equation

$$\phi(\alpha i) = i \cdot \phi\alpha. \tag{1}$$

By using these two properties, we can express $y = \phi(\alpha + \beta i)\delta$ as a rational function of $x = \phi\delta$ in the form

$$y = \frac{T}{S}, \tag{2}$$

where T and S denote polynomials of $x^4 = (\phi\delta)^4$ over the Gaussian number field. Then the equation

$$T = 0 \tag{3}$$

is the division equation of the periods of the lemniscatic function concerning the Gaussian integer $\alpha + \beta i$. If I may say so, this is an imaginary division equation. Under these situations, Abel showed that when $\alpha + \beta i$ is an odd prime number, this equation is cyclic just like an equation defining the division of a circle into an odd prime number of equal parts and so algebraically solvable over the Gaussian number field. This is the main point of Abel's theory.

Now back to the former case. According to Abel, the two imaginary division equations of the periods of the lemniscatic function concerning $\alpha + \beta i$ and $\alpha - \beta i$ are both algebraically solvable. By this, we can see that the equation defining the division of the periods of the lemniscatic function into $4\nu + 1$ equal parts is also algebraically solvable. This is a rough sketch of the division theory of a lemniscate by Abel. The idea of introducing an imaginary division equation is very creative and I want to emphasize here that this is the first step towards the discovery of the concept of an Abelian equation.

3   Singular Modular Equations

Abel's creative idea of an imaginary division equation is supported with the functional equation $\phi(\alpha i) = i \cdot \phi\alpha$. We will be able to see a sign of the theory of complex multiplication in this quite suggestive situation. It seems to me that Abel wanted to know a mathematical background of the equation $\phi(\alpha i) = i \cdot \phi\alpha$ and tried to find it in the transformation theory of elliptic integrals. Let's consider the differential equation

$$\frac{dy}{\sqrt{1-y^4}} = i\frac{dx}{\sqrt{1-x^4}}. \tag{4}$$

Then we have the equation

$$\int_0^y \frac{dy}{\sqrt{1-y^4}} = i \int_0^x \frac{dx}{\sqrt{1-x^4}} = i\alpha; \tag{5}$$

so, if we move to the inverse function of the lemniscatic integral, we have $y = \phi(\alpha i)$ and $x = \phi\alpha$; thus the functional equation $\phi(\alpha i) = i \cdot \phi\alpha$ is equal to an algebraic equation $y = ix$, which is an algebraic integral of (4). Abel started from this recognition. He proposed a general differential equation of the form

$$\frac{dy}{\sqrt{(1-y^2)(1+\mu y^2)}} = a \frac{dx}{\sqrt{(1-x^2)(1+\mu x^2)}}, \tag{6}$$

where a and μ denote constants, and he tried looking for the algebraically integrable conditions of this differential equation. It is clear the conditions depend on two constants a and μ, multiplier a and module μ. Here let's consider, following Abel's original idea, the inverse function of the elliptic integral of the first kind

$$\theta = \int_0^x \frac{dx}{\sqrt{(1-x^2)(1+\mu x^2)}}$$

and let's denote it by $x = \lambda\theta$; this is an elliptic function with module μ and the conditions we are looking for are the same as those for the elliptic function $\lambda\theta$ to have a multiplier a. They will be quite simple when a is real, but rather complicated when a is imaginary; therefore it is necessary to distinguish the two cases. Abel clearly understood the very core of the subject and so we can know that he had grasped the concept of complex multiplication. Abel obtained the two fundamental theorems for which he had no proof. He wrote:

> **Theorem I.** If a is real, and if the differential equation is algebraically integrable, then a must necessarily be a rational number.
>
> **Theorem II.** If a is *imaginary*, and if the differential equation is *algebraically* integrable, then a must necessarily be of the form $m \pm \sqrt{-1} \cdot \sqrt{n}$, where m and n are rational numbers. In this case the quantity μ is not arbitrary; it must satisfy an equation which has infinitely many real and imaginary roots. Each value of μ satisfies the question. ([Abel 2] p. 377)

We would like to pay attention in particular to the situation in Theorem II. On the premises that the elliptic function $\lambda\theta$ has complex multiplication[1], Theorem II says that every imaginary multiplier of $\lambda\theta$ must be an imaginary quadratic number and that the module of $\lambda\theta$ cannot be arbitrary. *Thus a module of an elliptic function with complex multiplication is not arbitrary;* so Kronecker later called it a *singular*

---

[1] If there exists an algebraic relation between two elliptic functions $\lambda(a\theta)$ and $\lambda\theta$, then we say that $\lambda\theta$ has a multiplier a. If in particular $\lambda\theta$ has an imaginary multiplier, then we say that $\lambda\theta$ has *complex multiplication*.

module. Theorem II further says that a singular module satisfies "an equation which has infinitely many real and imaginary roots." It is difficult to understand the precise meaning of this vague statement, but it seems to me Abel expresses here the recognition that a singular module is an algebraic quantity. In fact, using the transformation theory of elliptic integrals of lower degree, Abel presented the two ways of calculating a singular module and in both cases a singular module satisfies an algebraic equation which can be easily solves by radicals. Nevertheless Abel was not sure at this stage whether a singular module could always be expressed by radicals or not. Abel wrote:

> In the two cases we just considered, it was not difficult to find the value of the quantity e; but when the value of [degree of transformation] n is bigger, we will arrive at algebraic equations which might not be algebraically solved. ([Abel 1] p. 383)

Abel's hesitation completely disappeared in the next paper *Solution d'un problème général concernant la transformation des fonctions elliptiques* and Abel came to definitely say that a singular module is always expressed by radicals. He wrote:

> There is a remarkable case of the general problem; it is the case where we demand all the possible solutions of the equation
>
> $$\frac{dy}{\sqrt{(1 - c^2 y^2)(1 - e^2 y^2)}} = a \frac{dx}{\sqrt{(1 - c^2 x^2)(1 - e^2 x^2)}}.$$
>
> We will have in this regard the next theorem:
> If the preceding equation admits an algebraic solution with respect to x and y, where y is rational with respect to x or not, then the constant quantity a should necessarily have the form
>
> $$\mu' + \sqrt{-\mu},$$
>
> where μ' and μ denote two rational numbers and the latter is essentially *positive*. If we assign such a value to a, then we will be able to find infinitely many different values for e and c, which make the problem possible. All these values can be expressed by *radicals*. ([Abel 2] pp. 425–426)

According to Abel, a singular module always satisfies an irreducible algebraic equation over a domain of coefficients; so we can say that we have a general notion of a *singular modular equation*. Then a singular modular equation is always algebraically solvable. This is also Abel's discovery. Abel's indepth studies tell us almost perfect information about complex multiplication of elliptic functions; but they are slightly insufficient, for we don't know first the relation between a complex multiplier and a singular module, second the domain of coefficients of a singular modular equation and lastly the reason why a singular modular equation can be algebraically solved. It is Kronecker who succeeded Abel and made our last three questions clear. Let K be the imaginary quadratic number field $Q(a)$ generated by a complex multiplier a. Kronecker discovered that we have a singular modular equation over $K = Q(a)$ and that it is an Abelian equation (See [Kronecker 2]). Thus he completed Abel's studies and gave us a new kind of Abelian equation.

4   Kronecker's Youthful Dream

Now we are prepared to explain Kronecker's youthful dream . Thanks to Gauss, Abel and Kronecker, we know three kinds of Abelian equations in all, division equations of a circle over the rational number field,imaginary division equations of the periods of the lemniscatic function over the Gaussian number field and singular modular equations over an imaginary quadratic number field. Incidentally, as Abel suggested in a letter addressed to Crelle dated March 14,1826 (See [Abel 6] p. 266), every root of an algebraically solvable equation has a specific form of expression by radicals; so we can say that the three above-mentioned examples of Abelian equations are not accidental.This is a very subtle sign of something new ; with the intention of making it clear, Kronecker developed his own theory of algebraic equations under the deep influence of Abel's posthumous manuscript *Sur la résolution algébrique des équations* (See [Abel 4]) and tried to throw light on "the nature of algebraically solvable equations *themselves*" (See [Kronecker 1] p. 3).

Kronecker first discovered that "the roots of *any* Abelian equation with integral coefficients can be expressed as rational functions of roots of unity" (See [Kronecker 1] p. 10) This is the so-called Kronecker's theorem which says that an Abelian equation over the rational number field is nothing else than a division equation of a circle ,or that ,from a viewpoint of the structure of number fields, every Abelian number field is a cyclotomic field, or that, from a viewpoint of the way of construction of number fields, every Abelian number field can be generated by a special value of the exponential function. Next to Kronecker's theorem, Kronecker told the inverse situation of the division theory of a lemniscate by Abel and went further away to the most general situation which includes as a special case Kronecker's youthful dream. He wrote:

> There also exists a similar relation between the roots of an Abelian equation, whose coefficients include only integral complex numbers of the form $a + b\sqrt{-1}$, and the roots of the equation, which appears in case of the division of a lemniscate; and lastly we can further generalize the above result [Kronecker's theorem] for all the Abelian equations, whose coefficients include determined algebraic numerical irrationalities. ([Kronecker 1] p. 11)

The first half of this statement says that every Abelian equation over the Gaussian number field is an imaginary division equation of the periods of the lemniscatic function or that every relative Abelian number field over the Gaussian number field is generated by a special value of the lemniscatic function. The second half of the statement is not limited just to Kronecker's youthful dream; we will be able to consider it as clear evidence that Kronecker had already seen the the situation of Hilbert's twelfth problem (See [Hilbert 4] pp. 311–313). Here I would like to recall Kronecker's youthful dream. Let K be an arbitrary imaginary quadratic number field. As stated in the Introduction, Kronecker asked the method of constructing all the Abelian equations over K and conjectured that they might be exhausted by "transformation equation of elliptic functions with singular modules" (See [Kronecker 6] p. 455). Kronecker's conjecture says that *all the Abelian equations over K will be exhausted by two kinds of transformation equations of elliptic functions with complex multiplication in K, imaginary division equations of the periods concerning complex numbers of K and singular modular equations.*[2] This is the objective content of Kronecker's youthful dream or, if we like to say so, of the theory of complex multiplication. Kronecker found the key to this theory in Abel's two discoveries and questioned the inverse situations of what Abel observed in order to explain something universal hidden in Abel's discoveries; so, in my opinion, Kronecker's youthful dream should be called the *inverse problem of Abel*. This is the *theoretical character of the theory of complex multiplication*.

5   Origin of Class Field Theory

There is another remarkable statement of Kronecker's in the same letter addressed to Dedekind as I have often referred to. He said as follows:

> ... but for the proof of the theorem [Kronecker's youthful dream], which I have imagined and sought for a long time, an entirely different — I might say — philosophical insight into the nature of the remarkable equations for singular modules [singular modular equations] was necessary for me, and by means of it, it must be made clear *why* they give the just sufficient irrationalities which — according to Kummer's way of notation that I also used in 1857 in the report[3] — really expresses the ideal numbers for $a + b\sqrt{-D}$. ([Kronecker 6] p. 456)

This statement is extremely important, for we can find there the origin of class field theory. Let K be an imaginary quadratic number field and L a number field generated by a singular module of an elliptic function with complex multiplication in K. According to Abel and Kronecker, we have already found that L is a relative Abelian number field over K; and now Kronecker wondered at the fact that every

---

[2]Let f be an elliptic function. If the complex multipliers of f are in K, then we will say that f has complex multiplication in K. The word "transformation equation of f" is not so clear; but I think it should be understood to be various equations which accompany with transformations of f. As to an imaginary division equation of the periods of elliptic functions with complex multiplication, see [Abel 7] pp. 310–313, pp. 316–318.

[3]See [Kronecker 2].

ideal in K becomes principal in L. This is a very mysterious ability of singular modules which Kronecker tried to understand from a philosophical viewpoint in order to get the key to the proof of his youthful dream. Furthermore, Kronecker told a remarkable fact elsewhere that L is relatively non-ramified over K (See [Kronecker 3] p. 213). Thus *L is a relative Abelian and non-ramified number field over K in which every ideal in K becomes principal.* That is a prototype of Hilbert's class field, that is, absolute class field.

After the discovery of the absolute class field of an imaginary quadratic number field, Kronecker's path in mathematics branched in two; one is the path to the solution of the youthful dream and Kronecker devoted to it a long and massive series of papers *Zur Theorie der elliptischen Functionen I–XXII* (See [Kronecker 5]) in his last years; the other is the path to the construction of class field theory; it is with this intention that he developed his own theory of algebraic numbers in his big paper *Grundzüge einer arithmetischen Theorie der algebraischen Grössen* (See [Kronecker 4]). Thus began the class field theory. The discovery of the remarkable arithmetical properties of a singular module; that is the origin of the class field theory.

II   Construction of Class Field Theory — From Kummer to Hilbert

1   Kummer's Theory of Numbers

In 1825 and 1831 Gauss published two fundamental papers on the biquadratic reciprocity law (See [Gauss 4,5]), and a new branch of number theory opened: investigation of the reciprocity law of higher degrees. Many mathematicians, since then, such as Jacobi, Dirichlet, Eisenstein and Kummer etc., had tried to establish the reciprocity law of higher degrees. Amongst them it was Kummer who obtained the most excellent results. We must first review Kummer's studies on the reciprocity law in order to know how Hilbert was motivated to construct the class field theory, for Hilbert had extracted the notion of class field from Kronecker's arithmetical world and applied it to the reciprocity law with the intention of overcoming the difficulties that Kummer had faced.

Now let m be an add prime number and ζ_m an m-th root of unity; we denote by $K_m = Q(\zeta_m)$ the cyclotomic field generated by ζ_m. Then, according to Kummer (See [Kummer 2]), we can formulate the reciprocity law of degree m in K_m and give proof to it on the condition that m is regular, or, in other words, K_m is regular, where the condition means that the class number of K_m is not divisible by m.[4] There are two remarkable problems in this formulation. Kummer formulated

---

[4]From this way of defining a regular cyclotomic field, we can immediately see that Kummer's theory is founded on a deep knowledge of class numbers of cyclotomic fields. The class number formula for cyclotomic fields is especially remarkable; Kummer succeeded in establishing it under the influence of Dirichlet's research on the class number formula for quadratic number fields (See

his reciprocity law of higher degrees not in the rational number field but in a cyclotomic field. Our first problem, then, is why did he have to leave the rational number field. Next, a cyclotomic field that Kummer presented as a stage of the reciprocity law was not arbitrary; he imposed on it the condition to be regular. Our second problem is to find the mathematical meaning of this condition.

2   First Problem in Kummer's Theory of Numbers

We will be able to find the key to our first problem in Gauss's theory of the biquadratic reciprocity law. In the paper "*Theoria residuorum biquadraticorum. Commentatio secunda*" (See [Gauss 5]), Gauss wrote as to the extension of a domain of numbers:

> Already in 1805 we had begun to consider this theme and we soon came up to the conviction that we must seek the natural well of a general theory in an *extension of the fields* of arithmetic,... .

> That is to say, in the problems treated until now the higher arithmetic has to do only with integral real numbers.The theorems on the biquadratic residues appear in their perfect simplicity and natural beauty only when the field of arithmetic is extended to the *imaginary* numbers, so that the numbers of the form $a + bi$ make the object of arithmetic without restriction, where, as usual, i denotes the imaginary quantity $\sqrt{-1}$ and a, b denote all the indefinite integral real numbers between $-\infty$ and $+\infty$. We will name such numbers *integral complex numbers*... . ([Gauss 5] p. 102)

Today, an "integral complex number" is called a Gaussian integer; so, according to Gauss, if we want to get the biquadratic reciprocity law "in its perfect simplicity and natural beauty", we will have to leave the rational number field for the Gaussian number field. This statement is quite attractive but difficult to understand; however, Gauss put footnotes to it, which will help us to understand what he meant.

> Incidentally, we will remark here, at least, that such an extension of the field is specially adapted for the theory of biquadratic residues. The theory of cubic residues must be founded in a similar way on the consideration of numbers of the form $a + bh$, where h is an imaginary root of the equation $h^3 - 1 = 0$, for example $h = -\frac{1}{2} + \sqrt{\frac{3}{4}}i$, and similarly the theory of residues of higher powers needs other imaginary quantities. ([Gauss 5] p. 102)

[Kummer 1]).

This short message seems to be a small door to the mysterious world of Gauss's theory of numbers. The biquadratic reciprocity law needs the domain of Gaussian integers and the cubic reciprocity law needs the domain of numbers of the form $a + bh$; more generally we will have to introduce a suitable imaginary quantity and set up a proper domain of numbers for the reciprocity law of higher degrees. Thus *for any natural number n there exists a natural domain of existence of the reciprocity law of degree n; that is, an algebraic number field containing an n-th primitive root of unity.* Kummer exactly grasped what Gauss wanted to say and tried to establish the reciprocity law of odd prime degrees in a cyclotomic field.

Now we must go a step further and ask why the reciprocity law of higher degrees necessarily demands the extension of a domain of numbers. This question is the very core of our first problem and it is Gauss who indicated the direction in which we should proceed to answer it; we will soon find a very essential notion in number theory, the notion of prime ideal. The natural object of the reciprocity law of higher degrees is not arbitrary but *something prime* in a suitable sense. *Gauss noticed through the experience of seeking for the biquadratic reciprocity law that the "something prime" cannot be a usual odd prime rational integer and finally saw the fundamental principle that, in order to grasp the "something prime", we have to extend the domain of numbers in a proper way.* That is the essence of the extension of a domain of numbers.

3   Second Problem in Kummer's Theory of Numbers

We will be able to find the key to our second problem in the fact that Gauss gave the seven kinds of proofs to the quadratic reciprocity law; so we must return to Gauss again. In the paper *Theorematis fundamentalis in doctrina de residuis quadraticis demonstrationes et ampliationes novae* (See [Gauss 3]), Gauss expressed his true sentiments on the reason why he gave so many proofs to the quadratic reciprocity law. He wrote:

> ... when I began to do research on the theory of cubic and biquadratic residues in 1805, nearly the same destiny befell me, as that I once met in the theory of quadratic residues. Admittedly, the theorems [concerning the biquadratic residues], which completely settle these problems and in which there exists a mysterious analogy to the theorems with regard to the quadratic residues, were inductively found without any difficulty as soon as they were suitably sought, but all the attempts to arrive at their proofs, which are perfect in all aspects, remained barren for a long time. That was the impulse under which I tried hard to add other proofs to the already known proofs on the quadratic residues, in the hope that one of the many different methods was able to contribute to the elucidation of theme of the same class. ([Gauss 3] p. 50)

Thus it is clear that Gauss aimed at looking for a special kind of proof of the quadratic reciprocity law, which is applicable for the biquadratic reciprocity law.

Gauss's way of investigation became a guiding principle for studying the reciprocity law of higher degrees. For example, Dirichlet developed the theory of biquadratic forms and attempted to go deeper into the theory of biquadratic residues (See [Dirichlet]); the model of his study is Gauss's second proof of the quadratic reciprocity law, which is based on the theory of quadratic forms. Now Gauss's fourth, sixth and seventh proofs of the quadratic reciprocity law are closely connected with the division theory of a circle; so Kummer tried to generalize it with the aim of obtaining a proof of the reciprocity law of higher degrees. However, he said:

> ... and I finally had to abandon the way of generalization of the division of a circle, which I had followed until that time, and seek for another means. I turned my attention to Gauss's second proof of the fundamental theorem [the quadratic reciprocity law], which is based on the theory of quadratic forms. It seemed to me that, though its method remained limited to the quadratic residues until that time, this proof had in its principle the character of generality, which held out hope that it can be applied to the investigation of the residues of higher degrees; my expectation has been really satisfied. ([Kummer 2] pp. 709–710)

This statement definitely shows Kummer's determination to generalize the genus theory of quadratic forms, which is the essence of Gauss's second proof of the quadratic reciprocity law. His plan succeeded and he was able to establish the reciprocity law of odd prime degrees in a regular cyclotomic field. Now it is easy to understand what the condition "regular" means. Let K_m be a cyclotomic field. We know that, if it is regular, then every relative Kummer number field of degree m over K_m relatively ramified[5] and so there always exists a non-trivial relative discriminant which enable us to develop the genus theory in K_m.

However, if K_m is not regular, then in general there is a non-ramified relative Kummer number field over K_m and so we will be no longer able to consider the genus theory in K_m. Thus *the condition "regular" explicitly shows the very limit of application of the generalized genus theory*. This is the mathematical meaning of the condition "regular".

4 Hilbert's Theory of Numbers and Construction of Class Field Theory

Hilbert's theory of numbers is mainly composed of a series of three papers:

(A) *Zahlbericht: Die Theorie der algebraischen Zahlkörper* (See [Hilbert 1]),

(B) *Über die Theorie des relativquadratischen Zahlkörpers* (See [Hilbert 2]),

(C) *Über der ralativ-Abelschen Zahlkörper* (See [Hilbert 3]).

---

[5]Let k be an algebraic number field and l an odd prime number; we can show more generally that, if the class number of k is not divisible by l, then every relative cyclic number field of degree l over k is relatively ramified (See [Hilbert 1], Theorem 94, p. 155).

As to Hilbert's famous work *Zahlbericht*, Furtwängler, one of the Hilbert's successors in number theory, briefly wrote in the preface of his paper *"Über die Reziprozitätsgesetze zwischen l^{ten} Potenzresten in algebraischen Zahlkörpern, wenn l eine ungerade Primzahl bedeutet"* (See [Furtwängler 1]):

> ... [Hilbert] recognized the general idea with outstanding insight, which lies in Kummer's special calculations, and through their restructuring created auxiliary means, with which the most general reciprocity law among arbitrary power residues in arbitrary algebraic number fields will probably be approachable. ([Furtwängler 1] p. 1)

It seems to me that in this short statement Furtwängler told the essence not only of *Zahlbericht* but of Hilbert's whole arithmetical world. In fact, in the preface of the second paper (B), Hilbert himself wrote:

> The methods, which I have used in the following for the investigation of relative quadratic fields, are by a suitable generalization also useful with the same results in the theory of relative Abelian fields of arbitrary relative degree, and then lead especially to the general reciprocity law for arbitrary higher power residues in an arbitrary algebraic number domain. ([Hilbert 2] p. 370)

Hilbert clearly saw the general idea of Kummer's genus theory in regular cyclotomic fields and opened the way for moving the reciprocity law from cyclotomic fields to arbitrary algebraic number fields. To establish the reciprocity law of higher degrees in an arbitrary algebraic number field with a suitable root of unity is the guiding principle of Hilbert's number theory.

The subject of the paper (B) is to establish the quadratic reciprocity law in an arbitrary algebraic number field k. From the purely theoretical viewpoint, an algebraic number field which should be a natural domain of existence of the quadratic reciprocity law is completely arbitrary, for the square root of unity, $+1$ and -1, are contained in every algebraic number field; however, Hilbert's method of proof is based on the genus theory in a relative quadratic number field and so the base field k will be subjected to restriction which are caused by the application limit of the genus theory. In fact, Hilbert found the restrictions and expressed them as follows:

> 1. The field k of m-th degrees as well as all the conjugate fields $k', \ldots, k^{(m-1)}$ are imaginary;
> 2. The number h of the ideal classes in the field k is odd. (See [Hilbert 2] p. 393)

These two conditions guarantee that every relative quadratic number field over k is relatively ramified (See [Hilbert 2], Theorem 27, p. 413); so we can develop the genus theory which enables us to demonstrate the quadratic reciprocity law in k. Thus Hilbert succeeded in explicitly showing the application limit of the genus theory in the general study of the quadratic reciprocity law.

If we do not impose the above-mentioned conditions on the base field k, then we will no longer be able to have the genus theory in k; so we will have to create an entirely new method of demonstration in order to establish the quadratic algebraic number field: that is the class field theory. The third paper (C) looks like a sketch of class field theory. Hilbert relaxed the two conditions imposed on the base field k in the preceding paper (B) and observed the situations of appearance of non-ramified relative quadratic number field over k. It is in this way that Hilbert got the notion of non-ramified class field. Hilbert's idea of class field theory was, however, not limited to the theory of non-ramified class fields. At the end of the paper (C), he left these very impressive words:

> Finally in the last paragraph (§16) of the present paper I have given, as expectations, a series of general theorems for a relative Abelian number field of arbitrary relative degree and with the relative discriminant 1; these theorems are of wonderful simplicity and crystalline beauty, whose complete proofs and suitable generalization to the case of an arbitrary relative discriminant seem to me the final goal of the purely arithmetical theory of relative Abelian number fields. ([Hilbert 3] p. 484)

Class field theory was born when non-ramified class fields first appeared in the world of relative Abelian number fields. Furtwängler succeeded Hilbert and completed the theory of non-ramified class fields (See [Furtwängler 2]) by means of which he could establish the quadratic reciprocity law in a completely arbitrary algebraic number field (See [Furtwängler 5]) and the reciprocity law of odd prime degrees in an arbitrary algebraic number field with a suitable root of unity (See [Furtwängler 3, 4]). The notion of class field gradually accepted the generalization and finally filled the whole world of relative Abelian number fields. In fact, if we depend on the present concept of a class field obtained by Takagi (See [Takagi] p. 85), then we can consider every relative Abelian number field as a class field. It is at that time that Artin found his general reciprocity law (See [Artin] p. 361) with the aid of which we can complete the theory of reciprocity law among arbitrary power residues, that is, we can establish the reciprocity law of arbitrary degree in an arbitrary algebraic number field with a suitable root of unity. Thus it became clear, as Hilbert expected, that the class field theory is the decisive key to the study of reciprocity law.

III Complex Multiplication and Reciprocity Law

1 Demonstration of the Biquadratic Reciprocity Law by Eisenstein

Here I would like to return to Gauss's message (See I-1 of the present paper). According to Gauss, the principles on which the division theory of a circle is based can be applied not only to circular functions but also to other transcendental functions, for example the lemniscatic function. In fact, Abel discovered the concept of an Abelian equation in the division theory of the lemniscatic function and succeeded in explaining one of the principles to which Gauss referred, a principle of algebraic character. Then Eisenstein explained another one, a principle of arithmetical character. Let's give a short sketch of Eisenstein's theory.

Eisenstein started, just like Abel, from the differential equation

$$\frac{dy}{\sqrt{1-y^4}} = m\frac{dx}{\sqrt{1-x^4}}, \tag{7}$$

where $m = a + bi$ denotes a prime Gaussian integer. Thanks to Abel, we know that the algebraic integral of this equation (7) under the initial condition "$y = 0$ when $x = 0$" can be expressed in the form

$$y = x\frac{A_0 + A_1x^4 + A_2x^8 + \cdots + A_{\frac{1}{4}(p-1)}x^{p-1}}{B_0 + B_1x^4 + B_2x^8 + \cdots + B_{\frac{1}{4}(p-1)}x^{p-1}}, \tag{8}$$

where p denotes a natural number called the norm of m, that is, $p = a^2 + b^2$; this is nothing but the rational expression of $y = \phi(m\delta)$ by $x = \phi\delta$, where $\phi\delta$ denotes the lemniscatic function (See I-2, (2)). Under these situations, Eisenstein paid attention to the singular properties of coefficients of the rational function (8) and found "a theorem which seems to lay the foundation for all the applications of the division of a lemniscate to number theory" (See [Eisenstein] p. 301). He wrote:

> If m is a *two-term* complex *prime number* [a prime Gaussian integer which is not real], therefore if p is a real prime number $\equiv 1 \pmod 4$, then all the coefficients of the numerator U except the last one, and also all the coefficients of the denominator V except the first one are divisible by m. (Eisenstein] p. 301)

As Gauss showed (See, for example, [Gauss 2]), we can find a proof of the quadratic reciprocity law in the division theory of a circle by Gauss. This is the arithmetical aspect of Gauss's theory that Eisenstein succeeded.

2   Essential Meaning of the Theory of Complex Multiplication

Thus we can find two things in the differential equation (7); one is an Abelian equation over the Gaussian number field; the other is a proof of the biquadratic reciprocity law, which is of transcendental character. This situation suggests that these two things are not unrelated to each other; in fact, there is an intimate internal relation between them. According to Kronecker, we know that every relative Abelian number field over the Gaussian number field is generated by a special value of the lemniscatic function. This is why Eisenstein's theory succeeded, for, as the class field theory shows, we will be able to prove the biquadratic reciprocity law if we know all the relative Abelian number fields over the Gaussian number field. Kronecker tried looking for the reason and found it in Abel's theory of Abelian equations. This is the essence of Kronecker's theory on complex multiplication and at the same time we can clearly understand *why the class field theory, which is the key to the theory of reciprocity law, was born in the theory of complex multiplication.*

I think that Kronecker wanted further to know the general reason why the lemniscatic function is enough to generate all the relative Abelian number fields over the Gaussian number field; so *Kronecker's youthful dream shows why there can be a transcendental proof of the biquadratic reciprocity law.* This is the *essential meaning of the theory of complex multiplication.* From a purely theoretical viewpoint, Abel's theory is only one origin of the youthful dream; however, I would like to emphasize here that the essential meaning of the youthful dream can be found in Eisenstein's theory, not in Abel's theory.

When we go beyond the youthful dream, we will be in a quite vague situation; however, it seems to me that Kronecker had clearly known the path to follow as shown by his few words expressed at the end of the letter addressed to Dedekind in which Kronecker told his favorite youthful dream: "the hope to finish the kernel point to find analogues of singular modules for general complex numbers" (See [Kronecker 6] p. 457). This, I think, should be considered as the second piece of evidence that he had conceived a big plan to have the completely generalized theory of complex multiplication, that is, the theory suggested in Hilbert's twelfth problem (See [Hilbert 4] pp. 311–313). In my opinion, this suggests that there will be a transcendental proof of the reciprocity law among arbitrary power residues. This is the prospect of the theory of complex multiplication that I wish to put forth at the end of the present paper.

References

[N. H. Abel 1]   Recherches sur les fonctions elliptiques, Journal für die reine und angewandte Mathematik 2,3(1827, 1828), pp. 101–181, pp. 160–190; Œuvres I, pp. 263–388.

[N. H. Abel 2]   Solution d'un problème général concernant la transformation des fonctions ellilptiques, Astronomische Nachrichten 6, 138(1828), pp. 365–388;Œuvres I, pp. 403–428.

[N. H. Abel 3]   Mémoire sur une classe particulière d'équations résolubles algébriquement, Journal für die reine und angewandte Mathematik 4(1829), pp. 131–156; Œuvres I, pp. 478–507.

[N. H. Abel 4]   Sur la résolution algébrique des équations, Œuvres II, pp. 217–243.

[N. H. Abel 5]   Extraits de quelques lettres à Holmboe, Œuvres II, pp. 254–262.

[N. H. Abel 6]   Extraits de quelques lettres à Crelle, Œuvres II, pp. 266–270.

[N. H. Abel 7]   Notes aux mémoires du tome I, Œuvres II, pp. 290–323.

[E. Artin]   Beweis des allgemeinen Reziprozitätsgesetzes, Abhandlungen aus dem Mathematischen Seminar der Hamburgischen Universität 5(1927), pp. 353–141; Collected Papers, pp. 131–141.

[G. L. Dirichlet] Recherches sur les diviseurs premiers d'une classe de formules du quatrième degré, Journal für die reine und angewandte Mathematik 3(1828), pp. 35–69; Werke I, pp. 65–98.

[G. Eisenstein] Beiträge zur Theorie der elliptischen Functionen. I. Ableitung des biquadratischen Fundamentaltheorems aus der Theorie der Lemniscatenfunctionen, nebst Bemerkungen zu den Multiplications- und Transformationsformeln, (1845); Werke I, p. 299–324.

[F. Furtwängler 1] Über die Reziprozitätsgesetze zwischen 1ten Potenzresten in algebraischen Zahlkörpern, wenn l eine ungerade Primzahl beduetet, Mathematische Annalen 58(1903), pp. 1–50.

[F. Furtwängler 2] Allgemeiner Existenzbeweis für den Klassenkörper eines beliebigen algebraischen Zahlkörpers, Mathematische Annalen 63(1907), pp. 1–37.

[F. Furtwängler 3] Die Reziprozitätsgesetze für Potenzreste mit Primzahlexponenten in algebraischen Zahlkörpern (Erster Teil), Mathematische Annalen 67(1909), pp. 1–31.

[F. Furtwängler 4] Die Reziprozitätsgesetze für Potenzreste mit Primzahlexponenten in algebraischen Zahlkörpern (Zweiter Teil), Mathematische Annalen 72(1912), pp. 346–386.

[F. Furtwängler 5] Die Reziprozitätsgesetze für Potenzreste mit Primzahlexponenten in algebraischen Zahlkörpern (Dritter und letzter Teil), Mathematische Annalen 74(1913), pp. 413–429.

[C. F. Gauss 1] Disquisitiones Arithmeticae, Leipzig 1801; English Edition translated by Arthur A. Clarke, Springer-Verlag, 1986.

[C. F. Gauss 2] Summatio quarumdam serierum singularium, 1808; Werke II, pp. 9–45.

[C. F. Gauss 3] Theorematis fundamentalis in doctrina de residuis quadraticis demonstrationes et ampliationes novae, 1817; Werke II, pp. 47–64.

[C. F. Gauss 4] Theoria residuorum biquadraticorum. Commentatio prima, 1825; Werke II, pp. 65–92.

[C. F. Gauss 5] Theoria residuorum biquadraticorum. Commentatio secunda, 1831; Werke II, pp. 93–148.

[D. Hilbert 1] Zahlbericht: Die Theorie der algebraischen Zahlkörper, Jahresbericht der Deutschen Mathematiker-Vereinigung 4(1897), pp. 175–546; Gesammelte Abhandlungen I, pp. 63–363.

[D. Hilbert 2] Über die Theorie des relativquadratischen Zahlkörpers, Mathematische Annalen 51(1897), pp. 1–127; Gesammelte Abhandlungen I, pp. 370–482.

[D. Hilbert 3]   Über der relativ-Abelischen Zahlkörper, Nachrichten der Gesell-
                 schaft der Wissenschaften zu Göttingen (1898), pp. 370–399, or
                 Acta Mathematica 26(1902), pp. 99–132; Gesammelte Abhandlun-
                 gen I, pp. 483–509.

[D. Hilbert 4]   Mathematische Probleme, Vortrag, gehalten auf dem Interna-
                 tionalen Mathematikerkiongreß zu Paris 1900, Nachrichten der
                 Gesellschaft der Wissenschaften zu Göttingen (1900), pp. 253–297;
                 Gesammelte Abhandlungen III, pp. 290–329.

[L. Kronecker 1] Über die algebraisch auflösbaren Gleichungen I, Monatsberichte
                 der Königlich Preussischen Akademie der Wissenschaften zu Berlin
                 1853, pp. 365–374; Werke IV, pp. 3–11.

[L. Kronecker 2] Über die elliptischen Functionen, für welche complexe Multi-
                 plication stattfindet, Monatsberichte der Königlich Preussischen
                 Akademie der Wissenschaften zu Berlin 1857, pp. 455–460; Werke
                 IV, pp. 174–183.

[L. Kronecker 3] Über die complexe Multiplication der elliptische Funcitonen,
                 Monatsberichte der Königlich Preussischen Akademie der Wis-
                 senschaften zu Berlin 1862, pp. 363–372; Werke IV, pp. 209–217.

[L. Kronecker 4] Grundzüge einer arithmetischen Theorie der algebraischen
                 Grössen, Journal für die reine und angewandte Mathematik
                 92(1892), pp. 1–122; Werke II, pp. 237–387.

[L. Kronecker 5] Zur Theorie der elliptischen Functionen, I-XXII, Sitzungsberichte
                 der Königlich Preussischen Akademie der Wissenschaften zu Berlin
                 1883–1890; Werke IV, pp. 347–495 and V, pp. 3–132.

[L. Kronecker 6] Auszug aus einem Briefe von L. Kronecker an R. Dedekind,
                 Sitzungsberichte der Königlich Preussischen Akademie der Wis-
                 senschaften zu Berlin 1895, pp. 115–117; Werke V, pp. 455–457.

[E. Kummer 1]   Mémoire sur la théorie des nombres complexes composés de racines
                 de l'unité et de nombres entiers, Journal de mathématiques pures
                 et appliquées XVI(1851), pp. 377–498; Collected Papers I, pp. 363–
                 484.

[E. Kummer 2]   Über die allgemeinen Reciprocitätsgesetze unter den Resten und
                 Nichtresten der Potesten, deren Grad eine Primzahl ist, Mathe-
                 matische Abhandlungen der Königlichen Akademie der Wis-
                 senschaften zu Berlin (1859), pp. 19–159; Collected Papers I, pp.
                 699–839.

[T. Takagi]   Ueber eine Theorie des relativ-Abel'schen Zahlkörpers, Journal
                 of the College of Science, Imperial University of Tokyo 41, Art
                 9(1920), pp. 1–133; Collected Papers, pp. 73–166.

The Establishment of
the Takagi-Artin Class Field Theory

Miyake Katsuya

Introduction

In the preface of his book, "The Algebraic Theory of Numbers," (in Japanese; [T-1948]), Takagi states

"Abelian fields are none other than class fields" (in Japanese).

What he means by these words is more than just a mathematical fact in his theory. They are placed in a context which briefly describes the position of his theory in the research program on relative abelian fields proposed by Hilbert. The purpose of this paper is to give a close look at the development of class field theory mainly in the first 30 years of our century. Then we will see, in the sequel, what Takagi meant by the words.

Both of the concepts, "abelian fields" and "class fields," have their direct origin in Kronecker. The former is simple and clear as is seen in his paper [K-1853]; an abelian field over a fixed base number field is a finite Galois extension whose Galois group is abelian; (if we should strictly speak, Kronecker did not introduce an "abelian field" but an "abelian equation," and meant by it, at the beginning, one of those equations which have specific algebraic numbers as their coefficients and cyclic Galois groups.)

The latter, however, has a long history whose most important portion is our main subject. We are to study four items,

1. Evolution of concepts of "class fields,"

2. The existence of class fields — "Class fields are abelian fields,"

3. The first inequality $h \geq n$ for abelian fields — "Abelian fields are class fields," and

4. The canonical isomorphisms of ideal class groups onto the Galois groups of their class fields — Artin's reciprocity law.

1   Evolution of Concepts of "Class Fields"

1.1   Kronecker

An idea toward "class fields" was born in youthful dreams of Kronecker around 1857 (cf. [K-1857a and b]), and grew up to a mathematical concept after 25 years in his fundamental paper [K-1882]. His terminology for it was "the associated species (die zu associierenden Gattungen)" instead of "class fields". The main feature of it is a minimal Galois extension K of a fixed base field k in which every ideal complex number of k is realized as an actual number of K (the principal ideal theorem); the Galois group of K/k was supposed to be isomorphic to the absolute ideal class group of k; furthermore, it was strongly suggested that K should be unramified over k (cf. [K-1857a, -1862 and -1874]).

1.2   Weber I

It was Weber [W-1891] who first used the words "class field". In 1886, he succeeded in giving a proof to Kronecker's conjecture (stated first in [K-1853]),

Kronecker-Weber's Theorem. *The roots of every Abelian equation with (rational) integral coefficients are expressed as rational functions of roots of 1 with (rational) integral coefficients.*

Then he pursued the problem so called "Kronecker's youthful dream"; this is to obtain a version of this theorem for Abelian equations with coefficients in an imaginary quadratic number field along with singular moduli and special values of those elliptic functions, in place of roots of 1, for which complex multiplication by numbers of the quadratic field is admitted. In his book, "Elliptische Functionen und algebraische Zahlen" ([W-1891]), he used the term "class field" for the smallest extension of an imaginary quadratic number field which is obtained by the singular modulus of an elliptic function with complex multiplication by the quadratic numbers. This type of extensions has extensively been studied by Kronecker (e.g. [K-1857a, -1862, -1874 and -1882]), and was a fundamental source for him to formulate his grand research program of algebraic number theory. Frei [F-1989] even calls it "the Kronecker class field" of an imaginary quadratic field.

1.3   Weber II

In 1897, Weber introduced a concept of congruence ideal class groups in a general algebraic number field in a series of three papers entytled

"Über Zahlengruppen in algebraischen Körpern" ([W-1897]).

The concept was so general that, later, Takagi had only to add "prime divisors at infinity" to complete it for his class field theory.

Let k be an algebraic number field of finite degree, and fix an integral ideal m of (the maximal order, the ring of all integers of) k. Let A_m denote the multiplicative group consisting of those fractional ideals of k which do not contain any prime divisors of m in their reduced prime ideal decompositions. We take a subgroup of A_m consisting of certain principal ideals,

$$S_m = \{(a) \mid a \in k^\times, \ a \equiv 1 \mod m\}$$

and consider only a subgroup H_m of A_m which contains S_m; then the index $h :=$ $[A_m : H_m]$ is finite. Let us call such an H_m a congruence ideal group. (In case when m is trivial, i.e. $m = o$, the maximal order of k, S_o consists of all of the principal ideals of k; hence A_o/S_o is the ordinary ideal class group of k, and $h = h_k$ is its class number if we take S_o as H_m.) Weber's intension was to generalize and utilize the analytic method of Dirichlet; Kronecker had used it to determine the irreducibility of the class equation of an imaginary quadratic field in [K-1862]; (its roots consist of a collection of singular moduli of elliptic functions whose period modules belong to a set of h_k ideals representing all of the ideal classes of A_o/S_o of the quadratic field k). In [W-1897], Weber proved, among others, the following fact:

For a subgroup H_m of A_m of the above kind, suppose that there exists a finite extension K/k such that

1. *$n := [K : k] \geq h$, and*

2. *Prime ideals of absolute degree 1 in H_m and only they split completely into a product of prime ideals of degree 1 in K.*

Then K is uniquely determined by the conditions; in particular, we have $n = h$.

In case when k is an imaginary quadratic field, he had three types of H_m and K: one of them is an extension of k obtained by the singular modulus of an elliptic function with complex multiplication by numbers of k; he called such K an *order field* of k and the corresponding H_m an order group because A_m/H_m becomes "the ideal class group" of an order of k which is determined by the period module of the elliptic function with the chosen modulus. (An order of k is a subring with 1 whose **Z**-rank is equal to the degree $[k : \mathbf{Q}]$ of k over the rational number field **Q**.) He still used the term "class field" of k only for the minimal order field which corresponds to the maximal order of k. At the time he knew that order fields are abelian over k.

The second one is the genus field of an order field K/k, that is, the maximal abelian extension of **Q** contained in K. Weber determined the congruence ideal group corresponding to it.

The last is a *division field* which is obtained over the class field of k by the values of a Weierstrass $\wp$-function at various division points of the periods; it corresponds to S_m if all of m-division points are taken, and then is maximal for a fixed m. Its degree and the irreducible equation over the class field were determined, and so was its Galois group which is isomorphic to a subgroup of A_m/S_m consisting of all the classes of principal ideals in A_m. However, it was *not* proved that a division field is an abelian extension of the base quadratic number field.

1.4   Weber III

In his article, "Komplex Multiplikation," in Encyklopädie der Mathematischen Wissenschaften ([W-1900]), Weber did not use the words "order field" at all. He defined a "class field" for each class group of binary quadratic forms with a fixed negative discriminant. This actually covers all order fields of imaginary quadratic number fields. He states that a division field is abelian over the "class field" of the corresponding order of an imaginary quadratic field. Furthermore, he even writes

> "It is probable that the roots of this division equation $\Phi_m = 0$ are contained in the class fields (of other orders)"

(in German; [W-1900], p.731). Later in 1908, he is to give a corrected result for division by a power of a prime ideal.

His book, "Lehrbuch der Algebra III"([W-1908]), was published in 1908 as the second edition of "Elliptische Functionen und algebraische Zahlen"([W-1891]). Here he could show for division by a power of a prime ideal that a division field is abelian over the base quadratic field; he did not hesitate to give a definition that an extension K of an imaginary quadratic number field k is called the class field for a congruence ideal group H_m if the above condition (2) is satisfied. The "second inequality" $n \geq h$ holds for it by [W-1897]. (As for the "first inequality" $n \leq h$, it is obvious for order fields and division fields by definition.)

Remark. With regard to prime ideals of absolute degree 1 in an algebraic number field, Kronecker had an idea to associate certain "density" to the set of rational primes which split completely in a given finite Galois extension of $\mathbf{Q}$ (cf. [K-1880]). Then Frobenius picked it up and finally formulated his conjecture in [Fr-1896a] (cf. e.g. [M-1989]). This is to be proved by Tschebotareff in [Ts-1926]. And the method of his proof is to be utilized by Artin to prove his general reciprocity law in [A-1927]. Hasse chose the name, "Frobenius substitutions," in his Bericht [H-1926] because of the work [Fr-1896a] of Frobenius. (Cf. Section 4.5 below.)

1.5   Hilbert I

Hilbert first used the words "class field" in his Bericht, "Die Theorie der algebrais-chen Zahlkörpers"([Hi-1897]); in §58 he proved

Theorem 94. *Let k be an algebraic number field, and K be an unramified cyclic extension of k of a prime degree p. Then there exists a non-principal ideal j of k which becomes principal in K. Therefore j^p is principal in k; and the class number of k is divisible by p.*

He called an unramified cyclic extension K of k a class field because it corresponds to an ideal class of k in a specific way by the theorem. At this moment, therefore, his class fields were only a little more than a translation of Kronecker's concept "die zu associierenden Gattungen". He put, however, the property "unramification" in front.

In 1896, he had succeeded in giving a simple proof to Kronecker-Weber's Theorem ([Hi-1896]). It utilized his theory of ramification in algebraic number fields ([Hi-1894]), roughly speaking, as follows: Let K be an abelian extension of $\mathbf{Q}$; we may assume that $K/\mathbf{Q}$ is cyclic of degree p^n, a power of a prime. Let $p_1, \ldots, p_t, p_i \neq p$ for $1 \leq i \leq t - 1$, be the distinct primes which ramify in $K/\mathbf{Q}$. Then we have $p_i \equiv 1 \bmod p$ for $i = 1, \ldots, t - 1$, and also for $i = t$ if $p_t \neq p$. Let us consider the case $p_t = p$. Take a primitive p_1-th root of 1 and denote the cyclotomic field by F_1. Let K_1 be the inertia field of p_1 in the composite field $K \cdot F_1$. Then in $K_1/\mathbf{Q}$, only $p_2, \ldots, p_t$ ramify. Moreover, Hilbert's theory shows that K is contained in the composite field $K_1 \cdot F_1$. Next take a cyclotomic field F_2 for p_2, and denote the inertia field of p_2 in $K_1 \cdot F_2$ by K_2. Then in $K_2/\mathbf{Q}$, only $p_3, \ldots, p_t$ ramify, and K_1 is contained in $K_2 \cdot F_2$. We continue this process up to $i = t - 1$, and obtain K_{t-1} such that only $p = p_t$ ramifies in $K_{t-1}/\mathbf{Q}$ and K is contained in the conposite field $K_{t-1} \cdot F_1 \cdots F_{t-1}$. Finally, let K_t be the inertia field of p in $K_{t-1}/\mathbf{Q}$. Then $K_t/\mathbf{Q}$ is an unramified abelian extension; hence, by Minkowsky's result of geometry of numbers, we have $K_t = \mathbf{Q}$; thus, p is fully ramified in the abelian extension $K_{t-1}/\mathbf{Q}$. Therefore, again by Hilbert's theory, we can find a sufficiently large p-power root of 1 such that the cyclotomic field F_t generated by it contains K_{t-1}. Hence we see K lie in the cyclotomic field $F_1 \cdots F_{t-1} \cdot F_t$.

Now, what do we need to carry out a similar process for a general base field k ? At least we need precise informations on *the maximal unramified abelian extension* of k and a good collection of abelian extensions of k which can provide elements of ramifications over k.

1.6 Hilbert II

In a set of works [Hi-1898, -1899a and -1899b], Hilbert gave a concrete aspects on his class field theory accompanied with a proof of existence of his class field in case when the absolute class number of the base field is equal to 2. This is understood by Takagi (see [T-1948], Preface) as the first step of his research program for a theory of relative abelian fields. Hilbert may have seen some possible generality in the theory of Kummer fields in his Bericht, §125; (see [T-1948], Preface and p.217; Takagi succeeded in giving a proof to "Kronecker's youthful dream" for $\mathbf{Q}(\sqrt{-1})$ by adapting the above Hilbert's method ([T-1903]) while he stayed at Göttingen (see [T-1942] Appendix 1).)

Under the influence of Weber (cf. [F-1989] and its references), Hilbert presented his class fields with the following properties:

For an algebraic number field k, there will exist an abelian extension K/k such that

1. The Galois group is isomorphic to the ideal class group A_o/S_o of k;
2. K/k is unramified;
3. K contains every unramified abelian extension of k as a subfield;
4. For a prime ideal p of k, its relative degree in K/k coincides with the minimal exponent f such that p^f becomes a principal ideal of k;
5. Every ideal of k becomes principal in K; (the principal ideal theorem).

Remark. Hilbert did not give any direct comments on his class fields in his cele-brated 23 problems [Hi-1900]. However the 9th and 12th problems are related to the theory of relative abelian fields. The former is to generalize the reciprocity law for power residues; and he suggested to generalize his theories of Kummer fields ([Hi-1897], §125) and of relative quadratic number fields ([Hi-1989b]). The latter problem is to obtain similar results to Kronecker-Weber's theorem; the first thing to do should be to settle "Kronecker's youthful dream". (Apart from this, Hecke is to investigate Hilbert modular functions of two variables to construct abelian fields of real quadratic number fields in [He-1913].) Hilbert suggested that these two problems were closely related; to see the historical background, we have to go back to study at least some works of Gauss and Eisenstein on reciprocity laws. However it is out of our present concern.

1.7 Takagi

On the concept of congruence ideal groups, Takagi added only prime divisors at infinity to Weber's; they are indispensable to grasp all of abelian extensions of even degrees as class fields if a base field is not totally imaginary. He took, however, a different approach from Weber's to define a class field. As we see in what follows, his main concern was to characterize class fields among Galois extensions of a base field. For him in the year 1914 when he started his own investigations on the subject (cf. [T-1942], Appendix 1, and Iyanaga [Iy-1990], p.360), it was almost clear that the class fields for congruence ideal groups exist and are abelian fields.

Let K be a finite Galois extension of a fixed base field k, and take a positive divisor m of k which consists of certain powers of ramified primes in K/k (including infinite ones). Then K determines a subgroup $H_m(K/k)$ of $A_m = A_m(k)$ of k,

$$H_m(K/k) = \{N_{K/k}(A)|\ A \text{ is an ideal of } K, \text{and } (A, m) = 1\} \cdot S_m,$$

where $N_{K/k}$ denotes the norm map of K over k. Weber's method now gives us

The Second Inequality: $n := [K : k] \geq h := [A_m : H_m(K/k)]$.

Takagi defines that K is a class field of k if the equality $n = h$ holds ([T-1920a]). Of course, the condition must have been extracted as a technical criterion of class fields; and it is sufficient to check "the first inequality" $n \leq h$ to see that an extension K is a class field of k. He used this criterion to realize that a cyclic extension of prime degree is a class field, and then found this simple case be well utilized to show that all abelian fields are class fields.

2   The Existence of Class Fields — "Class Fields Are Abelian Fields."

2.1

At the end of the last century, "class fields" (of general base fields) meant the absolute ones in the present sense as we have studied. The Galois groups of them over their base fields were expected to be isomorphic to the ideal class groups; therefore they must be abelian fields. Hilbert could show the existence for those base fields whose class numbers were equal to 2. Weber had obtained more general types of Galois extensions over imaginary quadratic number fields which correspond to his congruence ideal groups, namely, order fields and division fields. He could show that the Galois groups of order fields were isomorphic to the corresponding ideal class groups; order fields are abelian fields of the base fields. However, he could only show for division fields, by the end of the last century, that they were abelian over the related order fields even though they corresponded to congruence ideal groups of the base imaginary quadratic number fields.

2.2

In 1907, Furtwängler made big progress in the subject; he succeeded in proving the existence of Hilbert's class field for a general base number field in his paper [Fw-1907]. He showed that the properties (1) and (2) given in Section 1.6 above determine Hilbert's class field among Galois extensions of the base field and proved the existence of an extension with the two properties. Though he could not prove the principal ideal theorem (the property (5)) at the time, he put stress on it in [Fw-1907]; he added the words (in German),

"all ideals of k are presented as actual numbers of K,"

to (5) when he explained Hilbert's concept of "class fields". This suggests that Furtwängler may have put a special emphasis on the absolute class fields by recalling Kronecker's view on them. Later he poses the class field tower problem. At the moment, he resumed his investigations on the 9th problem of Hilbert using this existence theorem (cf. [Fw-1904 and -1909]). Then in 1911, he gives a proof to the decomposition theorem of Hilbert's class fields, that is, the property (4) of Section 1.6 ([Fw-1911]). At any rate, the Hilbert-Furtwängler method to construct class fields as abelian fields was well established in [Fw-1907] even if it must be refined later by Takagi in [T-1915 and -1920a].

It had not been known, however, until these works of Takagi that the maximal unramified abelian extension of a base field coincides with its Hilbert's class field.

2.3

Independently from the work of Furtwängler, Weber showed in his book [W-1908] that, for division by powers of prime ideals, the division fields are abelian over the base imaginary quadratic fields; he could characterize those abelian extensions of the quadratic fields which were generated by roots of 1 over order fields, by associating certain congruence ideal groups to them; then by simple calculations on the groups, he showed that a division field lies in one of those abelian fields if the division of periods of related elliptic functions is done by a power of a prime ideal. As we have mentioned in Section 1.4, he is now ready to declare that a Galois extension of an imaginary quadratic field is *the class field for a congruence ideal group* if they are associated to each other through the condition (2) given in Section 1.3.

In the foreword of his book [W-1908], he wrote

"... Certainly are here not achieved all what I have set to myself as the ultimate aim. The performance of the theory of relative cyclic fields must have been postponed — hopefully just a while."

If his ultimate aim would have been to settle "Kronecker's youthful dream," what would have been the theory of relative cyclic fields in his mind, and what could he have obtained in his last five years till 1913 ? (His works [W-1909] are not intended to give an answer to these, but restricted to cyclic extensions of the rational number field.)

3   The First Inequality for Abelian Fields — "Abelian Fields Are Class Fields."

3.1

Appendix 1 to Takagi's book [T-1942] is taken from what he talked in the Colloquium of Mathematics of Imperial University of Tokyo on the 7th, December, 1940, and tytled "Reminiscences and Perspectives" (in Japanese; cf. also Iyanaga [Iy-1990], pp.357, 360–361); it appears to be intended especially for young mathematicians in the audience. If we take his words in it as they are, Takagi started his own serious investigations on class fields in 1914 when World War I began. He did because he could not expect the flow of academic books and journals from Germany anymore. Then, in a sequel, he was to face the fact that Abelian fields are class fields. He says, however,

"At the time, this was so unexpected that I thought I must have made mistakes."

(in Japanese; cf. [T-1942], p.184); in a while he even tried to find a counter example in vain, of course (cf. ibid. and Iyanaga [Iy-1990], pp.360–361). The results were reported to the Physico-Mathematical Society of Japan in May and in July, 1915 ([T-1915]). The full details were published in 1920 as [T-1920a]. The tytles of the papers are

"Zur Theorie der relativ-Abel'schen Zahlkörper, I, II," and
"Ueber eine Theorie des relativ Abel'schen Zahlkörpers."

As we see by them, Takagi intended to develope a theory of relative abelian fields by characterizing these as class fields. His fundamental theorem was, therefore, the following one (in a simplified form; [T-1920a], pp.102–103):

Theorem. *Every relative abelian field of prime degree p with the relative discriminant f^{p-1} is the class field of a class group for the module f.*

(In his book [T-1948], this intension of him is even clearer; the fundamental theorem in it is also the same one but for general relative abelian fields; cf. [T-1948], Chapter 13, whose title is "The Fundamental Theorem".) To prove this theorem, he showed the first inequality by means of genus theory. Then he generalized Furtwängler's existence theorem for his class fields, and utilized it to show the fundamental theorem for general abelian fields *by mathematical induction* starting with the cyclic case of prime degree.

It should be noted that Takagi established the connection of the relative discriminant and the module which defined the congruence ideal group. By this, we immediately see that the maximal unramified abelian extension of an algebraic number field is none other than its Hilbert's class field (the characteristic property (3) of Hilbert's class fields listed in Section 1.6).

As an application of his theory of relative abelian fields, Takagi also gave a complete answer to "Kronecker's youthful dream" in [T-1920a]. (As for Takagi's works in number theory, see Iwasawa [Iw-1990].)

3.2

In the foreword of [T-1920a], Takagi listed seven literatures as the basic references; they are Weber's [W-1897 and -1908], Hilbert's [Hi-1897, -1898 and -1899b], Furtwängler's [Fw-1907] and Fueter's [Fu-1914]. The last and newest of them appeared in the year 1914 when he started his own serious research on class fields, and had already been referred by him in [T-1915]. In the paper, Fueter made great progress for "Kronecker's youthful dream"; for example, he could state,

"Every abelian equation of odd degree in an imaginary quadratic field is solvable through the singular moduli of suitable elliptic functions and roots of unity" (in German).

He also gave a counter example to this result for an abelian equation of even degree. Takagi had been interested in the problem for long from his student days; his doctoral thesis [T-1903] supplied an affirmative answer to it for the Gaussian field $\mathbf{Q}(\sqrt{-1})$. It may then be supposed that it was his interest in "Kronecker's youthful dream" which lead him to "class fields" in the sense of Weber [W-1908] to disregard the unramification condition and motivated him to characterize those among abelian fields or even among Galois extensions of a base field.

By contrast, as was mentioned in Section 2.2, Furtwängler seemed strongly concerned with Hilbert's class fields (and also with the 9th problem of Hilbert, the reciprocity law for the power residue symbols).

However, both of them were naturally interested in "non-abelian cases". The latter was to raise the class field tower problem, and to work on metabelian groups to prove the principal ideal theorem ([Fw-1930]). The former closed his talk on his class field theory at the International Congress of Mathematicians held in 1920 at Strasbourg by putting a stress on the importance of generalizing his theory to non-abelian Galois extensions (cf. [T-1920b]); in a short paper [T-1920c], he gave a proof to a conjecture of Dedekind on arithmetic structure of the (absolute) class group of a non-abelian cubic field which was raised in [D-1900]. (This paper of Dedekind together with Takagi's [T-1920a] appeared to supply effective materials to young Artin; cf. [A-1923].)

4 The Canonical Isomorphisms of Ideal Class Groups onto the Galois Groups of their Class Fields — Artin's Reciprocity Law.

4.1

In [T-1920a], Takagi not only characterized relative abelian fields as class fields but also proved basic results which generalize the properties of Hilbert's class fields listed in Section 1.6 with the exception of the property (5):

Existence Theorem which was proved by an elaborate refinement of the Hilbert-Furtwängler method;

Uniqueness Theorem which determines a relative abelian field by the corresponding congruence ideal group;

Isomorphism Theorem which generalizes the first property (1) of Hilbert's class fields;

Conductor Theorem which gives the properties (2) and (3) of Hilbert's class fields;

Decomposition Theorem which naturally generalizes the property (4);

(for details, see Iwasawa [Iw-1990] for example).

Now, once the properties (1) to (4) have been confirmed and fully generalized, the remaining property (5) of Hilbert's class fields appears in the limelight; not only its validity but also how to formulate its generalization are the problem. At this stage, Isomorphism Theorem could not supply any reasonable explanation even with Decomposition Theorem. One may ask, after some serious consideration, why a "ray ideal group" (a basic congruence ideal group)

$$S_m = \{(a)|\ a \in k^\times, \ a \equiv 1 \bmod m\}$$

plays such an important role and what the structure of the finite abelian group A_m/S_m can supply, however naïve it would be.

Our concern here is to see a brief history of general reciprocity laws in the first 30 years of this century. The most significant event must be the establishment of Artin's general reciprocity law ([A-1927]) which clarified the arithmetic structure behind Isomorphism Theorem and Decomposition Theorem. One of its important consequences is the proof of the principal ideal theorem, the property (5) of Hilbert's class fields, given by the cooperation of Artin [A-1930] and Furtwängler [Fw-1930].

4.2

As was mentioned in Remark at the end of Section 1-6, Hilbert called attention to a general reciprocity law for power residue Legendre symbols in his talk at the International Congress of Mathematicians at Paris in 1900 (cf. [Hi-1900]). Furtwängler picked up the 9th problem of Hilbert and started his investigation with the odd prime power case in 1904 ([Fw-1904]).

After he established the existence of Hilbert's class fields in 1907 ([Fw-1907]), he could utilize it (and also his Decomposition Theorem for Hilbert's class fields ([Fw-1911])) to show the reciprocity law for the prime power case in a series of three papers [Fw-1909]. In the third paper, he treated the quadratic reciprocity law in the same way as he did for odd cases. Since he had Hilbert's class fields in hand, he could carry it out for a general algebraic number field without any assumptions while Hilbert must have worked in two cases separately, namely, for a totally imaginary number field with odd class number and for a number field with the class number equal to 2 ([Hi-1898 and -1899b]).

4.3

When he wrote his paper [T-1922], Takagi was at more advantageous position than Furtwängler had been because he now had his class fields at hand, and could show a clear and much simplified demonstration of the general reciprocity law for prime power residue symbols. He also put a stress on the following fact as the actual content of the law: Let k be an algebraic number field of finite degree and suppose that it contains a primitive q-th root of 1. For $a \in k^\times$ and an ideal x of k, let $\left(\frac{a}{x}\right)$ be the q-th power residue symbol. Then for each a, *the value $\left(\frac{a}{x}\right)$ depends only on the ideal class mod f to which an ideal x belongs, where f is the conductor of the class field $k(\sqrt[q]{a})$ of k.*

Inspired by this work of Takagi, Artin formulates his general reciprocity law in [A-1924] to give a canonical isomorphism of a congruence ideal class group onto the Galois group of the corresponding class field.

4.4

As the title, "Über eine neue Art von L-Reihen," suggests it, the main theme of the paper [A-1924] of Artin was to introduce "Artin's L-fuctions" for relative non-abelian Galois extensions. Such "L-functions" have been sought by Dedekind for long: let K/k be a (Galois) extension of algebraic number fields of finite degree, and $\zeta_K(s)$ and $\zeta_k(s)$ be Dedekind's zeta-functions of K and k, respectively. Then the problem is to decompose the quotient $\zeta_K(s)/\zeta_k(s)$ into a product of "L-functions" which are somehow analogous to Dirichlet's L-functions or Weber's generalization of them (in [W-1897]). Trying to find a class number formula for pure cubic fields, Dedekind gave non-abelian examples in [D-1900] which had been obtained in around 1873 (also cf. van der Waal [vW-1977]). He informed the case when K is abelian over $k = \mathbf{Q}$ to Frobenius in 1896 (see [D-1930], II, p.435; also cf. Frei [F-1989], Chapter 4). Naturally, he had also been concerned in finding a good theory of group characters of non-abelian finite groups; it is well known by Hawkins [Hk-1970 and -1974] that he was quite influential over Frobenius to creat the theory of group characters in [Fr-1896b]; (also cf. [M-1989] for the arithmetic background). Neither Dedekind nor Frobenius, however, introduced any L-functions with the group characters when Frobenius established the theory in 1986 and published his conjecture which explained an essential link between prime ideals and their "Frobenius substitutions" in the same year ([Fr-1896a and b]).

It should be noted that Hecke made important contributions on Dedekind's zeta- and Weber's L-functions in 1917; in [He-1917a and c], he proved the functional equations of these, and in [He-1917b], extracted some arithmetic informations of number fields from them; in [He-1917c], he also showed the fact, $L(1,\chi) \neq 0$ for $\chi \neq 1$, analytically, which not only was indispensable for class number formula but also assured the existence of infinitely many prime ideals in each congruence ideal class. (This was established algebraically by Takagi already in 1915 when he proved the existence of his class fields in [T-1915], II. However Hecke did not have any means to know it because of World War I.)

4.5

In [A-1923], Artin picked up the problem, whether the quotient $\zeta_K(s)/\zeta_k(s)$ is holomorphic or not, and referred two affirmative cases: namely, the case of Dedekind [D-1900] when K is a pure cubic field with $k = \mathbf{Q}$, and the case of relative abelian fields where we have

$$\zeta_K(s) = \zeta_k(s) \cdot \prod_{\chi \neq 1} L(s,\chi) \qquad (*)$$

with Weber's L-functions for non-principal characters of the congruence ideal class group of k associated with the abelian extension K/k by Takagi's class field theory [T-1920a]. Here Artin himself treated relative metacyclic extensions of square-free degrees and relative icosahedron extensions, and gave affirmative answers to the question in these cases; he could express the quotients of Dedekind's zeta-functions as products of power roots of Weber's L-functions of intermediate relative abelian fields by using Hecke's result in [He-1917b].

In [A-1924], Artin defines his L-function $L(s, \chi; k)$ for a relative Galois extension K/k with a group character of the Galois group $\mathrm{Gal}(K/k)$ in the following manner as an Euler product: let Γ be a linear representation of $\mathrm{Gal}(K/k)$ which gives the character χ; each prime ideal p of k which is not ramified in K/k, take a prime ideal P of K over it; then there exists an element σ_P of $\mathrm{Gal}(K/k)$ such that we have $A^{\sigma_P} \equiv A^{N_p} \bmod P$ for every integer A in K where N_p is the norm of the prime ideal p over $\mathbf{Q}$; (this is called the Frobenius substitution of P by Hasse [H-1926] because of Frobenius' work [Fr-1896a]). Now, let us denote the image $\Gamma(\sigma_P)$ simply by X_p; then Artin's L-function is defined by

$$L(s, \chi; k) := \prod_p \frac{1}{\det(E - N_p^{-s} X_p)}$$

where the product is taken through over all prime ideals of k unramified in K/k, and E is the unit matrix of size $\chi(1)$. (Note that the denominator of the Euler factor does not depend on the choice of P for p because another choice gives a conjugate of σ_P in $\mathrm{Gal}(K/k)$.) The definition is to be refined in [A-1931]. At this point, however, Artin could prove the equality,

$$\zeta_K(s) = \zeta_k(s) \cdot \prod_{\chi \neq 1} L(s, \chi)^{\chi(1)} \qquad (**)$$

where the product is taken over all non-principal absolutely irreducible characters of $\mathrm{Gal}(K/k)$. He also showed the functional equation for such $L(s, \chi)$ (assuming his general reciprocity law); however, its holomorphy has not yet been proved in general except a few cases, and is one of the most important conjectures of Artin.

If K/k is relatively abelian, in particular, then we have $\chi(1) = 1$ for every absolutely irreducible character χ; hence we have two decompositions (*) and (**) of $\zeta_K(s)/\zeta_k(s)$ in this case. To conclude that these are essentially identical, Artin claimed

Artin's General Reciprocity Law. *Let K/k be a relative abelian field and correspond to a congruence ideal class group A_m/H_m of k. Then (a) the Frobenius substitution σ_P depends only on the ideal class of A_m/H_m to which p belongs; (b) the assignment of the substitution to each class of A_m/H_m establishes an isomorphism of the class group onto $\mathrm{Gal}(K/k)$.*

If K/k is cyclic of degree q and if k contains a primitive q-th root of 1, then this is identical with the general reciprocity law for q-th power residue symbols as it is

seen by Takagi's work [T-1922] in the prime power case. This assertion also immediately implies Decomposition Theorem because the Frobenius substitution of an
unramified prime ideal generates the decomposition group of the ideal. However,
Artin used both of Isomorphism Theorem and Decomposition Theorem when he
proved his reciprocity law in 1927.

In 1926, Tschebotareff [Ts-1926] succeeded in proving his Density Theorem
which had been conjectured by Frobenius in [Fr-1896a]. The proof was fully analyzed by Schreier in [S-1927] who reconstructed a proof in no more than six pages;
a general case was reduced to the case of (well chosen) cyclotomic fields. In his
paper [A-1927], Artin could utilize the method in an ingenious manner to show his
reciprocity law for a general algebraic number field. The Takagi-Artin class field
theory was now established.

4.6

With his general reciprocity law, Artin investigated the property (5) of Hilbert's
class fields, the principal ideal theorem, and clarified the group theoretical mechanism behind it in Galois groups of certain metabelian extensions. The result was
informed to Furtwängler in the summer of 1927 (cf. [Fw-1930], Introduction); after
some efforts ("nach einiger Mühe") for a little more than one year, he could prove
a theorem on metabelian groups which implies the principal ideal theorem by the
work of Artin. Their results were published in 1930 in the same volume of Abh.
Math. Sem. Univ. Hamburg ([A-1930] and [Fw-1930]).

To understand the non-abelian nature and the generality of their works, let
us see the main result of [A-1930]. Let L/k be a finite Galois extension of algebraic
number fields, and K be an intermediate field of it. Denote the Galois groups by G
$= \mathrm{Gal}(L/k)$ and $H = \mathrm{Gal}(L/K)$. Then the maximal abelian extensions k' of k and
K' of K in L, respectively, correspond to the commutator subgroups $[G,G]$ of G
and $[H,H]$ of H. Let m be the discriminant of L/k. By class field theory, the abelian
extensions k'/k and K'/K are determined by ideal class groups $A_m(k)/H_m(L/k)$
of k and $A_m(K)/H_m(L/K)$ of K, respectively. Let us denote the isomorphisms of
Artin's reciprocity law by

$$\alpha_{k'/k} : A_m(k)/H_m(L/k) \to G/[G,G],$$

$$\alpha_{K'/K} : A_m(K)/H_m(L/K) \to H/[H,H].$$

Furthermore, we have a natural homomorphism

$$j_{K/k} : A_m(k)/H_m(L/k) \to A_m(K)/H_m(L/K)$$

which is defined by regarding an ideal of k naturally as that of K. Finally, let us
denote the transfer homomorphism of G to H by

$$V_{G \to H} : G/[G,G] \to H/[H,H]$$

(cf. e.g. Zassenhaus [Z-1958] for the definition). The content of the main result of
the paper [A-1930] is

Theorem(Artin). *Let the notation and the assumptions be as above. Then these four homomorphisms are linked together by the relation,*

$$\alpha_{K'/K} \circ j_{K/k} = V_{G \to H} \circ \alpha_{k'/k}.$$

In case when K is Hilbert's class field of k and L is that of K, i.e. the second class field of k, then G is metabelian and H is none other than $[G, G]$ which is abelian. Furthermore, we have the absolute ideal class groups of k and of K in this case. Hence the theorem shows that the principal ideal theorem is equivalent to the following assertion proved by Furtwängler in [Fw-1930]:

Theorem(Furtwängler). *The transfer map of a finite group to its commutator subgroup is trivial.*

It is clear that we easily obtain a generalization of the principal ideal theorem by proper combination of these two theorems whatever it may tell.

Up to now, there does not seem to have appeared yet anything which may indicate a new way of proving the principal ideal theorem other than by making use of these results of Artin and Furtwängler. This may be a strong support for Hasse to regard the theorem as a "non-abelian" fact (cf. [H-1926], II, p.2).

At any rate, all of the properties of Hilbert's class fields (including their existence) have been proved to be true, and fully generalized even beyond the horizon of the scope of Hilbert.

Epilogue

As we have seen it, Artin's general reciprocity law was established to give the canonical isomorphism of every congruence ideal class group onto the Galois group of the associated class field. It also gave a key to a proof to the principal ideal theorem. Even if we restrict ourselves to Hilbert's class fields, however, the structure of absolute ideal class groups does not reflect the mechanism of "capitulation of ideals" in Galois correspondence which may be suggested by Hilbert's Theorem 94 (see Section 1.5; also cf. Taussky [T-1929] and Scholz und Taussky [S&T-1934]). Moreover, we see a very general algebraic framework behind the principal ideal theorem as was explained in Section 4.6. Then we are again inclined to ask the naïve question on the big role of a "ray ideal group" S_m in the Takagi-Artin class field theory which was raised in Section 4.1. From this aspect, the idelic presentation of class field theory by Chevalley [Ch-1940] may appear more natural.

The concept of "ideles" of an algebraic number field had already appeared in an earlier work of him (cf. e.g. [Ch-1933]). It was, however, in his paper [Ch-1940] where Chevalley could give a beautiful description of class field theory with idele groups. Here we see all of the local class field theories at the places of an algebraic number field tied up harmoniously as the global one by the relations determined by the global numbers; the global numbers which first appeared as principal ideals in various S_m have been given a natural role distinguished clearly from that of ideals or divisors. This paper has one other important feature; here is the first completely arithmetic proof to class field theory; the second inequality is shown without use of analytic devices such as L-series.

It should also be pointed out that the result of Artin [A-1930] explained in the last Section 4.6 is naturally understood as one of the characteristic properties of Weil groups (cf. Weil [Wl-1951]).

References

[A-1923]   E. Artin, *Über die Zetafunctionen gewisser algebraischer Zahlkörper*, Math. Ann., **89** (1923), 147–156, = Collected Papers, pp. 95–104.

[A-1924]   _____, *Über eine neue Art von L-Reihen*, Abh. Math. Sem. Univ. Hamburg, **3** (1924), 89–108, = Collected Papers, pp. 105–124.

[A-1927]   _____, *Beweis des allgemeinen Reziprozitätsgesetzes*, Abh. Math. Sem. Univ. Hamburg, **5** (1927), 353–363, = Collected Papers, pp. 131–141.

[A-1930]   _____, *Idealklassen in Oberkörpern und allgemeines Reziprozitätsgesetz*, Abh. Math. Sem. Univ. Hamburg, **7** (1930),46–51, = Collected Papers, pp. 159–164.

[A-1931]   _____, *Zur Theorie der L-Reihen mit allgemeinen Gruppencharakteren*, Abh. Math. Sem. Univ. Hamburg, **8** (1931), 292–306, = Collected Papers, pp. 165–179.

[A-1965]   _____, *The Collected Papers*, Addison Wesley (1965).

[Ch-1933] C. Chevalley, *Sur la théotie du corps de classes dans les corps de nombres algebrique et dans les corps locaux*, J. Fac. Sci. Tokyo Imper. Univ., Sec.I-2, **1933**, 363–476.

[Ch-1940] _____, *La théorie du corps de classes*, Ann. of Math., **41** (1940), 394–418.

[D-1900]   R. Dedekind, *Über die Anzahl der Idealklassen in reinen kubischen Zahlkörpern*, J. reine angew. Math., **121** (1900), 40–123, = Werke II, pp. 148–233.

[D-1930]   _____, *Gesammelte Mathematische Werke* I, II, III, Braunschweig (1930–1932), reprinted by Chelsea, New York, 1969.

[E-1978]   W. and F. Ellison, *Théorie des nombres*, chapitre V, in Abrégé d'histoire des mathématiques 1700–1900, (J. Dieudonné, ed.), Hermann, Paris (1978).

[F-1989]   G. Frei, *Heinrich Weber and the Emergence of Class Field Theory*, in The History of Modern Mathematics, (D.E. Rowe and J. McCleary, eds.), Academic Press (1989), 425–450.

[Fr-1896a] G. Frobenius, *Über Beziehungen zwischen den Primidealen eines alge-braischen Körpers und den Substitutionen seiner Gruppe*, Sitzungsb. König. Preiss. Akad. Wiss. Berlin, **1896**, 689–703, = Ges. Abh. II, pp. 719–733.

[Fr-1896b] _____, *Über Gruppencharaktere*, Sitzungsb. König. Preiss. Akad. Wiss. Berlin, **1896**, 985–1021, = Ges. Abh. III, pp. 1–37.

[Fr-1968] _____, *Gesammelte Abhandlungen* I, II, III, Springer-Verlag (1968).

[Ft-1914] R. Fueter, *Abel'sche Gleichungen in quadratisch-imaginären Zahl-körpern*, Math. Ann., **75** (1914), 177–255.

[Fw-1904] Ph. Furtwängler, *Über die Reziprozitätsgesetze zwischen l-ten Potenz-resten in algebraischen Zahlkörpern, wenn l eine ungerade Primzahl bedeutet*, Math. Ann., **58** (1904), 1–50.

[Fw-1907] _____, *Allgemeiner Existenzbeweis für den Klassenkörper eines be-liebigen algebraischen Zahlkörpers*, Math. Ann., **63** (1907), 1–37.

[Fw-1909] _____, *Reziprozitätsgesetze für Potenzreste mit Primzahlexponen-ten in algebraischen Zahlkörpern*, Math. Ann., **67** (1909), 1–31; II, **72** (1912), 346–386; III, **74** (1913), 413–429.

[Fw-1911] _____, *Allgemeiner Beweis des Zerlegungssatzes für den Klassen-körper*, Nachr. Ges. Wiss. Göttingen, **1911**, 293–317.

[Fw-1930] _____, *Beweis des Hauptidealsatzes für die Klassenkörper alge-braischer Zahlkörper*, Abh. Math. Sem. Univ. Hamburg, **7** (1930), 14–36.

[H-1926] H. Hasse, *Bericht über neuere Untersuchungen und Probleme aus der Theorie der algebraischen Zahlkörper* I, Jber. Deutschen Math.-Ver., **1926**; Ia, **1927**, 1–134; II, **1930**, 1–204.

[Hk-1970] Th. Hawkins, *The origin of the theory of group characters*, Arch. His-tory Exact Sci., **7** (1970/71), 142–170.

[Hk-1974] _____, *New light on Frobenius' creation of the theory of group char-acters*, Arch. History Exact Sci., **12** (1974), 217–243.

[He-1913] E. Hecke, *Über die Konstruktion relativ-Abelscher Zahlkörper durch Modulfunktionen von zwei Variablen*, Math. Ann., **74** (1913), 465–510, = Werke, pp. 69–114.

[He-1917a] _____, *Über die Zetafunktion beliebiger algebraischer Zahlkörper*, Nachr. Akad. Wiss. Göttingen, **1917**, 77–89, = Werke, pp. 159–171.

[He-1917b] _____, *Über eine neue Anwendung der Zetafunktionen auf die Arith-metik der Zahlkörper*, Nachr. Akad. Wiss. Göttingen, **1917**, 90–95, = Werke, pp. 172–177.

[He-1917c] _____, *Über die L-Funktionen und die Dirichletschen Primzahlsatz für einen beliebigen Zahlkörper*, Nachr. Akad. Wiss. Göttingen, **1917**, 299–318, = Werke, pp. 178–197.

[He-1959]   _____, *Mathematische Werke*, Vandenhoeck & Ruprecht, Göttingen (1959), 2.Auflage, 1970.

[Hi-1894]   D. Hilbert, *Grundzüge einer Theorie des Galoisschen Zahlkörpers*, Nachr. Akad. Wiss. Göttingen, **1894**, 224–238, = Ges. Abh. I, pp. 13–27.

[Hi-1896]   _____, *Ein neuer Beweis des Kroneckerschen Fundamentalsatzes über Abelsche Zahlkörper*, Nachr. Akad. Wiss. Göttingen, **1896**, 29–39, = Ges. Abh. I, pp. 53–62.

[Hi-1897]   _____, Bericht: *Die Theorie der algebraischen Zahlkörper*, Jber. Deutschen Math.-Ver., **4** (1897), 175–546, = Ges. Abh. I, pp. 63–363.

[Hi-1898]   _____, *Über die Theorie der relativ-Abelschen Zahlkörper*, Nachr. Akad. Wiss. Göttingen, **5** (1898), 377–399, = Acta Math. Stochh., **26** (1902), 99–132, = Ges. Abh. I, pp. 483–509.

[Hi-1899a]   _____, *Über die Theorie der relativ-quadratischen Zahlkörper*, Jber. Deutschen Math.-Ver., **6** (1899), 88–94, = Ges. Abh. I, pp. 364–369.

[Hi-1899b]   _____, *Über die Theorie des relativ-quadratischen Zahlkörpers*, Math. Ann., **51** (1899), 1–127, = Ges. Abh. I, pp. 370–482.

[Hi-1900]   _____, *Mathematische Problem. Vortrag auf dem internationalen Mathematiker Kongresse in Paris 1900*, Nachr. Akad. Wiss. Göttingen, **7** (1900), 253–297, = Ges. Abh. III, pp. 290–329.

[Hi-1932]   _____, *Gesammelte Abhandlungen* I, II, III, Berlin (1932–1935), reprinted by Chelsea, New York, 1965.

[Iw-1990]   K. Iwasawa, *On papers of Takagi in number theory*, Appendix I to [T-1973, 2nd edit.] (1990), pp. 342–351.

[Iy-1990]   S. Iyanaga, *On the life and works of Teiji Takagi*, Appendix III to [T-1973, 2nd edit.] (1990), pp. 354–371.

[K-1853]   L. Kronecker, *Über die algebraisch auflösbaren Gleichungen*, Monatsber. König. Preuss. Acad. Wiss. Berlin, **1853**, 365–374, = Werke IV, pp. 1–11.

[K-1857a]   _____, *Über die elliptische Functionen, für welche complex Multiplication stattfindet*, Monatsb. König. Preuss. Acad. Wiss. Berlin, **1857**, 455–460, = Werke IV, pp. 177–183.

[K-1857b]   _____, *Brief an G. L. Dirichlet vom 17 Mai 1857*, Nachr. Akad. Wiss. Göttingen, **1885**, 253–297 = Werke V, pp. 418–421.

[K-1862]   _____, *Über die complex Multiplication der elliptischen Functionen*, Monatsb. König. Preuss. Acad. Wiss. Berlin, **1862**, 363–372, = Werke IV, pp. 207–217.

[K-1874]   _____, *Über die Congruenten Transformation der Biliniear Formen*, Monatsb. König. Preuss. Acad. Wiss. Berlin, **1874**, 397–447, = Werke I, pp. 423–483.

[K-1880]   _____, *Über die Irreductibilität von Gleichungen*, Monatsb. König. Preuss. Acad. Wiss. Berlin, **1880**, 155–162, = Werke II, pp. 85–93.

[K-1882]   _____, *Grundzüge einer arithmetschen Theorie der algebraischen Grössen*, J. reine angew. Math., **92** (1882), 1–122, = Werke II, pp. 237–388.

[K-1895]   _____, *Mathematische Werke* I–V, Leipzig(1895–1930), reprinted by Chelsea, New York, 1968.

[M-1989]   K. Miyake, *A note on the arithmetic background to Frobenius' theory of group characters*, Expo. Math., **7** (1989), 347–358.

[S&T-1934] A. Scholz und O. Taussky, *Die Hauptideale der kubischen Klassenkörper imaginärquadratischer Körper: ihre rechnerische Bestimmung und ihr Einfluss auf den Klassenkörperturm*, J. reine angew. Math., **171** (1934), 19–41.

[S-1927]   O. Schreier, *Über eine Arbeit von Herrn Tschebotareff*, Abh. Math. Sem. Univ. Hamburg, **5** (1927), 1–6.

[T-1903]   T. Takagi, *Über die im Bereiche der rationalen komplexen Zahlen Abelscher Zahlkörper*, J. Coll. Sci. Tokyo, **19** (1903), 1–42, = Collected Papers, pp. 13–39.

[T-1915]   _____, *Zur Theorie der relativ-Abel'schen Zahlkörper* I, Proc. Phys.-Math. Soc. Japan, Ser. II, **8** (1915), 154–162; II *243–254*, = Collected Papers, pp. 43–50, 51–60.

[T-1920a]   _____, *Ueber eine Theorie des relativ Abel'schen Zahlkörpers*, J. Coll. Sci. Tokyo, **41** (1920), 1–133, = Collected Papers, pp. 73–167.

[T-1920b]   _____, *Sur quelques théorèmes généraux de la théorie des nombres algébriques*, Comptes rendus du congrès internat. math., Strasbourg, **1920**, 185–188, = Collected Papers, pp. 168–171.

[T-1920c]   _____, *Sur les corps résolubles algebriquement*, Comptes rendus hebdomadaires des séances de l'Académie des Sciences Paris, **171** (1920), 1202–1205, = Collected Papers, pp. 172–174.

[T-1922]   _____, *Ueber das Reciprocitätsgesetz in einem beliebigen algebraischen Zahlkörper*, J. Coll. Sci. Tokyo, **44** (1922),1–50, = Collected Papers, pp. 179–216.

[T-1942]   _____, *Topics from the History of Mathematics of the 19th Century (Japanese)*, Augumented edit., Tokyo (1942).

[T-1948]   _____, *The Algebraic Theory of Numbers (Japanese)*, Iwanami Shoten, Tokyo (1948).

[T-1973]   _____, *The Collected Papers*, Iwanami Shoten (1973); the 2nd edit., Springer-Verlag (1990).

[Ta-1929]   O. Taussky, *Über eine Versch ärfung des Hauptidealsatzes für alge-braische Zahlkörper*, Diss. Wien (1929), = J. reine angew. Math., **168** (1932), 193–201.

[Ts-1926]   N. Tschebotareff, *Die Bestimmung der Dichtigkeit einer Menge von Primzahlen, welche zu einer gegebenen Substitutionsklasse gehören*, Math. Ann., **95** (1926), 191–228.

[vW-1977]   R. van der Waal, *Holomorphy of Quotients of Zeta-Functions*, in Algebraic Number Fields, (A. Fröhlich, ed.), Academic Press (1977), 649–662.

[W-1886]   H. Weber, *Theorie der Abelschen Zahlkörper* I, Acta Math. Stokh., **8** (1886), 193–263; II, **9** *(1887), 105–130*.

[W-1891]   _____, *Elliptische Functionen und algebraische Zahlen*, Braunschweig (1891).

[W-1897]   _____, *Über Zahlengruppen in algebraischen Körpern* I, Math. Ann., **48** (1897), 433–473; II, **49** (1897), 83–100; III, **50** (1898), 1–26.

[W-1900]   _____, *Komplexe Multiplikation*, in ENCYKLOPÄEADIE der MATHEMATISCHEN WISSENSCHAFTEN, Leibzig (1900–1904), I.2.C. Zahlentheorie, 6, pp. 718–732.

[W-1908]   _____, *Lehrbuch der Algebra*, Vol.III, Braunschweig (1908), (the second edition of [W-1891]).

[W-1909]   _____, *Zur Theorie der zyklischen Zahlkörper* (I), Math. Ann., **67** (1909), 32–60; II, Math. Ann., **70** (1911), 459–470.

[Wl-1951]   A. Weil, *Sur la théorie du corps de classes*, J. Math. Soc. Japan, **3** (1951), 1–35, = Collected Papers I, pp. 487–521.

[Wl-1979]   _____, *Collected Papers* I, II, III, Springer-Verlag (1979).

[Z-1958]   H. Zassenhaus, *The Theory of Groups second edit.*, Chelsea, New York (1958).

Where Did Twentieth-Century Mathematics Go Wrong?

Chandler Davis

Abstract

Mathematics in the first half of the twentieth century saw two striking tendencies. Dominant mathematical schools turned sharply away from interest in applications; and during the same years a "naïve platonism" became the dominant metaphysics among working mathematicians.

These two apparently related tendencies are so familiar we may forget how strange they are. The rejection of applications is singular in being so extreme, in comparison with parallel moves within other sciences. "Naïve platonism" is bizarre in that (i) most mathematicians insist on living by it but don't really believe it; (ii) it followed an earnest examination of foundations which did not in any logical way lead to it; (iii) it is out of tune with the notions of scientific theory-building and theory-testing which developed during the same decades.

Explanation of such a peculiar phenomenon is called for. It should be sought in the social context of science and perhaps also in the state of physical theories. Only some very preliminary suggestions toward such an explanation are made in this paper. Specifically, two "folk explanations" are evaluated and found pertinent but insufficient; and it is ventured that a run of successes in applications, partly fortuitous, lengthened the sway of a paradigm internal to mathematics by making mathematics relatively exempt from criticism from outside.

The concluding section comments cursorily on the prospects for rapprochement between the prevailing mode of thought of academic mathematics and that of applied mathematics.

My subject is the nature and origin of a curious malaise, afflicting mathematics preferentially among the sciences, and appearing rather suddenly around the turn of the century — no earlier than Georg Cantor, and no later than Emil

Artin and Emmy Noether.[1] Hence my title.

I took up the same topic, under the same title, at the international Science and Technology meetings in Toronto in 1980. At that time the title begged a question, for the notion that 20th-century mathematics *had* gone wrong was not widely held. That has changed. Many more have come to share my perception since then, and yet there is little analysis of it; so that I may hope that this venture is even more timely now than a decade ago.

I   As Others See Us

Let us begin with the view of mathematics from the outside. Specifically, how do we mathematics professors as a group look to most professors of engineering?

Most engineering professors, when they are not downright hostile, are at least impatient with us. We think only of rigor, they say. They pronounce our term "existence proof" with a distinctly derisive tone. If we want to study solutions of differential equations, they say, why don't we look for them, instead of debating their existence? And when an answer to a problem has been proposed, they can't see why we should want to linger over the proof of its correctness instead of getting on to a fresh problem.

They might prefer to ignore us, but they can't because we teach the undergraduate engineering students calculus. The engineering professors allege that their undergraduates are forbidden in our classes to use a method of computation (even a method like the Laplace transform that has been tested by centuries of use) until after enduring our proof of convergence.

And these are just their strictures on the *useful* parts of what we study. They are convinced that most of what we spend our time on is of no practical concern whatever. Some of them denote all of this part of mathematics by the word "topology", again quite sarcastically pronounced.

You know, there's something to it. The engineering professors know us pretty well. They have no motive to be forgiving, for they compete with us (very successfully) for university resources and government research funds. I am not saying their appraisal is unbiased. I do say that, whatever its exaggerations, it provides a terse statement of the problem I want to address.

---

[1] I am perhaps referring to the same historical currents as Richard Courant in the introduction to Courant and Hilbert's *Methoden der Mathematischen Physik* in 1924, which he still thought timely when the English edition appeared in 1953.

> ... mathematicians, turning away from the roots of mathematics in intuition, have concentrated on refinement and emphasized the postulational side of mathematics, and at times have overlooked the unity of their science with physics and other fields.... This rift is unquestionably a serious threat to science as a whole.

Yet A. Douady writing just last year in the *Gazette des Mathématiciens* put the split between physicists and mathematicians as occurring about 1930. No later, surely.

II   Head and Hand

Are we dealing simply with a matter of pure versus applied science? Is the pure-versus-applied polarity, in turn, just a special case of the dialectic of theory versus practice, or "head and hand" — the sons of Mary and the sons of Martha, in Rudyard Kipling's memorable image? Surely this is part of the story.

What we would expect to find, on this basis, is a division of mathematical sciences into several specialties, some of them motivated by intellectual or esthetic quests, some of them by the needs of the economy and public policy. We would expect to find the second sort, the more applied scientists, securing funds more readily, communicating more easily with the rulers and policy-makers of society; we would expect to find the first sort, the theoretical scientists, convinced of the greater importance of their problems (which after all were chosen freely and not imposed by any customer or client) and perhaps even of their own individual superiority as scientists.

We would expect to find a repeated problem of isolation. All sub-domains of science are subject to bouts of isolation: there are stable periods where one remains content with the local definitions of the problems of investigation; but then the field is given a new burst of activity by influences from other fields and progresses rapidly in some new direction. And if fields are isolated from one another not only by having different local concerns but also by difference in the degree of "purity," then it will be that much harder to hear across the boundaries, the difficulty of breaking through boundaries may be increased and the periods of inward-looking may be longer.

We do find all of this. We find it in every part of science, and in particular we find it in mathematical parts. If a department of physics is deciding whether to appoint a new professor of cosmology or of meteorology, all the familiar problems of theoretical versus practical will come into the discussion. They will stand out more clearly yet if a department of electrical engineering is choosing a professor in signal processing and has one candidate sophisticated in Fourier analysis and another skilled in use of modern hardware. The proponents of the more directly applied alternative will justify it by its utility but may also speak of its intellectual respectability; the proponents of the other alternative will concentrate on the need to support basic research but may also contemplate potential applications in a more distant future.

Although the themes are found throughout science in our times, mathematics is singular. First, because on the scale of most to least applicable, it ordinarily sits by itself far at the inapplicable end. So much so that within the mathematical community, work is regarded as applied not only if it can serve agriculture or transportation, but if it can serve any other science! Applications to cosmology are sufficient to make a piece of mathematics applied. Thus in mathematics the pure-versus-applied debates (though they may resemble those in other disciplines) are between opponents both of which would in another discipline be far to the non-applied extreme.

This far, the uniqueness of mathematics is perfectly natural. The distinguishing characteristic of mathematics (as the word is used in this century) is not its concern with numbers but its concern with deduction. Reasoning which can be carried out without frequent reference to experience is said to be mathematical; and though many others reason mathematically, mathematicians in particular do so. It is only to be expected that mathematics, the field where applications *can* be forgotten, should turn out to be the field where they most often *are* forgotten.

Second, it is almost unheard of outside of mathematics for scientists to deny that their work has material subject-matter. A purely theoretical research topic in biological sciences will of course not concern potential medical therapies; but it will ordinarily concern some living organisms. Theoretical biology may deal with hypothetical organisms, but the practitioner would not deny that behavior of existing organisms had relevance to the problem. Theoretical biology at its most extreme is called mathematical biology, and counts as *applied* mathematics. Since the early 20th century, on the other hand, mathematics (at least if not applied) has been held normally not to be about existing things.

When the leading mathematician was K. F. Gauss, who devoted as much effort to calculating terrestrial magnetic fields and planetary motions as to any other aspect of science, no doubt was raised that mathematics referred to the material world. A century later, Bertrand Russell and Ludwig Wittgenstein alike took it as certain that it did not. The word and concept of "existence" were left free to reassign. Within mathematics, existence could be allowed to mean something else.

This philosophical (or at least terminological) peculiarity has become part of the content of the struggle around applications. If a department of mathematics is deciding between building its fluid mechanics and building its differential geometry (two fields which Gauss practiced brilliantly and which to him were closely related), metaphysics will enter the debate. The fact that fluid mechanics is generally taken to have referents in the world we live in will be held *against* it: in the mentality of 20th-century mathematicians, differential geometry seems more soundly based than fluid mechanics because less doubt arises that it can be derived from axioms without physical referents.[2]

So we are seeing more than a mere rejection of applications. The contemporary "pure" mathematician does not say only, "I am too noble to get my hands dirty on mechanical problems like you mere engineers," but something even more hostile in its defensiveness: something like, "You are destroying my true science if you entangle me with your reality."

The pure-versus-applied tension is aggravated among mathematicians (and slanted so that most resolutions are in favor of the pures) by this metaphysical peculiarity. Most 20th-century mathematicians talk as if they had a subject-matter outside of time and space. No wonder they seem snooty to others! How did they get that way?

---

[2]Paul Halmos assembles in "Applied mathematics is bad mathematics" (in *Mathematics Tomorrow*, ed. L.A. Steen, Springer, New York, 1981) a number of polemical points, including this one.

III "Naïve Platonism"

When the number theorist asserts the existence of an integer with a stated property, all fellow mathematicians believe that something has been asserted, but it really is not always clear what. If the integer has never been written down in any explicit way, then the notion of existence has departed already from that appropriate to mountains (or even elementary particles). It may depart much farther. There may be no known bound on the integer, or even if there is, there may be no known bound on the length of an algorithm for specifying it. Then the existence assertion fails the constructivist's criterion: it is not a statement about the outcome of any finite computation. I have spoken about existence of integers because this seems, of all existence theorems, the sort most likely to be concrete. An assertion of existence of a function or a set with desired properties is that much harder to make sense of. Yet such assertions are our normal focus of attention. However little I know of a fellow mathematician's specialty, I can count on the usage of the term "existence".

Let me use "naïve platonism" as an informal term for the doctrine that every properly defined mathematical object exists. No need to be precise in stating this: it is an imprecise doctrine.[3] My readers doubtless know what the doctrine is by long exposure to it, the same way I do. I am stating it just precisely enough to invite you to share my appreciation of how preposterous it is.[4]

Yet the contemporary mathematician conventionally plays the platonist, pretending to believe in an ideal world peopled by all that is imaginable. One of my colleagues answered my indignant sputterings with a ringing affirmation, "Yes, it exists, it all exists! The only criterion is consistency." I said, "So if it is not contradictory for a giraffe to be blue, then there exists a blue giraffe?" "Yes," he cried gleefully, "even the blue giraffe." (Easily said! He wasn't promising to *show* me a blue giraffe.) This departs even farther from common-sense metaphysics than merely attributing existence to products of our imagination, doesn't it? It no longer has even analogic resemblance to common-sense existence; if it did, we might say that *in*consistency of a giraffe's blueness *dis*proved the existence of a blue giraffe, but not the converse.

My colleague was kidding, but he meant it anyway. His stubborn platonism is not atypical, only his humorous openness in avowing it. Nor is this a case of the adaptation of a common-language word to use in a defined (or informal undefined) specialized technical sense. The word "consistency", to be sure, is consciously felt to be a technical term, and this is salutary, for it invites exploration of the

---

[3]An indication of wherein my informal statement of it is imprecise: All specific definitions comprise a countable set, perhaps, and the real numbers are said to be a more-than-countable set, yet all real numbers are said to exist.

[4]Those who do not find it preposterous may find further polemic elsewhere. See for example Errett Bishop, *Foundations of Constructive Analysis*, McGraw-Hill, New York, 1967; *New Directions in the Philosophy of Mathematics* (ed. Thomas Tymoczko), Birkhäuser, 1985, especially Part I; and my own "Criticisms of the usual rationale for validity in mathematics", in *Physicalism in Mathematics* (ed. A. Irvine), Kluwer, Dordrecht, 1989, pp. 343–356.

significance of sentences incorporating it. But the word "existence" is most often handled naïvely. If we know that an operator A commutes with some compact operator, then by Lomonosov's theorem A has a non-trivial invariant subspace, so the mathematician says, "Let M be any such subspace," and proceeds as if M had been *specified*. (This despite the fact that Lomonosov's proof gives no indication of how such a subspace might be located in general.)

It seems to me, despite the 20th-century mathematician's treating this mathematical existence as an article of faith, to be defended with indignation against the infidel, it is not really believed in any ontological sense. (Presumptuous of me, when my colleagues say they are believers, to deny the genuineness of their belief. But they seem to recognize by their behavior that their ideal world is under human control. I am trying to give them the benefit of the doubt: to ascribe to them no more absurdity than I have to.)

Did this ideology arise, perhaps, from the great debate on foundations of mathematics?[5] It surely can not be said to have *won* the debate, for it was not a serious contender in it. It is as if logicism, formalism, and intuitionism had battled each other into exhaustion, and the field had been left to naïve platonism by default. If it is surely not a logical consequence of the foundational debate, it may be in part a historical consequence. Subtle thinkers with complicated opinions on foundational questions abandoned this debate after the twenties and joined the uninvolved in tacitly accepting a philosophy of mathematics which nobody claimed to have found adequate.[6]

We may reconstruct what made this concept of existence, so bizarre to everyone else, acceptable to mathematicians. To us, theorem-provers, it *is* useful to be told that "there exists" a continuous function whose Fourier series does not converge everywhere. If one has a function (perhaps thoroughly constructible) whose Fourier series one would like to prove everywhere convergent, one may be glad of the warning that one can not settle the matter by proving the function continuous. The counterexample performs this favor without any of its values being displayed, because the essence is not the example itself, the essence is the warning that certain implications do not hold. For this it is not necessary that the counterexample exist in any customary sense, only that it be imaginable. In summary of this point — as it is true that mathematics is largely reasoning about what *may be* calculated or proved, it has come to seem that that is its only concern, that we think only about *what we might think*.

Another process was at work. At the same time that they were being staggered by the "pathological" examples in real analysis, nineteenth-century mathe-

---

[5]I am referring only to the first "crisis in foundations", from Cantor and Weierstrass to Russell and Hilbert and Brouwer. The "crisis" to which Kurt Gödel was central came later, and did not reinforce naïve platonism so far as I can tell.

[6]Hermann Weyl, to take the most striking example. See his *Philosophy of Mathematics and Natural Science*, Princeton 1949, especially p. 234. I discuss this in III of my "Materialist mathematics" in *For Dirk Struik* (ed. R.S. Cohen, J. Stachel, M. Wartofsky), Boston Studies in the Philosophy of Science XV, 1974.

maticians were absorbing from non-euclidean geometry the lesson that one must beware of inadvertently importing hidden assumptions from common experience into one's deductions. Formalism may not have won the great debate about foundations, but it became unchallengeable that mathematics should be kept clean by sweeping away the associations of our concepts. The mathematician's triangles should be free not only of any margin of error in the magnitudes of sides and angles, but of any properties whatever which merely arose in the mind by analogy with everyday triangular things. The purely imagined mathematical object is not only sufficient for service as a counterexample, and simpler, it is actually superior in avoiding fallacious proofs.

I have suggested two internal motivations for naïve platonism. I am not saying that it has proved a useful fiction over the decades of its ascendancy. It has surely hurt.

In the first place, if it was motivated partly by the fact that a counterexample needs only to be possible in order to have interest, naïve platonism has overdone this service to the point where it is no longer useful. Too many things are said to exist. My favorite example, but not atypical, is the non-measurable subset of the real line. This is generally regarded as a significant concept. So it would be, if in studying real functions and integrals we risked defining sets or functions which escaped from the theory. Such a risk existed prior to Borel, Baire, and Lebesgue; but it does not exist in Lebesgue theory. The non-measurable sets which Vitali proved to "exist" do not occur in mathematical practice, and there is no danger that they will.

A second drawback to the rule of naïve platonism is that it contributes to aimlessness. I recall asking a colleague why study the structure of the Calkin algebra. He smiled: "Because it's there." Now that phrase may or may not be an adequate justification for trying to climb a conspicuous mountain, and it surely is a good though partial justification for studying (say) bryophytes. As a justification for studying the structure of the Calkin algebra, it is flimsy. The Calkin algebra is *not* there in any ontological sense; it is a construct which facilitated understanding something else whose study was plainly justified, and that makes it part of the mathematical universe. Did my friend then want to infer that *everything* we can deduce from its definition (however alien to the original motivation) is worthy of investigation? Physicists, however theoretical, don't do that. It becomes clear to every student of quantum mechanics that the operators and vectors dealt with have certain aspects which are crucial to their role in describing phenomena and that the rest are not to be insisted upon.

To support studying everything definable — everything which in the naïve platonist sense exists — is too uncritically generous. For such uncharted drifting we should not be thankful. We need to orient our exploration better than that, and from time to time we remember that we do. If platonism gives us excuses to evade the question where we are headed, I hold that against it. But I fear that to many mathematicians this is one of its main attractions.

How comforting to have a subject-matter inaccessible to scrutiny except on your own terms — especially if you can convince yourself and others of its importance! While N-rays were rejected by science because no-one other than their discoverer and his direct disciples could observe them, von Neumann algebras — though just as invisible — roll merrily on. This invulnerability makes the mathematician feel secure against the engineers' exasperation at not getting their "real-world" questions answered. If the subject-matter of mathematics is outside of space and time, why should we let the ephemeral concerns of the world determine our choice of a problem, let alone the way we treat it?

The issue is not only one of closing doors to voices from physics. I recall when I was a student telling one of my teachers (with conscious temerity) that I was interested in physics. He was perfectly forgiving: "Oh, yes, physics is all right. It's a fine source of problems." Of problems, you see, to be enjoyed on our terms; not a source of direction. Perhaps he will remember the conversation now and see how at cross purposes we were.

Albert Einstein said to a mathematician, "Your job is easier than mine. What you do only has to be correct, but what I do has to be both correct and right."[7] We do not have the test of experimental truth to guide us. What then?

A natural answer was available in the early 20th century, natural at any rate to the heritors of Mach and Wittgenstein. We could have continued to think of mathematical theories in isolation from observation, for the sake of keeping our deductions clean, yet have tried to test them as Einstein would by seeking a match with observation. This level of reference to experience would have followed a trend in the thinking of the times. The 20th-century mathematician chose instead to dig a deeper moat between the ivory tower and the world. The attitude I say was natural then was not seen so: it was seen as not permissible, as non-mathematical. Revived decades later as "mathematical modelling," it remains a stepchild in most mathematicians' view.

IV   The Function We Perform

Evidently I am describing an attitude of insularity. The individuals who hold it can't really withdraw. Even the most platonist of them, whatever independence from the temporal world they might attribute to their spaces, their fields, and their rings, would hardly claim equal immutability for their own persons. They live in society, their attitudes came from social sources, and we must ask how they come to feel this way and what social function their odd behavior serves. We must seek answers specific to a time and a culture, since what we are explaining is localized. And for this phenomenon, on which the rulers of society have apparently smiled for decades, we must ask in particular how it serves the rulers.

---

[7]Quoted by J.L. Kelley, "Once over lightly", in *A Century of Mathematics in America*, Part III (ed. P. Duren, R.A. Askey, H.M. Edwards, U.C. Merzbach), Providence, American Mathematical Society, 1989, pp. 471–493, see p. 483.

One need not expect the answer to be simple: it is hard to doubt that the engineer who decries our existence theorems and our "topology" enjoys at least as warm a blessing from the captains of industry as we do. Industry needs mathematical skilled labor as well as manual and managerial. If the engineer wants the mathematician to supply it and the mathematician is recalcitrant, first appearances are that the mathematician is a rebel, or at least an obstacle, against the rulers of industry. Why then have they tolerated mathematics in its modern ethereal guise, and indeed continued to award it especially high prestige?

I will mention two answers which have been made to this, in which I see considerable truth, but not definitiveness.

It is sometimes said that the main function of mathematicians in advanced industrial capitalist society is the maintenance of social stratification. Our courses, it is said (and this part of the explanation is especially well confirmed by our experience), are the feature of formal education that performs the most decisive winnowing of students. Society tells the student, even the working-class student, you may be a dentist if you pass the test, you may be a military officer if you pass the test... and the decisive test is in math.

For this purpose, 20th-century mathematics may be peculiarly well adapted. The incomprehensibility of our research problems and their lack of plain motivation — together, of course, with their difficulty — invest us with a prestige more mysterious than that of the conventionally successful. The aridity of our courses, their remoteness from students' human concerns — together, of course, with their difficulty — make them especially forbidding, hence especially good as selectors of students with superior capacity for self-discipline (sometimes called repression). At the same time the apparent objectivity of our criteria certifies the impartiality of our grading. In these respects, 20th-century mathematics surpasses Latin grammar, which was the sieve used to determine admission to the 19th-century British élite, and far surpasses the authentically applied mathematics which maintains its interfluence with other sciences and engineering.

Not only those who want to have entrants selected will be happy with such a system. The testers get something out of it too. The testers, in this case the mathematicians, are given the prestige of mystery and the authority to determine young people's lives: rewards which have attracted priestly castes in many societies.

This explanation seems to me valid but insufficient; many aspects of the phenomenon escape it. We do serve this function, but many of us do so unwillingly. Also the account seems off-target historically, if I am right in dating the phenomenon to early in the century, for the great importance of mathematics as a selection gate does not begin until around 1940. We want at least some of our explanation to cover the entire period, say, 1910–1980.

A second explanation would subsume the mathematicians' turn toward cloud nine under a more general turn toward abstraction. During the same period as mathematics, other arms of the cultural superstructure in Europe and America went abstract. Indeed there is clear affinity between the "modern" mathematics of the category, the scheme, and the topos, the "modern" music of the row and the cluster, and the "modern" painting and "modern" poetry of multiple isms; there is the same defiant pride in incomprehensibility. More than coincidence, one feels. And the modernist mystique is clearly one more readily attached to mathematics than to the other sciences.[8] As to the social roots of this modernism, that is an outstanding puzzle which many have addressed. Perhaps one should take historians' accounts of the roots of modernism in humanist fields, and see whether they account for modern mathematics.[9] Or perhaps one should hunt not for parallel developments but for lateral influence: modernism may have been picked up by mathematicians from musicians and artists and poets.[10]

This explanation again seems persuasive but insufficient. In particular, it would predict that the modern tendency to cut ties with tradition would have gone even farther in mathematics than it has. When Edgar Varèse and John Cage strain to avoid building their music on the familiar, I have to say that they have succeeded very well, alas. Now, do we find a comparable rejection of tradition by modernists in mathematics? The Bourbaki provide a clearly central example on which to test the matter. How completely do they repudiate their antecedents? To be sure, they affect a style which sounds like indifference to the past. But on closer examination we find that they are familiar with Gauss, Abel, Riemann, and Schur, and care quite seriously about a continuity which their style denies.

There seems to be more in the phenomenon than can be dealt with by these explanations.

V   A Shield against Detractors

Twentieth-century mathematics turned toward a "pure" ideal as many sciences have, and claimed unassailability for its own definitions and criteria as many sub-cultures have; the event has much in common with other such events. We will be helping answer our question if we find factors providing mathematics of this era with some immunity from the forces countering its isolation. Let me try to suggest such a factor.

It is what Eugene Wigner called the unreasonable effectiveness of mathematics in the natural sciences.[11] I do not now refer to mathematics as defined by departmental lines within universities, nor to mathematics as defined elsewhere in this essay, but more narrowly to the classical mathematics of real numbers and

---

[8] In saying this I am asserting something about modernism. Medieval and Renaissance scientific mystique attached to other sciences as readily as to mathematics.

[9] A theoretically ambitious attempt is Peter Bürger, *Theorie der Avantgarde*, Suhrkamp, Frankfurt, 1974; and explanatory suggestions are scattered through social histories like Carl Schorske, *Fin-de-siécle Vienna*, Knopf, New York, 1980.

[10] Some mutual influence between different modernisms is seen in such movements as pataphysics. See *Oulipo: A Primer of Potential Literature* (ed. W.F. Motte), University of Nebraska Press, Lincoln & London, 1986. But this seems quite peripheral to me.

[11] His article so titled is in *Commun. Pure Appl. Math.* 13(1960), 1–14. A disclaimer: some of the important queries made by Wigner fall outside the scope of my analysis here.

manifolds, differentiable functions, and differential equations with solutions. More of experience is susceptible of expression in these terms than could have been anticipated in 1870 by any but the unimaginative optimist.

How do I as a materialist take this observation? As evidence that evolution has built into humanity a mathematical sense, pre-adapted to construct successful physical theories? Or that everyone's childhood experience of 3-space gives individual pre-adaptation? Hardly: if it were so, the Newtonian revelation would not have waited till the 17th century. I take the great success of mathematical analysis as a fact about the physical universe. Whether or not it can be called accidental in any metaphysical way, it is therefore, from the view of the 19th-century physicists who elaborated the approach, a historical accident.

Some scientific theories prove valid in no conditions much beyond those in which they were originally recognized; others, such as electromagnetic theory, are luckier. Even when they are tested unreasonably far beyond their original range, they still find confirmation. I regard the presumption that physical things can be described in terms of differentiable manifolds as a physical theory of a higher order of generality; this super-theory also has turned out to be valid far beyond the range of situations which first led it to be developed.

This was good fortune in the sense of permitting rapid progress. Its consequences for the context of our work were ironic. Just as mathematicians were learning not to take for granted that the Euclidean geometry in our minds existed out there, this run of luck with analysis-based theories deceived them by making naïve platonism less manifestly absurd than it would otherwise have appeared.

First, continual success puts one in a position to tolerate waste. Mathematical analysis, over the last 150 years, generated so much precious output that it could afford to generate a lot of junk. Carried along free were some excrescences which may have had a much less essential role than the pilots thought. I return to the example of measurability. It has long seemed to me (even before I had heard of Robert Solovay's work on alternate consistent models of analysis) that the proofs of measurability of functions which analysts have dutifully put into so many papers may in 21st-century hindsight be judged totally superfluous. Even if I am right about this, it means only that some honest mathematical toilers wasted some hours,[12] or at worst got discouraged from approaches that would have worked if they had persisted. Our lemmas asserting measurability, even if not needed to *validate* our theorems, never *in*validated them. It was just a matter of excess baggage. Analysis could bear the burden.

Second, a theory which is repeatedly confirmed begins to seem permanent and universal, inviting us to erect it into an a priori principle. This is clear in general. From the instance of Euclidean geometry, whose non-universality has been well appreciated by now, we should take warning. It will be well to keep watch for the limits of validity of the high-level theory of manifold models, lest we overstep them.

---

[12]As Kurt Friedrichs said in irritation, "In analysis the phrase 'everywhere dense' is everywhere dense."

In the great transformation of physics, Euclidean geometry was thrown open to question by special relativity, but without opening linearity of space to doubt; then it was shown by general relativity that one might not want to take it as self-evident even that space-time was linear, yet the presumption that a model of space-time would be a manifold survived unscathed.

But the transition from classical to quantum mechanics? I think the only fair answer is that *this* transition took us to domains where, behold! events no longer transpire in a space resembling a manifold; that our theory of theories saying they must was refuted by experiments some decades ago and we didn't notice; and that theory from Bohr to Gell-Mann and beyond has been patching over. The patches have been made of larger helpings of real analysis than ever entered into the older, more literally interpretable mechanics; granted. This may be in the nature of the gap to be bridged between our world and the world of subatomic distances and high energies. And the theories I am calling patchwork have sometimes been amazingly successful, granted; I do not mean to be scornful of their successes. I suggest that there is a size gap, with something qualitatively different on the other side. This attitude is informally very widespread. Maybe we should recognize it more squarely as a refutation of our thitherto invincible super-theory.

Patchwork between parameter ranges where different theories apply is respectable in some parts of physics. It should be acceptable even as to the super-theory.

I have a further conclusion about real analysis as a super-theory. It succeeds in more than one way, not only in the way which the now-conventional rigor rationalizes.

(1) A differential equation, say $y' = f(y, t)$, often describes situations very well in which the derivative can not be understood as a limit of $\Delta y/\Delta t$ as $\Delta t \to 0$. The usual example is exponential growth of a population with a finite minimum generation time; but maybe this is a superficial discrepancy, to be overcome by switching to difference equations. A more striking example is the equations of elasticity governing vibrations of subatomic amplitude (seismic oscillations, for example). Then our theory's Δy can not get down even to the order of magnitude of y itself without entering the range where classical mechanics is false! The differential equation nonetheless applies very well — better, therefore, than does the story we conventionally tell to "justify" it.

(2) If the only way for our imagined Euclidean space $\mathbf{R}^2$ or $\mathbf{R}^3$ to correspond to reality were the straightforward way — for us to put functions or sets into it which approximate point-for-point functions or sets we can measure — then the intuitionist criticism would demolish pathological sets like the Cantor set and the Koch snowflake curve. After all, how are we to declare after examining finitely many digits of a ternary expansion that *none* of its digits is 1? Nevertheless Benoît Mandelbrot argues very persuasively that some of these bizarre objects are usable models in applied mathematics. He evades the intuitionist argument by setting up a slightly different sort of correspondence between the model and the object modelled. The correspondence, involving real numbers, must inevitably be approximate

since we can only verify finitely many data; for Mandelbrot, the approximation is not to deal point-by-point with a mathematical set but to replace it by one of the sequence of computed sets involved in its definition.

The two kinds of correspondence are as different as the p-adic numbers are from the reals. It is only the prejudice for the independent existence of the reals which blinds us to this complication.

Our imagined sets show versatility, not only in the kinds of physical systems described but in the sense in which the describing is done.

This "unreasonable" success of analysis was a fortuitous circumstance, a conjuncture between a stage of theory building and the realities of the universe. It gave mathematics an exceptional claim to be left alone. The sort of economic power that insisted engineers change from studying vacuum tubes to studying semi-conductors did not insist that mathematicians lay off subjects of past concern (like the Riemann Hypothesis or the Hauptvermutung) just because new needs arose. It didn't need to, because the goose was still laying golden eggs.

This, then, is my extension of the explanation for the strange turn of 20th-century mathematics. The circumstances of progress, though they did not *oblige* us to seclude ourselves (and some of us refused to), left the option available. The turn toward the recondite was doubtless impelled by a variety of motives, including those adduced in the two explanations above; but special circumstances in mathematics created the possibility to take it with exceptional impunity.

VI Escaping Obsolescence

Time has run out. Thousands of engineers, physicists, and computer scientists have laid aside invective, they have learned to take from us whatever ideas they can use and use them. We have no monopoly on anything any more. An independent tradition is beginning to coalesce out of the proto-planetary fragments now orbiting different user disciplines. We can concede most of mathematics to them and so preserve our cherished fantasies and our intellectual isolation (and our genteel poverty). I hope we are not so timid. Conferences and journals now often imbed our papers among those of engineers or chemists. The excitement of the theory of untangling DNA, for example, is greater than the sum of the excitement on the topological side and that on the molecular biology side. The comfort of the ivory tower is outweighed by the rewards of breaking out of it.

We must enter the new communication with the "other" mathematics without illusions of superiority. The mystique of rigor is not only specious but also creates specious pecking orders. But we can enter without doubt that we have something distinctive to offer. Some things we have associated with our platonism may remain and serve well when it is abandoned. The 19th-century rigorizing of calculus teaches something if the pretentiousness is stripped away; the "global" approach in algebra and topology sometimes does jobs the engineer couldn't do without it.

More and more we will be patching between domains of different approaches.

I have given an alleged example in basic physics. My favorite example is the Eudoxus-Dedekind derivation of the real numbers: it is not needed to legitimate the reals (I would be willing to think about them anyway, for I see their like in the world), but to clarify their relation to the natural numbers (which I also like to think about). I have expressed a similar view of epsilonics in calculus.

I have been pleading for years for a patchwork conception of rigor. What canon of rigor is appropriate may vary with the context and the intended application. As I have noted elsewhere,[13] many an existence proof ought to be taken as an exhortation to construct the object. And I noted above that in some contexts non-constructive objects are natural. Anyway, it is hard to put a wall around constructivism; for example, as Marian Pour-El and Ian Richards have explained, a differential equation with constructive data may have non-constructive solution. For some purposes, what is really at issue is not existence or constructiveness in accustomed senses but computability in polynomial time. Much of what is called theoretical computer science today is a branch of mathematics largely concerned with patching between domains governed by what we could call different standards of rigor.

The change I am describing is under way. There is no question of the direction of change, but the extent of the change is minor compared to what I have advocated. When Pierre Cartier exclaimed after the International Congress of 1986 that the ivory tower had crumbled,[14] he was surely overstating. And though we now see openness to outsiders in many quarters, it is much more openness to their problems than to their criteria. Still we are headed toward breaking out of our isolation.

Let me follow Cartier with some comments from the International Congress of 1990. Two of the plenary speakers on the last day were Yakov Sinai and Vaughan Jones. Sinai observed with satisfaction that ten of the thirteen plenary talks had been related to applications,[15] and declared, "We are witnessing the reappearance of natural philosophy." Jones closed his talk with a plea to fellow mathematicians not to reject Witten's formula for counting knotted topological graphs just because it hadn't been rigorously proved. Notice that he wasn't asking us to regard the formula only as a conjecture. Witten, a mathematical physicist, considers it established; Jones was asking us to give some recognition to a different standard of rigor from the usual.

Twentieth-century mathematics is beginning to recover from its aberration.

---

[13] "Criticisms of the usual rationale for validity in mathematics", in *Physicalism in Mathematics* (ed. A. Irvine), Kluwer, Dordrecht, 1989, pp. 343–356; see Sec. 4(iii).

[14] *Gazette des Mathématiciens* 32 (1987), 20–26.

[15] I counted fewer. But he might have noted that of the five medals awarded at the Congress (four Fields medals and one Nevanlinna prize) at least three went to rather application-motivated work.

Indian Mathematics in Arabic[*]

Roshdi Rashed

Although speculations abound concerning the transmission of Indian Mathematics into Arabic, there is little convincing evidence that many methods and results were known to Arab mathematicians. Many texts are either lost or not yet published, or carelessly edited and analyzed. No important work either, as far as I know, is devoted to the Indian legacy in Arabic, and to translation from Sanskrit to Arabic. Consequently, it is not yet possible to draw a synthetic picture of the transmission of Indian mathematics or grasp its general characteristics. The only conjecture that I dare suggest is that the Indian mathematics transmitted were mainly computational, dependent on Astronomy. My concern in this talk is rather modest; I should like to present a hitherto unknown source[1] which provides further confirmation of the previous conjecture — partially at least — and which illustrates why and how a sector of Indian mathematics was transmitted and diffused among Arab mathematicians. It is necessary however to recall at first — very briefly — the mathematical context in which this Indian text took place.

In the second half of the 10th century, Arab mathematicians were already highly interested in numerical analysis. The origin of this interest is to be found in other topics: algebra and observational astronomy. We know better now that mathematicians like al-Karajī were developing algebraic calculus. Many of the algebraic activities were about the new algorisms, necessary for polynomial algebra, the numerical resolution as to say "pure" and "affected" equations. One of the mathematicians who had contributed to this research at the time of al-Karajī, was Abū al-Rayḥān al-Bīrūnī. He wrote a one hundred folios in length book — now lost — on the extraction of the nth root of an integer. During the same period, astronomers were studying interpolation methods required to undertake the composition of their trigonometrical tables and their astronomical *zījs*. A good illustration of this activity at that time are the works of Ibn Yūnus[950–1009] in Egypt, and al-Khāzin — mid 10th century — in Western Iran. Both of them invented and applied a quadratic interpolation. This interpolation, given by Ibn Yūnus in his *zījs* for instance, may be written in modern notations as follows:

$$y = y_{-1} + \left[\frac{x - x_{-1}}{d} \right] \left[\frac{1}{2}(\Delta y_{-1} + \Delta y_0) + \frac{1}{2} \left[\frac{x - x_{-1}}{d} \right] \Delta^2 y_{-1} \right],$$

---

[*]This paper is summarized version of "Al-Samaw'al, al-Bīrūnī et Brahmagupta: les méthodes d'interpolation" in *Arabic Sciences and Philosophy: A Historical Journal*, **1** (1991): 101–60.

[1]This source has been edited and translated into French, *ibid.*

with $d = x_i - x_{i-1}(i = 0, 1, 2, \ldots)$ Δ first order difference, Δ^2 second order difference.

It is well known for more than half a century[2] that with his work on Astronomy, *al-Qānūn al-Mas'ūdī*, al-Bīrūnī contributed to the advancement of this research on interpolation methods. We should note before going further that al-Bīrūnī had contributed to the two topics of numerical analysis at that time. In this respect, he had been followed by other mathematicians, al-Samaw'al in the next century, for example. The combination of both activities must be kept in mind because it influenced how to conceive mathematical research on algorisms.

This is, briefly, the mathematical context, when al-Bīrūnī became interested in the *Khaṇḍakhādayka* of Brahmagupta. Not only al-Bīrūnī quoted the Brahmagupta's zīj in his writings, notably in *al-Qānūn al-Mas'ūdī*, but, from a list of his works written by him when he was sixty-three years old, we know that he revised an Arabic translation of it. He wrote: "I corrected the zīj of al-Arkand and translated it into my own words because the extant translation is incomprehensible and the Indian terms were left as such." Al-Bīrūnī tells us also that then he was writing the final version of his Astronomy, *al-Qānūn al-Mas'ūdī*. The historian can not avoid a topic of some interest to the history of Indian mathematics in Arabic. Brahmagupta in his zīj did apply a formula of parabolic interpolation. However, there is no trace of this formula in *al-Qānūn al-Mas'ūdī*, but, on the contrary, we find another quadratic interpolation. Al-Bīrūnī own words confirm that he was very well acquainted with Brahmagupta's zīj. This situation has puzzled historians of science, and E. Kennedy, a specialist of al-Bīrūnī, wrote recently: "the curious fact that a book of Brahmagupta which was well known to al-Bīrūnī contains a legitimate parabolic interpolation ignored by him although he mentions the book many times in his writings."[3] This reaction was the only one possible as long as the unique source of our knowledge about al-Bīrūnī's contribution to this topic remained his *al-Qānūn al-Mas'ūdī*.

Now the discovered source will modify this unlikely conclusion, and will help us to prove that:

1. al-Bīrūnī and his successors were fully aware of Brahmagupta's methods;

2. al-Bīrūnī was, moreover, well informed about another Indian parabolic method, whose origin I cannot trace;

3. al-Bīrūnī and his successors did not apply these methods — Indian and their own — indiscriminantly. They compared them with one another to choose the best for the given function. In order to realize these comparisons, they combined empirical research and theoretical considerations.

---

[2]M. A. Kazim, "Al-Bīrūnī and trigonometry" in *Al-Bīrūnī Commemoration volume* (Calcutta, 1951), pp. 161–70.

[3]E. S. Kennedy, "The Motivation of al-Bīrūnī's Second Order Interpolation Scheme" in *Proceedings of the First International Symposium for the History of Arabic Sciences* (Aleppo, 1978), pp. 67–71 (p. 67).

The new source which provides this information is a chapter never studied before from the book of the 12th century mathematician: al-Samaw'al. This is chapter seven of his "On the discovery of the defects of Astrologers and their errors in most of their procedures and predictions," written in 561/1165-6, i.e. six years before his very important book in which he presented the so-called Ruffini-Horner method, and the first theory of decimal fractions, and eight years before his death. The chapter itself is entirely devoted to a tract of al-Bīrūnī. Although its title is omitted, it is very likely that this tract is the one mentioned in al-Bīrūnī's list as "*the Sankalt of numbers*." Al-Samaw'al began by a long quotation from al-Bīrūnī, in which the 11th century mathematician deals with Brahmagupta's method, and the other Indian method. Al-Bīrūnī explicitly attributed the Brahmagupta's method to its author, to which he preferred another method he called: "the sankalt method." This choice was severely criticized by al-Samaw'al.

Now, the word *sankalt* is an Arabic transcription of a sanskrit word. In another work on astrology by al-Bīrūnī, *al-Tafhīm*, the word is used as a technical term in mathematics which means "the sum of a series." This meaning is confirmed: the word means "add together, heap together." As a technical term in mathematics, *sankalita* (the past passive particle) means "the sum of a series." However, the translation of the word by al-Bīrūnī, the Arabic word "*al-mufrad*" which means as a mathematical term "monomial." The method will therefore be called "the monomial method." There is no doubt, I believe, as to the Indian origin of this method, in spite of the fact that al-Bīrūnī did not attribute it to an Indian author.

Let us take now the interpolation methods explained in al-Bīrūnī's fragment — the linear interpolation designated by the Astronomers' method — to which we add al-Bīrūnī's method presented in his *al-Qānūn al-Mas'ūdī*. These methods could be written in another notation for

$$x_{-1} < x < x_0, \qquad d \;=\; x_0 - x_{-1} = x_i = x_{i-1} \quad (i = 1, 2, \ldots)$$

$$(\alpha) \qquad\qquad y \;=\; y_{-1} + \left(\frac{x - x_{-1}}{d}\right) \Delta y_{-1},$$

the Astronomers' method;

$$(\beta) \qquad\qquad y = y_{-1} + \left(\frac{x - x_{-1}}{d}\right) \left[\Delta y_{-2} \left(\frac{x - x_{-1}}{d}\right) \Delta^2 y_{-2}\right],$$

al-Bīrūnī's formula; its application requires the calculus of $\Delta y_{-1}, \Delta^2 y_{-2}$ and $x_{-1} > d$.

$$(\gamma) \qquad\qquad y = y_0 + \left(\frac{x - x_0}{d}\right) \left[\frac{\Delta y_{-1} + \Delta y_0}{2} + \frac{1}{2}\left(\frac{x - x_0}{d}\right) \Delta^2 y_{-1}\right].$$

According to al-Bīrūnī, this method supposes $x < x_0$ and would be written

$$y = y_0 + \left(\frac{x_0 - x}{d}\right) \left[\frac{\Delta y_{-1} + \Delta y_0}{2} + \frac{1}{2}\left(\frac{x_0 - x}{d}\right) \Delta^2 y_{-1}\right],$$

with $(x_0 - x) > 0$, $\Delta y_{-1} < 0$, $\Delta y_0 < 0$ and $\Delta^2 y_{-1} > 0$.

In other terms, according to al-Bīrūnī, Brahmagupta's method supposes $x < x_0$, and the correction is additive. In a sense, there is a slight difference with Brahmagupta's expression, at least in Sengupta's translation: "Multiply the residual arc left after division by 900′ [d] (i.e. by 15°), by half the difference of the tabular difference passed over and that to be passed over and divide by 900′ (i.e. by 15°); by the result increase or decrease, as the case may be, half the sum of the same two tabular differences; the result which is whether less or greater than the tabular difference to be passed, is the true tabular difference to be passed over" [The *Khaṇḍakhādayka*, p.141].

Finally, the monomial method:

$$(\delta) \qquad\qquad y = y_0 - \frac{(x_0 - x)(x_0 - x + 1)}{d(d+1)} \Delta y_{-1}$$

which proceeds by calculating the increments from x_i to x_{i-1}.

It is of some interest for the history of numerical analysis to answer the question about the origin of these methods. There is a partial answer by al-Bīrūnī himself, indicated but not justified. For him quadratic methods, and particularly Brahmagupta's one, were invented to improve linear interpolation. Now, if we go back to the cotangent tables considered by al-Bīrūnī— without diminishing whatsoever the generality of the discussion — we see that the values obtained by the linear interpolation are by excess, while the variation of the first order differences is not uniform. Al-Bīrūnī's interpretation is that Brahmagupta, conscious of this fact, had searched the second order interpolation. Let us take anew the conjecture of al-Bīrūnī for all three quadratic interpolations, and replace in (α) Δy_0 by a Δ dependent on x. Start from Δy_{-1} by linear interpolation for $x = x_0$, $\Delta = \Delta y_{-1}$ and for $x = x_1$, $\Delta = \Delta y_0$. On $[x_0, x_1]$ this interpolation gives

$$\Delta \;=\; \Delta y_{-1} + \left(\frac{x - x_0}{d}\right)(\Delta y_0 - \Delta y_{-1})$$

$$\text{hence}\quad y \;=\; y_0 + \left(\frac{x - x_0}{d}\right)\Delta = y_0 + \left(\frac{x - x_0}{d}\right)\left[\Delta y_{-1} + \left(\frac{x - x_0}{d}\right)\Delta^2 y_{-1}\right].$$

If we take again this interpolation on $[x_{-1}, x_0]$, we obtain the (β) formula. Let us now consider the linear interpolation on $[x_{-1}, x_0]$, we obtain

$$\Delta = \Delta y_{-1} + \frac{1}{2}\left[\frac{x - x_{-1}}{d}\right]\Delta^2 y_{-1} = \Delta y_{-1} + \frac{1}{2}\left[\frac{x - x_0}{d}\right]\Delta^2 y_{-1} + \frac{1}{2}(\Delta y_0 - \Delta y_{-1})$$

$$\text{hence}\quad \Delta = \left[\frac{\Delta y_{-1} + \Delta y_0}{2}\right] + \frac{1}{2}\left[\frac{x - x_0}{d}\right]\Delta^2 y_{-1} \qquad \text{hence}$$

$$y = y_0 + \left(\frac{x - x_0}{d}\right)\Delta = y_0 + \left(\frac{x - x_0}{d}\right)\left[\left(\frac{\Delta y_{-1} + \Delta y_0}{2}\right) + \frac{1}{2}\left(\frac{x - x_0}{d}\right)\Delta^2 y_{-1}\right]$$

which is Brahmagupta's formula.

Next we come to the monomial method. Divide the interval $[x_0, x_1]$ into d equal parts. It may be evident from the cotangent table that the function decreases more rapidly near x_0 than near x_1. Let us consider cumulative increments from x_1 to x_0

$$\epsilon + 2\epsilon + \cdots + d\epsilon = \frac{d(d+1)}{2}\epsilon = |\Delta y_0| \qquad \text{hence} \qquad \epsilon = \frac{2|\Delta y_0|}{d(d+1)}.$$

The correction to be made on y_1 is additive and corresponds to a cumulative increment on $(x_1 - x)$, where $(x_1 - x)$ is an integer; then

$$y = y_1 + c$$

where

$$c = \frac{(x_1 - x)(x_1 - x + 1)}{2}\epsilon = \frac{(x_1 - x)(x_1 - x + 1)}{d(d+1)}|\Delta y_0|$$

and

$$y = y_1 - \frac{(x_1 - x)(x_1 - x + 1)}{d(d+1)}\Delta y_0$$

But $x_1 - x = x_1 - x_0 + x_0 - x = d + x_0 - x$,

$$y = y_1 + \left(\frac{x - x_0}{d} - 1\right)\left(1 - \frac{x - x_0}{d + 1}\right)\Delta y_0$$

$$\text{hence} \quad y = y_0 + \frac{x - x_0}{d}\left[\frac{2d + 1}{d + 1}\Delta y_0 - \left(\frac{x - x_0}{d}\right)\frac{d}{d + 1}\Delta y_0\right]$$

relation of the form

$$y = y_0 + \frac{x - x_0}{d}\Delta$$

where Δ is first degree relatively to $\dfrac{x - x_0}{d}$.

Finally, we obtain the monomial formula if we consider the interpolation on $[x_{-1}, x_0]$.

Al-Bīrūnī was definitely right when he affirmed that the three quadratic methods are three different procedures to improve linear interpolation.

The second topic which is interesting for the history of these methods is why al-Bīrūnī did not apply his own method in his tract, and why he preferred the monomial formula to that of Brahmagupta. As al-Bīrūnī did not give any indication, whatsoever, these historical question may only have a mathematical answer. The answer, in this case, is quite probable, but not completely certain. We compared the different interpolations given by each of the three formulas, and the cotangent curve on the different intervals considered by al-Bīrūnī. In this way, we can prove that for the interval $[2°, 5°]$ he considered, and for other intervals of 3 degrees amplitude, $[4°, 7°]$, $[5°, 8°]$, $[6°, 9°]$, the monomial method gives a better interpolation than Brahmagupta's one. For all the other intervals of 3 degrees amplitude, and the intervals of smaller amplitude, however, the last one gives a better interpolation. Moreover the Brahmagupta method in general has greater possibilities of approaching better the cotangent function. Al-Bīrūnī method (β) is less suitable than the two others for this function.

In his commentary on al-Bīrūnī text, al-Samaw'al not only criticized his preference for the monomial method, but also attempted to explain it. In his opinion, al-Bīrūnī might have generalized the result obtained for particular interval $[2°, 5°]$ to all intervals. This criticism could be justified if al-Bīrūnī had generally preferred the monomial formula to Brahmagupta's formula, and not only for certain intervals where the interpolation of the cotangent function is not easy, because we are in the neighbourhood of the pole. In any case, it is difficult to imagine the true intention of al-Bīrūnī from this fragment alone.

More important than the defense of Brahmagupta's formula by al-Samaw'al against the Sankalt formula, is his own contribution to improve all the methods proposed. His main idea was to find convenient ponderations. This idea is indeed interesting when first order differences are important. This is quite true for the intervals $[2, 3]$, $[3, 4]$, $[2, 5]$. But it is less interesting when the cotangent decreases less rapidly. In this case, the ponderated mean is nearer and nearer the arithmetical mean.

Without going into greater detail, let us now conclude. Thanks to al-Bīrūnī's fragment and to al-Samaw'al's commentary, we have a better understanding of the contributions of these authors to the advancement of research on interpolation methods as well as their Indian predecessors'. Unlike what has been asserted until now, al-Bīrūnī knew three quadratic methods, two of them are of Indian origin; his successors were not only his followers, but also his critics. All of them were obviously trying to improve linear interpolation. They followed two directions: substitute linear interpolation for a parabolic one, or find convenient ponderations. The increasing number of these methods led the mathematicians to ask questions, which they had not yet the conceptual means to answer. Al-Samaw'al, for instance, was asking about the speed of the differences. Nevertheless, to test the performance of each method, they determined the approximative magnitude of the error for different intervals. A simple examination of these empirical determinations confirms the fact that they were guided by a certain knowledge of the studied function, here the cotangent. This knowledge however is only presented in the text in numerical form. Nevertheless this knowledge determined the choice of this method rather than another for a certain interval. This is, I believe, al-Bīrūnī's motive.

Al-Bīrūnī's example illustrates the topicality of Indian mathematics in Arabic. Although the picture of Indian mathematics in Arabic still cannot be complete, the extant documents confirm that transmission was obviously motived by the research taking place at that time, and also suggest a conjecture: Arab mathematicians had translated only results and methods of applied mathematics connected with astronomy, or, more generally, computational techniques. Once translated, these results and methods were submitted to a new treatment and acquired a new signification.

The *Tetsujutsu Sankei* (1722), an 18th Century Treatise on the Methods of Investigation in Mathematics

Annick Horiuchi

Introduction

The mathematics which developed in Japan during the Edo period (1600–1868) may be considered as a prolongation or a late rebirth of the Chinese mathematical tradition. The former retained the latter's basic features as a science of calculation (*sangaku, suanfa* in Chinese), that is a science aimed at calculating quantities and elaborating general algorithms through which these quantities could be determined.[1]

The scarcity of comments on methodology and the extreme concision of mathematical texts have often been an impediment to the understanding of the way ancient Chinese mathematicians constructed and formulated their algorithms. Although written in Japan in a late period, Takebe Katahiro's[2] *Tetsujutsu sankei* appears as an invaluable guide to the way mathematics was thought and practised in the Chinese tradition. Furthermore Takebe's book is a marvelous illustration of the high standard of Japanese Mathematics at the beginning of the 18th century and informs us of the new reflections prompted by the latest developments.

To get an overall view of Takebe's work and to understand its significance against its historical background, I will consider first some aspects of his life.[3]

---

[1] The general framework of Chinese mathematics as a science of calculation was constituted by the Han period and it remained almost unchanged until the Jesuit's introduction of Western mathematics in the 17th century. During the Song and Yuan periods (11th to 13th centuries), new powerful procedures were added to the corpus, which gave a significant impetus to the mathematical research. The *tianyuan* procedure for solving problems which historians generally designate by "algebra," although it differs from the Arabic tradition in many aspects, played a key role in this renewal [Li Yan, Du Shiran 1987], pp. 109–174.

This new trend in mathematics didn't last long and a good part of 13th achievements fell into oblivion during the Ming dynasty (14th-17th centuries). Japanese mathematicians had access to ancient treatises in the 17th century only because a number of them were preserved in Korea where they were used as official textbooks. [A. Kodama 1966].

[2] Following the Japanese custom, Japanese names will be given with family names in the first position.

[3] The reader will find a thorough account of his life and his work in [A. Horiuchi, 1990].

The *Tetsujutsu sankei*, a work written sometime before 1722, crowns Takebe's brilliant career as a scientist. His work and thought in mathematics was strongly influenced by Seki Takakazu whom he met in 1676, when he was still a young man. By that time, Seki was already famous for his achievements in various fields and especially in the field of algebra where he succeeded in extending the *tianyuan*[4] procedure for solving problems in such a way that a very wide range of problems, including all the difficult problems of the time, could be easily solved. Takebe entered Seki's school in order to receive his teaching about the problem-solving methods.[5]

Between 1680 and Seki's death in 1708, there was a close collaboration between Seki and the Takebe brothers (Katahiro's elder brother Kataaki was also mathematician) and a number of important treatises were produced during this period. Their collaboration was so close that historians sometimes find it difficult to identify the genuine author of the treatises.[6]

Seki could hardly have found a better associate than the young Takebe. Their mathematical approaches were quite complementary: Seki was certainly the most imaginative of the two, opening up new fields in mathematical research. During Seki's lifetime, Takebe concentrated on refining and rectifying Seki's intuitions and on improving numerical methods.[7]

Takebe's work was most innovative in the mathematical treatment of problems raised by calendrical astronomy. By Seki's time, Japanese mathematicians were especially interested in the search for a procedure for computing the length of an arc of the circle a in terms of the sagitta s and the diameter d [see Fig. 1]. This problem was crucial for Chinese calendrical astronomy in the same way as trigonometry was indispensable to Western classical astronomy.[8]

Seki opened up a new path by using the algebraic tool in the calculation but he did not find any solution which completely satisfied his successors. The *Tetsujutsu sankei* appears as the first treatise in which a definitive solution is given as an infinite series or, to use Takebe's own terms, an "inexhaustible" procedure. It will become clear below that Takebe's reflections on mathematical science were precipitated by this experience and were supported by the confidence he thus gained in his method of investigation.

---

[4]The *tianyuan* (*tengen* in Japanese) procedure spread among Japanese mathematicians in the middle of 17th century after the publication of a Japanese edition of the *Suanxue qimeng*, a 13th century treatise of Zhu Shijie. For an outline of this work, see [Lam Lay Yong 1979].

[5][A. Horiuchi, 1992].

[6]This is especially true for the *Taisei sankei* [The Great Accomplished Classic of Calculation], which was planned in 1683 by the three mathematicians but was achieved by the Takebe brothers.

[7][A. Horiuchi, 1990], pp. 399–435.

[8]Concerning the relations with the calendrical astronomy, see [A. Horiuchi, 1987].

Fig. 1

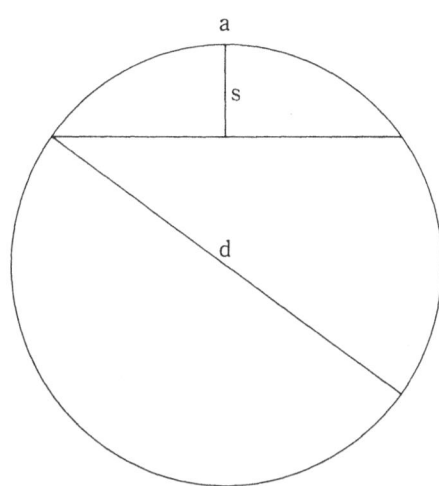

I Preliminary Remarks

The *Tetsujutsu sankei* consists of a foreword, twelve sections composing the main body of the treatise and a short postscript entitled "Discussing one's own attributes." Each section consists of a detailed description of the investigating process used for a well known mathematical problem.

Takebe's treatise is an unusual attempt to articulate the actual process of investigation, which was rarely recounted in the mathematical textbooks. The specific terminology Takebe used is especially of great interest.

When dealing with this terminology, one should be aware that mathematical concepts are not necessarily valued in the same way in different mathematical traditions. One rudimentary concept in Western mathematics may correspond to several refined concepts in Takebe's vocabulary. Conversely, an elaborate analysis found in Western mathematics may have no counterpart in the Chinese and Japanese mathematics. For example, terms like *hō*, *jutsu*, *hōjutsu*, or *waza* are all connected to the idea of method, technique or device. But clearly the Japanese terms (most of which are of Chinese origin) are not interchangeable and we ought to consider the differences between them. Similarly, Takebe juggles with many words to express the process of *seeking* and *finding* a mathematical object: *motomeru* (the verb can be written in two ways using different characters), *sassu*, *saguru*, *kaisu*, *uru*, *satoru*, *tsukusu*.... It is a difficult task to know if by translating them into determine, discern, seek, grasp, obtain, realize, exhaust, we retain their genuine meaning. Conversely, the Japanese mathematician merely comments on the deductive process and the terminology related to it is very rudimentary.

Most of the terms used by Takebe were not alien to his Chinese or Japanese predecessors. But Takebe made a remarkable effort to define the context in which each term was to be used and to create a refined tool of analysis. Of importance is the fact that many of the terms, particularly the ones referring to the abstract part of mathematical research, have a Neo-Confucian[9] origin. This gives evidence of the significant role played by Neo-confucian thought in the formation of Japanese science in the early Edo period.

II   The Purpose of Mathematics

[The science] of calculation deals with the establishment of rules, the clarification of the principle of procedures and the calculation of numerical values.[10]

In this single sentence, Takebe summarizes the three basic duties and aims he assigns to mathematics. They are connected with three important mathematical objects in the Japanese tradition, namely the rules (*hōsoku*), the procedures (*jutsu*) and the numerical values (*sū*).[11]

---

[9]Neo-Confucianism was a trend of renewal in philosophy which appeared in China during the Song period among educated elite. They especially developed a new metaphysical discourse as a response to the immense power of Buddhism and Taoism. Neo-Confucianism was raised to the status of a national orthodoxy during the Yuan dynasty (1279–1367) and kept this role until the end of the Qing (1662–1911). In Korea and Japan too, Neo-Confucianism was in great favour with educated people and at Takebe's time, "learning" was almost synonymous of studying Neo-Confucian classics.

[10][K. Takebe, 1722], foreword.

[11]The *Tetsujutsu sankei* is divided into three parts, each of which consisting of four chapters.

I. Investigation of rules

   1. Multiplication and division

   2. *Tengen* [*tianyuan* in Chinese]

   3. Simplification of fractions

   4. *Shōsa* [literally the "determination of differences" i.e. the Chinese interpolation method using finite differences).

II. Investigation of the principle of procedures

   5. Procedure of repeated exchanges between weavers

   6. Procedure for finding the maximal volume of a parallelepiped

   7. *Sandatsu* [a mathematical problem of congruences stemming from the problem known in the West as the Josephus problem]

   8. Procedure for calculating the spherical area

III. Investigation of numerical quantities

   9. Numerical quantities stemming from the pulverization

   10. Numerical quantities obtained through square-root extraction

   11. Numerical quantities related to the circle

   12. Numerical quantities related to the arc of the circle

To understand the scope of this definition, it is worth recalling some basic features of Chinese and Japanese mathematics. Mathematical problems in most of Chinese treatises followed a conventional scheme consisting of the terms of the problem, the numerical answer and the "procedure."[12] The procedure, the *jutsu* (*shu* in Chinese), was the list of operations to be performed on the appropriate instrument of computation (the abacus or the counting-rods) in order to obtain the desired quantity. The operations were expressed in general terms and were, in most cases, similar to what today's mathematicians call algorithms.[13] Although this general scheme went through important changes during Seki's time, Takebe's definition should be considered against this general background.

For Takebe too, the "calculation of numerical values" (the term *sū*, usually translated as "numbers," designates all the numerical values which can be sought in mathematical problems) is the raison d'être of the mathematical investigation. The procedures show the way which leads to numerical results and the rules are the tools necessary to obtain the procedures.

Concerning the procedures, Takebe especially stresses the fact that their principle (*ri*, *li* in Chinese) should be clarified.

Principle, in Neo-Confucian metaphysics,[14] meant something close to a pattern, an order or a logic inherent to each existing thing in the universe. This order itself was seen as the manifestation of an all-encompassing order alternatively called *li* or *dao* (the way) which gave the whole meaning to the life in the present world. The existence of the inherent logic within existing things and the possibility to know it was an important claim of Neo-Confucian philosophers. The

---

[12]The term "procedure" is used by several historians and especially by K. Chemla who studied in detail the Chinese Algorithmic Tradition. See [K. Chemla, 1991].

[13]Here is a typical example of a problem solved by using the *tengen/tianyuan* method. Let us give before the general pattern of the solution.

First, a quantity is chosen as the unknown. Then two different algebraic expressions equal to the same quantity are set up. The equation is derived by subtracting one of the expressions from the other. Finally, the numerical value of the unknown is determined by extracting the root digit by digit. One basic feature of the *tianyuan* method was the use of counting rods to carry out the polynomial calculation as well as the extraction of a root. For convenience, the usual representations with counting-rods are replaced in the following translation with modern algebraic notations.

"Suppose we now have a rectangular area of 180 *ho*, the sum of the width and the length being 27 *ho*. The length and the width are required. The answer is: 12 *ho* as the width and 15 *ho* as the length. The procedure is: take one rod and put it as the width (x). Subtract it from the sum of the width and the length. The remainder gives the length ($27 - x$). Multiply it by the width. That gives the rectangular area ($0 + 27x - x^2$). Keep it on the left. Set the rectangular area. Subtract it from [the rods] kept on the left. The following configuration is obtained ($180 - 27x + x^2$). Open it like a square [i.e. solve the second-degree equation]. 12 *ho* is obtained for the width. When subtracted from the sum, we obtain the length 15 *ho*." [K. Takebe, 1722], Chapter 2. The equation satisfied by x is $180 - 27x + x^2 = 0$ which is solved using the classical procedure of extraction. For further details, see [Li Yan, Du Shiran, 1987], pp. 135–140.

[14]See [Th. De Bary, Wing-Tsit Chan, B. Watson 1960], pp.455–502.

clarification of the principle of things took on a primary importance in this context insofar as the very understanding of the meaning of the universe was engaged in this investigation.

Interestingly, Takebe distinguished between the *formulation* of the procedure and the clarification of its principle. The investigation would be completed only when the principle has been obtained, that is when a certain level of understanding of the procedure has been reached. A more detailed study of the book will precise the point.

Let us now turn to the aim Takebe quoted the first, that is the elaboration of rules (*hōsoku*). The meaning of this term which was not of a current use among mathematicians, is close to norm, law or fixed model. Takebe designates by this name rules which can be carried out in various contexts such as rules of multiplication, division, simplification of fractions or the algebraic method for solving problems.

A category of universal tools was thus defined which was to play a prominent role in the process of elaborating procedures, as expressed by the following remark:

> To establish the rules is the basis of carrying out the procedures.[15]

Let us now turn to a more specific purpose that Takebe had in mind when writing the *Tetsujutsu sankei*.

III   The *Tetsujutsu*: The Art of Linking

The *tetsujutsu*, Takebe wrote in the foreword, consists in linking, investigating and thus finding out the "principle" of procedures.[16]

Takebe further specified:

> If investigating a case is not sufficient for finding out the principle of the procedure, investigate another case. If two cases are not enough, investigate a third case. Even though the principle is deeply buried, if one keeps investigating several times, a point of maturation will be reached where it is impossible not to find [the principle].[17]

The first difficulty here is to clarify the meaning of the central concept of *tetsujutsu*. The word *tetsujutsu* splits into *tetsu*, which means to link up, to gather, to combine, to bind up and even to write and *jutsu*, which means procedure or technique. The concept of *tetsujutsu* (*zhuishu* in Chinese), has quite a long history in Chinese astronomy. Takebe's only explicit reference in his foreword was to the work of the mathematician Zu Chongzhi (5th century AD) which was also entitled *zhuishu*. But this book had been lost long before Takebe's time and he had little access to Zu's idea of *zhuishu*. Takebe only knew from the *Suishu* (the History of Sui

---

[15] [K. Takebe 1722], chap. 1.
[16] *Op. cit.*, Foreword.
[17] *Op. cit.*, Foreword.

dynasty) that there was some connection between the *zhuishu* and Zu Chongzhi's works on circle measurement.

We can find several other instances of the use of this term between the 5th century and Takebe's time. In most cases, it referred to the interpolation method used by Chinese calendar makers. They are not of great help here insofar as Takebe clearly resorted to it as a means of legitimazing a concept of his own.[18]

Now, what can be said of Takebe's *tetsujutsu*? The last sentence of the foregoing quotation provides us with a clue. Takebe says that the pursuit of particular researches in a given area finally leads to the knowledge of the principle and that *tetsujutsu* is no more than the process of investigating and finding a pattern common to a set of cases. Takebe's statement brings to mind the so-called incomplete induction consisting in generalizing a mathematical relation verified by a great number of numerical values. Yet, the *tetsujutsu* differs from the induction in one respect. It dose not include any idea of *generalization* and emphasizes mainly the cumulative aspect of mathematical research and the need to link up several experiences to obtain the result. Rather than stressing the creative or constructive aspect of mathematical research, mathematics is viewed here as a task of familiarization with tangible elements such as numbers leading to the knowledge of the pattern underlying them.

This interpretation is supported by Takebe's conclusion to the foreword.

Having been led to use the term *tetsu* of this author, I cannot help thinking that Zu Chongzhi was a genius of Antiquity. Isn't the truth (*shinjitsu*) a thing that we cannot reach either through education or through contemplation?[19]

The *tetsujutsu* is described here as a cognitive method which could neither be communicated orally nor be reached by a direct apprehension. This suggests that it could only be obtained through a material or a physical approach, that is through practice. This significance of Takebe's statement will be better appreciated when examining more closely his concept of mathematical investigation.

---

[18] As regards this point, a striking coincidence should be pointed out: some of the mathematical treatises produced in China under the Jesuits' supervision also refered to the concept of *zhuishu* as a term encompassing the whole science of mathematics. In the present state of research, there is no evidence of contacts between Chinese and Japanese mathematicians in the 17th century. Because of important discrepancies in the way the term was used in the two countries, we are inclined to think that the term was familiar enough to the calendrical astronomers to be used independently by Takebe in Japan and by the Jesuits in China.

[19] *Op. cit.*, foreword.

IV The Distinction between "Investigation Relying on Principle" (*Ri ni yotte saguri motomeru*) and "Investigation Relying on Numbers" (*Sū ni yotte saguri motomeru*)

Takebe's analysis relies on a very important distinction between two types of investigation: investigation relying on principle and investigation relying on numbers which are also designated as "conforming"(*jun*) and "contrary"(*geki*).

> Mathematics deals with the establishment of rules, the clarification of the principle of procedures and the calculation of quantities. This [process] is "conforming" if the principle is discerned, the procedure unfolded and the quantities are obtained by using this procedure. It is "contrary" if the procedure is tested according to quantities and the principle is sought by using the procedure. The conforming and the contrary are unified in the *tetsujutsu*.[20]

The distinction lies in the way numbers (that is numerical values) are used in the investigation. In the former case, the principle of the procedure is grasped at once and the mathematician does not need to calculate any numerical value to obtain the procedure. In the latter, the procedure is guessed and constructed with the help of numerical values.

a) Investigation Relying on Principle

The notion of principle is not seen in this context as a pattern or an inherent logic that one tries to grasp (as in most Neo-Confucian texts and mathematical texts before Takebe) but as an object of knowledge which can play an active part in the mathematical investigation. What sort of thing can this principle be?

There is no precise answer to this question in the *Tetsujutsu sankei*. It seems that Takebe did not consider the principle as a thing which could be expressed in words. Thus, there is no other way to grasp the outline of the investigation relying on principle than compare it with the investigation relying on numbers.

The distinction between the two types of investigation is best exemplified in chapter 8 where Takebe considers two different ways of obtaining the procedure for calculating the area of a sphere with a diameter of 1*shaku* [the *shaku* is a unit of length. 1*shaku* = 10*sun*].

The first solution, which was also Takebe's own, ran as follows. He calculated the difference between two spherical volumes V_1 and V_0 corresponding respectively to the spheres of diameters $d_1 = 1.001shaku$ and $d_0 = 1shaku$ and divided it by the thickness $e_1 = 0.0005$ to obtain the approximate area: 314*sun*.473529344 [the

---

[20] *Op. cit.*, foreword.

sun in this case is a unit of area]. The same calculations were carried out with the diameters d_1' and d_1'', respectively equal to 1.0001 and to 1.0000001.

He obtained: $(V_1' - V_0)/e_1' = 314 \, sun \, 16240 \, 6962$ and $(V_1'' - V_0)/e_1'' = 314 \, sun$ 15929 6775. He then derived from the three values, using a procedure called *sonyaku no jutsu*,[21] a further value of the area equal to: 314 *sun* 15926 5359. He guessed from this that the procedure of calculating the area was: $S = \Pi d^2$ (where Π designates the ratio of the circumference to the radius of a circle).

The second method that Takebe ascribed to Seki consisted in viewing the sphere as a circular cone having the center of the sphere as its vertex, the radius as its heights, the area as its base. The volume of the cone $V = (1/3)S \cdot d/2$ was thus identified to the spherical volume $V' = 1/6 \, \Pi \, d^3$. $S = \Pi d^2$ could then be derived easily. Takebe concluded this study by the following remarks:

> As regards the previous procedures for determining the spherical area, the former consisted in investigating "slices," determining numbers, and investigating the procedure with the help of numbers. In the latter, there was no investigation either of numbers or of procedure. The principle was directly discerned and the procedure was also directly obtained.[22]

The distinction between the two types of investigation emphasized above all Takebe's faith in the power of suggestion of approximate numerical values. General expressions of volumes or areas, Takebe claimed, could be grasped, as was exemplified here, through the study of particular numerical values provided they were chosen adequately. This type of numerical approach is given a prominent role in Takebe's concept of mathematics. In contrast, one is struck by the total absence of comments on the way principle should be handled. Takebe only mentioned briefly how the success of the investigation relying on principle can be secured:

> In the investigation relying on principle, even though we do not resort to the determination of numbers, as long as our way of using our mind is right we will certainly find out [the result].[23]

He also stressed the point that the algebraic method of solving problems was the most usual way to carry out investigations relying on principle. He wrote:

> As regards [problems] where investigation relies on principle, there is the rule of *tengen* which unifies all the procedures.[24]

Takebe's statement reveals his general perception of the *tengen* method as a norm in the investigation of procedures. This is also true for Seki's improvements of

---

[21][A. Horiuchi, 1990], p. 411.
[22][K. Takebe, 1722], Chapter 8.
[23][K. Takebe, 1722], Chapter 4.
[24][K. Takebe, 1722], Chapter 9.

the *tengen* rule which Takebe regarded equally as "marvelous."[25] We thus see that algebraic techniques were highly valued by Takebe and that they symbolized for him a kind of perfection in mathematics. Takebe regarded the algebraic solution as a transparent process in which all the steps were perfectly clear and reflected the genuine order of nature (the *ri*). No further justification was then necessary in this case.

b) Investigation Relying on Numbers

Comparing to the investigation relying on principle that Takebe often qualified as "easy," the investigation relying on numbers was regarded as a blind and painstaking approach to the problem.[26] The understanding of the procedure was not expected from the outset. The investigation aimed at finding out the form of the procedure by calculating particular values and testing them. The success of this task depended on the mathematician's ability to choose the suitable values and to carry out the calculation in an effective way.

According to Takebe's commentaries, many significant results had been discovered in this way, such as the rule of simplification of fractions, the rule of finite differences, the calculation of the ratio of the circumference to the diameter or the calculation of the length of an arc. Takebe's examples show that this type of investigation included the process which is usually referred to as the incomplete mathematical induction. But Takebe was also taking into account a more sophisticated numerical investigation which he described in the following terms:

> The method of investigation consists in this: in general, at the extreme points of saturation and exhaustion, there is either increase or decrease; by pulverizing or by slicing, numerical values will be determined, and according to their variations, a basis for the principle or for the numbers will be found. And from this basis, the rule or the procedure will be formed by trying innumerable modifications.[27]

This description is clearly modelled on the path he actually followed while investigating the procedure for calculating the length of the arc in terms of the sagitta. We know from the detailed analysis of his method Takebe gave in the corresponding chapter[28] that he obtained the expression of the arc as an infinite

---

[25] [K. Takebe, 1722], Chapter 2.

[26] [K. Takebe, 1722], Chapter 8.

[27] [K. Takebe, 1722], Chapter 8.

[28] Let us sum up Takebe's method. Taking the diameter to be 10, an approximate value of the square of the arc length a_0 corresponding to the sagitta $so = 10$ was first calculated with more than sixty digits by using a method of successive approximations.

$$(a_0/2)^2 = 0.000\ 100\ 000\ 033\ 333\ 351\ 111\ 122\ 253\ 969\ 066\ 667\ 282\ 347\ 769\ 479\ 595\ 875$$

The initial arc was divided into two equal arcs, and each of these arcs was divided again into two other arcs etc. Approximate values were then obtained by adding up the chords corresponding to these equal parts of the arc. Takebe then examined the sequence of digits in the approximate

power series through observing variations (*increase* or *decrease*) of the numerical values of arc lengths when they approached the arc of length zero, that is when the arc was at the *extreme point of exhaustion.*

The technical term of *pulverization* refers to the process of dividing a curved line into infinitesimal segments similar to the one he used for approximating the length of the arc. Likewise, *slicing* refers to the process of dividing an area or a volume into infinitesimal parts. Takebe emphasized on many occasions the significance of these two processes in the investigation relying on numbers.[29]

The importance Takebe conferred to the last example and more generally to the investigation relying on numbers indicates how eager he was to demonstrate the far-reaching possibilities of this type of investigation. Takebe's main argument was that, in some cases, procedures could not be obtained otherwise.

> As far as circle and arc are concerned, neither the procedure nor the numerical values can be obtained as long as investigation relies only on principle. It is only by using numbers that they can be obtained. This is due to the attribute (*shitsu*) of the circle and the arc.[30]

The notion of attribute (*shitsu*) of mathematical objects is a keystone of Takebe's argument in the *Tetsujutsu sankei*. In other places, Takebe combined it with another Chinese character which means shape (*katachi*). Both words belong to Neo-confucian terminology and express the properties or characteristics of existing things. *Shitsu* must be understood as a tangible or a physical characteristic of things rather than an abstract quality. In a Chinese treatise of botany influenced by Neo-Confucian philosophy, the concept of *shitsu* was used in the sense of tangible morphology or texture of things.[31]

Takebe's following comment suggests a understanding of the term.

> This is a procedure which exhausts the natural attribute of the arc. We thus found out that the attribute of arcs and circles is the inexhaustible and therefore that the corresponding procedures must be sought with the help of the inexhaustible.

> Truly, among numbers, there are the exhaustible ones and the inexhaustible ones. Among procedures, there are the exhaustible ones and

---

value and reconstructed the expressions of the first terms of the following infinite power series.

$$(a/2)^2 = sd + \frac{1}{3}s^2 + X_1\frac{8}{15}\frac{s}{d} + X_2\frac{9}{14}\frac{s}{d} + X_3\frac{32}{45}\frac{s}{d} + X_4\frac{25}{33}\frac{s}{d} + X_5\frac{72}{91}\frac{s}{d} + \cdots$$

where $X_1, X_2, X_3, X_4, X_5 \ldots$ designate respectively the second, third, fourth, fifth, sixth... terms of the series; a, f and d are respectively the arc length, the sagitta and the diameter.

Once the first terms of the series had been found out, he managed to find out the general expression of the successive coefficients. [K. Takebe 1722], Chapter 12. For further details, see [A. Horiuchi 1990].

[29][K. Takebe 1722], Chapter 9.

[30]*Op. cit.*, chapter 12.

[31][Needham 1986], p. 315

the inexhaustible ones. Among attributes, there are the exhaustible
and the inexhaustible.[32]

We see that the concept of attribute was introduced in connection with the
finite or infinite character of numbers or procedures. For Takebe, numbers or procedures had either attribute: the exhaustible or the inexhaustible.[33]. One ought
to respect this attribute when elaborating the procedure. Moreover, Takebe was
convinced that the only way to have an insight into the attribute of a procedure
was to carry out a numerical investigation.

V   A Mathematical Investigation Oriented towards the Discovery of a Natural Order.

From the foregoing, we see that Takebe's concept of mathematics fundamentally
rested on the idea that the mathematician was always dealing with a tangible order
which was apprehended sometimes directly and sometimes by means of numbers.
The mathematician also had to make sure that his search was in accordance with
this order and was not departing from the *principle*. Takebe warned his readers
against the danger of making up artificial procedures or rules and applying them.

> Numbers, principles of procedures and rules are all arranged according
> to a natural order. To find them, it is not necessary to make up a new
> way. It is enough to follow the Way of Nature.[34]

Takebe further claimed that trouble in mathematics often stemmed from the
fact that one did not reflect beforehand upon the possibility of getting the result
and was consequently subjected to doubt. Feeling doubtful about the success of
one's undertaking, one usually could not make much effort. Conversely, the confor-
mity with the attribute of numbers was a source of confidence and the investigation
was secured by the feeling of following the right way.[35]

The concern with the attribute of numbers was intimately related in Takebe's
mind to the idea that human intelligence had limitations and that they should be
taken into account. He wrote:

> Alas! One's own attributes of purity and shortcomings remain such as
> we received them at our birth. Even if we study thoroughly, they will
> not increase further. Conversely, we may forget things, they will not
> decrease.[36]

---

[32] [K. Takebe 1722], Chapter 12.

[33] By inexhaustible numbers, Takebe referred to numbers with an infinite decimal development.
[A. Horiuchi 1990], p. 436

[34] [K. Takebe 1722], Chapter 8.

[35] [K. Takebe 1722], "Discussing one's own attributes."

[36] *Ibid.*

In other words, there is no human endowed at birth with a boundless intelligence. Mathematics must be practised according to one's own capacities. A prerequisite for an optimal practice of this art is to know oneself thoroughly and especially to know one's own limits. Takebe blamed all those who pretended to grasp the whole field of mathematics.

> I truly think that if one takes an art and puts it into one's mind, then, because this art splits up into the thinkable and the unthinkable, it will submit as far as the thinkable is concerned, but the unthinkable will not submit.[37]

According to Takebe, there is always an area of mathematics which lies beyond the scope of human understanding because of the imperfection of human nature. When one chooses a purely mental or a contemplative way of approach (Takebe used the expression: "to put the art into one's mind"), one is inevitably led to miss the part of mathematics which is unthinkable. Takebe advocated to *go into* mathematics as a solution of this paradox.

> I say that if one goes thoroughly into mathematics in such a way that there is not the slightest opposition of one's personality, then one's mind will form one body with the Way and the thinkable will submit as thinkable and the unthinkable will submit as unthinkable. This is one aspect of the identification with the Way.[38]

The fundamental meaning of Takebe's advocation was the respect for natural order (the Way). It would lead at times to a length of painstaking research but this research would be supported by the confidence in the success of the task. Through this kind of research, he hoped to reach both the thinkable and the unthinkable fields of mathematics.

The *tetsujutsu* or the technique of linking which is only briefly mentioned in the foreword might just be another way of expressing this thought. By *tetsujutsu*, Takebe would then be referring to the method consisting in *going into* mathematics by accumulating experiences.

VI Conclusion

To conclude, let us summarize the main features of Takebe's argument.

The most striking aspect of Takebe's treatise is the utmost importance he conferred to the use of numerical values in the investigation. This was not in contradiction with the long practice of seeking the intuition of mathematical relations in numerical values. Takebe's originality resided in his acknowledgment of this investigation as an invaluable heuristic method.

---

[37][K. Takebe 1722], "Discussing one's own attributes."
[38][K. Takebe 1722], "Discussing one's own attributes."

An other striking point is the fact that the determination of a procedure and the clarification of its meaning were distinguished and that the latter was almost identified to the knowledge of the algebraic construction of the procedure. If the procedure had been reached through algebraic calculations, its meaning was considered as perfectly clear. If it had been obtained by intuition through numerical values, its meaning was yet to be found. Regarding this last point, it is interesting to know that some time after the *Tetsujutsu sankei*, Takebe found an algebraic justification of the procedure for calculating the arc length.[39]

One could also note the little importance Takebe attached to logical reasoning or to deduction. This should probably be linked with the kind of mathematics Takebe was dealing with, a science of calculation oriented towards the discovery of a natural order. Such a vision which can be seen as an outcome of Neo-Confucian thought and of Takebe's growing engagement in the calendrical science (where he had to translate the regularities of celestial movements into mathematics), prompted him to consider the mathematical research in an optimistic perspective.

We can finally mention Takebe's emphasis on the necessity of respecting one's own capacities and of going into mathematics. All these remarks, which are developed in the last section, suggest that Takebe considered mathematical activity in the same way as he regarded the scholarly formation in general, that is as a fundamentally moral undertaking.[40]

References

[Th. De Bary, Wing-Tsit Chan, B. Watson 1960] *Sources of Chinese Tradition*, vol. 1. New York: Columbia University Press.

[K. Chemla 1991]   "Theoretical Aspects of the Chinese Algorithmic Tradition (first to third century)," *Historia Scientiarum*, **42**, 75–98.

[A. Horiuchi 1987]   "La science calendérique de Takebe Katahiro," *Historia Scientiarum*, **33**, 3–24.

[A. Horiuchi 1989]   "Sur un point de rupture entre les traditions chinoise et japonaise des mathématiques," *Revue d'histoire des sciences*, XLII-4, 375–390.

[A. Horiuchi 1990]   *Etude de Seki Takakazu (3?–1708) et Takebe Katahiro (1664–1739), deux mathématiciens de l'époque d'Edo*. Thèse de doctorat de l'Université Paris 7, 1990.

[A. Horiuchi 1992]   "The Development of Algebraic Methods of Problem-solving in Japan in the Late Seventeenth and the Early

---

[39][K. Takebe].

[40]I wish to thank Kim Orun, Lowell Skar and Catherine Jami for their helpful comments on earlier versions of this paper.

Eighteenth Centuries," *Proceedings of the International Congress of Mathematicians 1990*, vol. 2 Tokyo, Berlin, Heidelberg: Springer-Verlag, 1639–1649.

[A. Kodama 1966]   *Jūgoseiki Chōsenkan dōkatsujiban sūgakusho* (Mathematical treatises printed with bronze movable type characters in the 15th century Korea). Tokyo.

[Lam Lay Yong 1979]   "Chu Shih-chieh's Suan-hsüeh ch'i-meng (Introduction to Mathematical Studies), *Archive for the History of Exact Sciences*, **21**, 1–31.

[Li Yan, Du Shiran 1987] *Chinese Mathematics, A Concise History*. Oxford: Clarendon Press.

[J. Needham 1986]   *Science and Civilisation in China*, Vol. VI.1. Cambridge: Cambridge University Press.

[K. Takebe 1722]   *Tetsujutsu sankei* (A Mathematical treatise on *tetsujutsu*), Manuscript of Kokuritsu Kōbun Shokan Naikaku Bunko.

[K. Takebe]   *Enri kohaijutsu* (The Procedure for Calculating the Arc Length according to the Principle of the Circle), Manuscript of Nihon Gakushiin.

Glossary of Japanese and Chinese Terms

geki   逆

hō   法

hōjutsu   法術

hōsoku   法則

jishitsusetsu   自質説

jun   順

jutsu   術

kaisu   会ス

motomeru   求メル／索メル

ri／li   理

saguru   探ル

sanpō／suanfa   算法

sangaku   算学

sassu   察ス

Seki Takakazu   関 孝和

shitsu   質

sū   数

Takebe Katahiro   建部賢弘

tengen／tianyuan   天元

tetsujutsu／zhuishu   綴術

Tetsujutsu sankei   綴術算經

tsukusu   尽ス

uru   得ル

waza   技

Zu Chongzhi   祖沖之

The Adoption of Western Mathematics in Meiji Japan, 1853–1903

Sasaki Chikara

Introduction

Those mathematicians and historians of mathematics who have participated in the 21st International Congress of Mathematicians in Kyoto and the Tokyo History of Mathematics Symposium 1990 may be interested in the following problem: Japan is now one of the big powers in mathematics, as participants may have observed at the Kyoto Congress. Why was Japan able to introduce Western mathematics so rapidly? I propose the following answer to this problem: The old order of Japan already possessed a very sophisticated traditional mathematics called 'wasan', and new Japan was in a great hurry to reform educational systems drastically and to institutionalize military academies, a mathematical society, and universities, which all facilitated Japan's adoption of Western mathematics.

I will discuss the problem of the adoption of Western mathematics in Japan during the half century from 1853 to 1903. In 1853, Japan was shocked by the arrival of the black ships of Commodore Perry of the United States, and began to open its closed door to the West. The year 1903 is significant as the date Takagi Teiji, an eminent Japanese mathematician, published a paper which marked the establishment of an independent research tradition of Western originated mathematics in Japan.

Today, we cannot talk about world history without attention to non-Western civilizations. Europe is only one area of the world, not only geographically but also culturally. For this reason, it is interesting to study episodes in East-West relations which took place in later nineteenth century Japan.

The modernization process in Japan is often dated from the Meiji Restoration of 1868. Events which occured during this period interested many foreign visitors as well as the Japanese themselves. They often left documents which reported how rapidly and drastically the process of change was carried out. For instance, Erwin Bälz (1849–1913), a German doctor who taught at the Medical School of the University of Tokyo for more than a quarter century, wrote to his family in Germany as follows:

> Tokyo, October 25, 1876
> I am feeling very well, and propose today to tell you a little more

about the conditions here.

They are, in fact, remarkable in many ways. No doubt you have read a good deal about them. You may not have realized, however, that they are of altogether exceptional interest to a student of the history of civilization.

To understand the situation you have to realize that less than ten years ago the Japanese were living under conditions like those of our chivalric age and the feudal system of the Middle Ages, with its monasteries, guilds, Church universal, and so on; but that betwixt night and morning, one might almost say, and with one great leap, Japan is trying to traverse the stages of five centuries of European development, and to assimilate in the twinkling of an eye all the latest achievements of Western civilization. The country is thus undergoing an immense cultural revolution — for the term "evolution" is inapplicable to a change so rapid and so fundamental. I feel myself lucky to be an eye-witness of so interesting an experiment.[1]

Of course, some Japanese were also aware of how important this modernization was for the future of Japan. Fukuzawa Yukichi (1835–1901), champion of the enlightenment movement of this period, stated in the Foreword to his *Bunmeiron no Gairyaku* (An Outline of a Theory of Civilization), published in 1875:

> Contemporary Japanese culture is undergoing a transformation in essence, like the transformation of fire into water, like the transition from non-being to being. The suddenness of the change defies description in terms of either reformation or creation. Even to discuss it is extremely difficult.
>
> I trust that we present-day scholars will measure up to this challenge. But let me point out that in addition we have an accidental opportunity for greatness thrust upon us. Since the opening of the ports Japanese scholars have been assiduous in mastering Western Learning. Though the results have been sketchy and limited so far, we have been able to get some idea of Western civilization. Yet just twenty years ago we scholars were steeped in a purely Japanese civilization; there is little danger of our falling into vague inferences when discussing the past. We also have the advantage of being able directly to contrast our own personal pre-Meiji experience with Western civilization. Here we have an advantage over our Western counterparts, who, locked within an already matured civilization, have to make conjectures about conditions in other countries, while we can attest to the changes of history through the more reliable witness of personal experience. This actual experience of pre-Meiji Japan is the accidental windfall we scholars of the

---

[1] *Awakening Japan: The Diary of a German Doctor: Erwin Baelz*, edited by his son, Toku Baelz (New York, 1932; Bloomington-London: Indiana University Press, 21974), pp. 15–16.

present day enjoy. Since this kind of living memory of our generation will never be repeated again, we have an especially important opportunity to make our mark. Consider how all of today's scholars of Western Learning were, but a few years back, scholars of Chinese Learning, or of Shinto or Buddhism. We were all either from feudal samurai families or were feudal subjects. We have lived two lives, as it were; we unite in ourselves to completely different patterns of experience.[2]

By studying the transition of mathematical thought in Meiji Japan, we come to know not only the situation of science in one particular non-Western country but also various aspects of Western science, both good and bad, for Japan was, then, a melting pot, so to speak, of European scientific thought of the latter half of the nineteenth century. The European culture which Commodore Perry brought to Japan in 1853 on his four black ships had experienced the Scientific Revolution, the Enlightenment, and the Industrial Revolution. And it was, sometimes, British science, which was in general liberal and loosely organized, and, sometimes, German science, which can be characterized as idealistic and authoritarian, that Japanese scientists tried to adopt after having decided to accept Western science.

I will argue why and how the Japanese adopted Western mathematics rapidly, for convenience, in comparison with the case of China. To do this, I will have to characterize the old Japanese mathematics before the adoption of Western mathematics. Next, I will discuss the study of Western mathematics before the Meiji Restoration, and show that the first students of the subject were people related to the Japanese navy. I will then briefly mention that an important enlightenment movement led by Fukuzawa Yukichi played a decisive role in the later development of Japanese culture. Under the influence of this movement, in the educational reform of 1872, Western arithmetic was accepted as an official subject in elementary schools. In 1877 the first learned society, the Tokyo Mathematical Society, and in 1884 its successor, the Tokyo Mathematico-Physical Society were established. Through the establishment of these societies, I will show how the old type *wasan* mathematicians disappeared from the mainstream of the history of mathematics. And finally, I will touch on the beginning of the research tradition of Western originated mathematics at Japanese universities.

Throughout my report, the names of Japanese appear in the normal Japanese order, surname first and given name second, except when they are referred to as authors of writings in Western languages.

1   Pre-Modern Japanese Mathematics — *Wasan*

To study the main characteristics of the process of the modernization in Japan, it is reasonable to provide ourselves with some knowledge on the final stage of

---

[2]Fukuzawa Yukichi, *An Outline of a Theory of Civilization*, translated by David A. Dilworth and G. Cameron Hurst (Tokyo: Sophia University, 1973), pp. 2–3.

the old order, that is, the Tokugawa period from 1600 to 1868. The culture of
Japan is usually considered a subculture of China. During most of the Tokugawa
period Confucianism enjoyed a position similar to that of a state religion. It can
be said that, at the beginning of the seventeenth century, these two countries of
East Asia had quite similar settings. It seems no accident that Professor Marius
B. Jansen of Princeton University began his consideration on the conditions of
Japan's modernization with the eighteenth century treating China and Japan as
if they had been at almost the same position, intellectually and socially.[3] The
same conception seems applicable to mathematics and other disciplines. According
to Max Weber's simplified characterization in his "Confucianism and Taoism,"
one chapter of his *Gesammelte Aufsätze zur Religionssozioligie*, in the intellectual
milieu of Confucianism, systematic and naturalist thought failed to mature.[4] Even
those who cannot totally agree with Weber have to admit that the old orders of
both China and Japan did not creat mathematics and natural sciences in the
modern European sense.

At this point the following problem presents itself. Between these two countries, why could only Japan succeed in adopting modern mathematics and sciences
so rapidly? To answer this question, we will confine our problem to mathematics,
comparing the adoption process of Japan with that of China.

As I have mentioned, Confucianism flourished in Tokugawa Japan. Confucianism is essentially an anthropocentric moral and political philosophy, although Neo-Confucianism, reformed by Zhū Xī (1130–1200), modified its thought-
structure so as to be able to discuss metaphysics and philosophy of nature. Therefore, the students of politics, law, or philosophy in Meiji Japan found most of the
conceptual distinctions in the European models already familiar and easily expressed in their own vocabulary. On the other hand, the new chemists of the Meiji
period had to develop a whole set of new words to express new ways of classifying
matter. The new physicists at first used the vocabulary of Neo-Confucian philosophy of nature, but later gradually coined new words which could exactly express
Western scientific concepts.

What about mathematics? Before the Meiji Restoration, Japan had its own
traditional mathematics, extremely sophisticated and difficult for non-specialists
to understand. This constitutes a crucial difference between the history of mathematics and that of physics, or natural philosophy, in pre-modern Japan.[5] For this
reason, we must study the Japanese adoption of Western mathematics independently of other natural sciences.

---

[3]See Marius B. Jansen, "Changing Japanese Attitudes Toward Modernization," in Jansen,
ed., *Changing Japanese Attitudes Toward Modernization* (Princeton: Princeton University Press,
1965), pp. 43–89.

[4]Max Weber, *The Religion of China: Confucianism and Taoism*, translated and edited by
Hans H. Gerth with an Introduction by C. K. Yang (New York-London: Free Press, 1968). See
especially Chapter VI, "The Confucian Life Orientation," pp. 142–172.

[5]For a well-written survey of the history of physics in Meiji Japan, see Kenkichiro Koizumi,
"The Emergence of Japan's First Physicists: 1868–1900," *Historical Studies in the Physical Sciences*, **6** (1975), pp. 1–108.

Before considering Western mathematics in Japan, we have to look at the content and institutional setting of the Japanese traditional mathematics.[6] This mathematics has been called 'wasan', a word compounded from 'wa', an ancient name for Japan, and 'san', meaning calculation. However, wasan was not completely original, but was based on the mathematical techniques imported from China, as was the other learning of Japan. Through the reintroduction of Chinese mathematics in the seventeenth century, the Japanese accomplished a great reform in both mathematical techniques and thought. It was after this reintroduction of the Chinese mathematics that the Japanese became really observant of mathematical studies. The name of most important reformer is Seki Takakazu (?–1708). Chinese mathematics performed operations fundamentally by means of particular calculating rods, and therefore could not treat any algebraic expressions except those with one variable in numerical coefficients. This art of calculation was called 'tiān yuán shù' (technique of the celestial element). Seki introduced Chinese ideographs and wrote them to the right of a vertical line — such as ｜甲 . For instance, ｜甲｜乙 stands for $ab + (a + b)x - x^2$ in today's notation. He called his method 'endan jutsu', which literally means the 'method of explanations'. Endan jutsu enabled him to represent known and unknown quantities by Chinese ideographs and led him to form equations with literal coefficients of any degree and with several variables. This reform closely resembles that of François Viète in Western mathematics.[7] After the establishment of the endan algebra, Seki pursued his studies further and arrived at the analytic system of Japanese mathematics, namely 'kigen seihō'. This literally means 'a method or art by which the hidden or relations of things may be made clear.' In short, Seki's kigen seihō was an algebraic system in the writing style. This system was later named 'tenzan jutsu', which means 'an art of rectifying words'.

In the algebra of the tenzan method the considerations were all carried out in writing. The analysis leading to the final results was considered as constituting a part of available knowledge. The subjects studied under the common heading of the tenzan algebra covered a wide scope, including the theory of equations, the solution of indeterminate equations, and algebraic treatment of geometrical relations. So it is noteworthy that in the world history of mathematics, the notion of determinants was first employed by Seki. He used it before 1683, while Leibniz

---

[6] For an introduction to the history of traditional Japanese mathematics, see Yoshio Mikami, *The Development of Mathematics in China and Japan* (First published in 1913; New York: Chelsea, 2 1974) and David E. Smith and Yoshio Mikami, *A History of Japanese Mathematics* (Chiacago: Open Court, 1914). These two books are partly out of date. The best work on the history of *wasan* is no doubt: The Japan Academy, *Meiji-zen Nippon Sūgaku-shi* (History of Mathematics in the Pre-Meiji Era), 5 vols. (Tokyo: Iwanami Shoten, 1954–1960), actually written by Fujiwara Matsusaburō (1881–1946).

[7] As to main charactersitics of Viète's algebra, see Michael S. Mahoney, "The Beginnings of Algebraic Thought in the Seventeenth Century," in Stephan W. Gaukroger, ed., *Descartes: Philosophy, Mathematics and Physics* (Brighton, Sussex: The Harvester Press-Totowa, New Jersey: Barnes & Noble Books, 1980), pp. 141–155.

introduced a similar notion in 1693. Of course, such a comparison is dangerous because the foundations of the two were quite different. But the most important contribution of Seki and his followers, *'enri'* (principle of the circle), which consists of the rectification of the circumference of a circle, curvature of a sphere, and so forth, has been thought to be very similar to a rudimentary form of Western infinitesimal analysis.

Despite such excellence, *wasan* can be characterized in comparison with Western mathematics by the following technical shortcomings: (1) Absence of axiomatic method and synthetic proof. In his *In artem analyticem isagoge* (1591), Viète created a new algebra dependent on the whole heritage of Greek geometrical thought. He could, therefore, establish the axiomatic system of analytic art; (2) No relationship with areas other than the calculation of the calendar. Western mathematics enjoyed interrelationships with other sciences, particularly rational mechanics. We should remember that strictly speaking, in eighteenth century Europe, it seems doubtful there existed the concept of 'pure mathematics' in today's sense; (3) Lack of the concept of function. Newton and Leibniz, particularly the latter, established the general calculus on the basis of the concept of function; (4) Lack of a coordinate system; (5) No reform of symbolism; (6) Neglect of the concept of angle.

Further, *wasan* flourished in a Confucian society which had adopted much of Chinese culture. In such a cultural milieu, as Weber pointed out, the specialist or expert has no truly positive dignity, no matter what his social usefulness. The "cultured man" (gentleman) was "not a tool"[8]; that is, in his adjustment to the world and in his self-perfection, he was an end unto himself, not a means to any functional end. Chinese mathematics, upon which the Japanese is based, is thus of a character fundamentally different from the Greek tradition. Since European mathematics after Viète was founded on the basis of Greek mathematics, we can contrast Chinese mathematics with Greek mathematics most effectively. In this regard Joseph Needham's observation is very interesting:

> 'On account of its more abstract and systematic character' — so came the words of themselves to the keys of the machine. Systematic, yes, there no doubt is possible, but abstract — was that wholly an advantage? Historians of science are beginning to question whether the predilection of Greek science and mathematics for 'the abstract, the deductive and the pure, over the concrete, the empirical and the applied' was wholly a gain. [...] In the flight from practice into the realms of the pure intellect, Chinese mathematics did not participate. [...]
>
> For the 1st century B. C., the time of Lohsia Hung and Liu Hsin, the *Chiu Chang Suan Shu* (Nine Chapters of the Mathematical Art) was a splendid body of knowledge. It dominated the practice of Chinese reckoning-clerks for more than a millennium. Yet in its social origins

---

[8]Weber, *Op. cit.*(n. 4), p. 160.

it was closely bound up with the bureaucratic govenment system, and devoted to the problems which the ruling officials had to solve (or persuade others to solve). Land measuring and survey, granary dimensions, the making of dykes and canals, taxation, rates of exchange — these were practical matters which seemed all-important. *Of mathematics 'for the sake of mathematics' there was extremely little.* This does not mean the Chinese calculators were not interested in truth, but it was not that abstract systematized academic truth after which sought the Greeks.[9] [The emphasis is added.]

The advantages which Needham points out are, at the same time, the shortcomings of Chinese mathematics. However, we encounter a question. Is Needham's thesis applicable to Japanese mathematics, *wasan*? Our answer is "no". One of the most eminent historians of Japanese mathematics, Mikami Yoshio (1875–1950), stated that in *wasan* mathematical rules or methods were treated as a kind of art, not as a science. The old Japanese had a concept of mathematics for its own sake under the general influence of the well-known manner of doing arts (*geidō*), which might be compared to '*l'art pour l'art.*'[10] The setting of *wasan* seems to have resembled traditional Japanese arts much more than European sciences. *Wasan* masters built their guilds and gave their students licenses which corresponded to the students' skill just as did the masters of the tea ceremony and flower arranging.

Wasan mathematicians were competitive, with controversies occurring between the different guilds or schools. Mikami notes that "as the mathematical knowledge of those times lacked any scientific basis, there was nothing that could be taken for a standard of reference, and so it came that frequently the accusation and the defence both failed in precision. Mathematics being cultivated as art, not as science, the old Japanese in pursuing their studies of the subject must have lost much time and labour in useless entanglement."[11]

But within the cliquish guilds the arts could become elaborate and sophisticated, so that *wasan* mathematicians proposed to other mathematicians increasingly difficult problems which were often useless, in order to demonstrate the excellence of their skills. In this regard Weber's suggestion is again quite interesting. Weber claims, in his *Wirtshaft und Gesellshaft*, that among Japanese vassals, in an artistic style of life games occupied a most serious and important position

---

[9]Joseph Needham, *Science and Civilization in China*, Vol. 3: *Mathematics and the Sciences of the Heavens and the Earth* (Cambridge: Cambridge University Press, 1959), pp. 151–153. As to an up-to-date introduction to the history of Chinese mathematics, see Lǐ Yǎn and Dù Shírán, *Chinese Mathematics: A Concise History*, translated by John N. Crosseley and Anthony W.-C. Lun (Oxford: Oxford University Press, 1987) and Jean-Claude Martzloff, *Histoire des Mathématiques Chinoises* (Paris-Milan-Barcelone-Mexico: Masson, 1988).

[10]This aspect was emphasized in Mikami Yoshio's *Bunkashi-jo yori Mitaru Nippon no Sūgaku* (Japanese Mathematics from the View-point of Cultural History), ed., by Hirayama Akira, Ōya Shin'ichi, and Shimodaira Kazuo (Tokyo: Kōseisha-Kōseikaku, 1984). I claim that Mikami's this work is one of the most insightful monographs ever written on the old Japanese mathematics.

[11]Mikami, *The Development of Mathematics in China and Japan* (n. 6), p. 169.

in the life of the knightly strata.[12] Knowledge for its own sake is certainly not compatible with orthodox Confucianism. But the Japanese had a conception of '*l'art pour l'art.*'

Some historians of Japan regard the popularity of education in the Tokugawa period as one of the causes of the successful modernization. In the case of mathematics, schools of the feudal clans began to accept *wasan* arithmetic into their curricula from the mid-eighteenth century. Weber thinks that in China, under the influence of Confucianism, all remained sublimated empiricism. This might be true. However, because of the technical reform and the 'formalist' spirit of art, *wasan* produced mathematical results of a very high standard. Originally, the reformed algebra of *wasan* was certainly connected with calendrical calculations, but especially during the later period was advanced as pure mathematics to be pursued for its own sake. This aspect nourished a kind of mathematical spirit which may be called 'formalistic'. This was quite different from China. This fact seems to have been an important difference between the Chinese and the Japanese when they introduced European modern science.

Wasan mathematicians were so proud that, although from the end of the eighteenth century they imported numerous treatises on astronomy and mathematics from Holland, they thought that there were no foreign books treating mathematics of advanced nature. The students of medicine and translators of treatises of physics and chemistry could learn much from the Occidental studies. But *wasan* mathematicians considered their 'pure' mathematics much better than that of the West. We quote the following passage from a writing of 1811 of a *wasan* mathematician:

> Although the Europeans highly excel in all matters relating to astronomy and the calendar, nevertheless their mathematical theories are inferior to those that we have so accurately developed. From this we can conclude that foreign mathematics is not on so high a plane as the mathematics of our own country.[13]

Similar statements can be cited from books of 1877, ten years after the Meiji Restoration.

This is the situation of Japanese mathematics at the eve of the Meiji Restoration. How and why was "the most excellent mathematics in the world" superseded by Western mathematics during the process of modernization? To state my conclusion in advance, it was not mainly because of internal or technical characteristics of the two mathematics, but for external or environmental reasons.

---

[12] Max Weber, *Economy and Society: An Outline of Interpretive Sociology*, edited by Guenther Roth and Claus Wittich, Vol. 2 (Berkeley-Los Angeles-London: University of California Press, 1978), pp. 1105–1106.

[13] Furukawa Ujikazu (1783–1837), *Sanwa Zuihitsu* (Essays on the Story of Mathematics, 1811): Quoted in Ogura Kinnosuke, *Kindai Nippon no Sūgaku* (Mathematics in Modern Japan), *Chosaku-shū* (Selected Works), Vol. 2 (Tokyo: Keisō Shobō, 1973), p. 7.

2   The Acceptance of Western Mathematics as Military Science, 1853–1868

A shocking event happened in Edo Bay in 1853. Japan had closed its door to the West, permitting only a few Dutch and Chinese merchants who were closely watched at Nagasaki. On July 8, 1853 Commodore Matthew Calbraith Perry's two steamers and two sailing vessels, the first modern Western squadron to reach Japan, anchored in Edo Bay. The traditional Japanese feudal rulers had begun to feel the strength of the West when its parent country, China, was defeated by Britain at the so-called Opium War and Britain acquired Hong Kong by the treaty settement of 1842–43. The fear of colonization became a real concern on this occasion.

Perry's visit to Japan caused two historically important consequences. One is the end of Japanese seclusion. The other is the establishment of a program to build a Japanese navy to guard the independence of Japan. Consequently, a naval training school with Dutch instructors was founded at Nagasaki in 1855, the *Nagasaki Kaigun Denshū-sho* (Nagasaki Naval Academy). In the same year, the *Yōgaku-sho* (Office for Occidental Learning; in the following year, under the resistance of Confucianist scholars, renamed the *Bansho Shirabe-sho* (Office for the Investigation of Barbarian Books)) was also established mainly to study Western books, especially those on ship-building and armaments. These two institutes are extremely important in the history of mathematics in Japan, because it was there that the Japanese first studied Western mathematics systematically. In these institutes, the Japanese students — officials of the central and local feudal governments — learned from the Dutch instructors such practical knowledge as the arts of navigation and surveying, trigonometry, logarithms, geography, and artillery. We can see in a notebook of these students, for instance, a trigonometric formula: $\sin(a+b) = \sin a \cos b + \sin b \cos a$.[14] It should be noted that Western arithmetic and algebra were taught, but they were closely connected with the practical purposes of the Japanese. It is known today that the books which the *Bansho Shirabe-sho* collected are still extant. The kind of subjects they are concerned with is interesting: The books were mostly related to military science. We can conclude that Western mathematics was at the beginning imported for military use. The object of study was to protect national independence. The adoption of Western learning and science, especially its military aspects, was a matter of vital importance to the military leaders of pre-Restoration Japan.

Another important aspect in the process of adoption of Western mathematics can be seen in first two elementary books on the European methods of calculation published in 1857. One was Fukuda Riken (1815–1889)'s *Seisan Sokuchi* (A Short Course in Western Arithmetic), and the other was Yanagawa Shunsan (1832–1870)'s *Yōsan Yōhō* (Methods of Western Arithmetic). Fukuda Riken was a *wasan*

---

[14]The Japan Mathematical Society, *Nippon no Sūgaku Hyakunen-shi* (100 Year History of Japanese Mathematics), Vol. 1 (Tokyo: Iwanami Shoten, 1983), p. 25.

mathematician and specialist in the Chinese calendar. It is impressive that in his book he treated addition, subtraction, multiplication, and division of integers, but all numerals and symbols were written in the Japanese way. So while his book had the title *A Short Course in Western Arithmetic*, actually it was Japanese arithmetic influenced by Western methods. The author of the other book, Yanagawa Shunsan was a scholar of Dutch learning and later became a journalist. His book was very elementary but written in a completely European way. Shortly after the Meiji Restoration he began publishing a magazine titled *Seiyō Zasshi* (Western Magazine) and a newspaper *Chūgai Shimbun* (Domestic and Foreign Newspaper) in 1868. This marked the beginning of the magazine and newspaper business in Japan.

It is also worth describing the chair of mathematics in the *Bansho Shirabe-sho*. The *Bansho Shirabe-sho* was renamed *Kaisei-jo* (Office for Reclamation) in 1862, and was a forerunner of the University of Tokyo, established as the first national university in 1877. In February, 1862 Kanda Takahira (1830–1898) was made professor of mathematics of the *Kaisei-jo* institute. As far as I know, he was the first Japanese instructor of mathematics, whether Japanese or European style in any institute sponsored by the central government. Kanda was a scholar of Dutch learning. Before his nomination he served at the *Bansho Shirabe-sho* as a librarian of Western books. He seems to have been made instructor of mathematics not because of his mathematical skill but because of his knowledge of the Dutch language. It is supposed that he only taught elementary arithmetic and algebra, which were useful for the art of navigation.

To summarize, the Western mathematics which was studied before the Restoration was for the most part related to military sciences and was very elementay. Further, both students and teachers of mathematics were amateurs in Western mathematics, and they were almost all ignorant of the 'higher' Japanese mathematics. Very few *wasan* mathematicians intended to study Western mathematics before the Restoration. Such was the situation in the adoption of Western mathematics in the crucial period of pre-Restoration Japan.

3   The Enlightenment Movement and Educational Reform, 1868–1877

Drastic changes took place with the Meiji Restoration of 1868. Before then, the acceptance of Western learning and science was made in most cases under the popular slogan *"wakon yōsai"*: "The West is superior only in technical arts, but Japan is superior in basic spirit or moral philosophy." But the important shift in the general attitude toward learning could be observed with the deepening of the Restoration process. To discuss this shift we must talk briefly about the spirit of enlightenment at the beginning of Meiji Japan. The champion of the Japanese enlightenment movement was Fukuzawa Yukichi, mentioned above. Fukuzawa has been called a Japanese Voltaire. It is not too much to say that to talk about the enlighenment of Japan is to talk about him.

Fukuzawa was born of a samurai family. He learned the Dutch language in Nagasaki and then the Occidental sciences at a private school in Osaka. In 1858 he opened his own Dutch language school in downtown Edo (now Tokyo) and later learned English by himself. He visited the United States and Europe several times before the Restoration. His books, *Seiyō Jijō* (Things Western, 1866–70), *Gakumon no Susume* (An Encouragement of Learning, 1872–76), and *Bunmeiron no Gairyaku* (An Outline of a Theory of Civilization, 1875), were the best sellers at the beginning period of the Restoration. He attacked the traditional Chinese learning which had been the official doctrine of the Tokugawa period, while advocating the spirit of the Occidental sciences. As Weber stated, the 'reason' of Confucianism was a rationalism of order. Confucianism, like Buddhism, consisted only of ethics. However, in sharp contrast to Buddhism, Confucianism exclusively represented an inner-worldly morality of the commoner. Confucianism meant adjustment to the world, to its order and conventions. Further, the order of the world was considered fixed and inviolate and the order of society was but a special case of this.

How did Fukuzawa conceive of the differences between the traditional Oriental learning and the newly arriving Occidental learning? In his autobiography, he states this creed in the section under the subtitle "My fundamental principle in education":

> In my interpretation of education, I try to be guided by the laws of nature in man and the universe, and I try to co-ordinate all the physical actions of human beings by the very simple laws of "number and reason".[15]

The hierarchical order of the universe and of the society which was strongly protected by Confucianism was destroyed by Fukuzawa's principles which were closely connected with modern sciences. He further continues his argument:

> From my observations in both the Occidental and Oriental civilizations, I find that each has certain strong points and weak points bound up in its moral teachings and scientific theories. But when I compare the two in a general way as to wealth, armament and the greatest happiness for the greatest number, I have to put the Orient below the Occident. Granted that a nation's destiny depends upon the education of its people, there must be some fundamental differences in the education of Western and Eastern peoples.
>
> In the education of the East, so often saturated with Confucian teaching, I find two things lacking; that is to say, a lack of studies in number and reason in material culture, and a lack of the idea of independence in spiritual culture. But in the West I think I see why

---

[15] *The Autobiography of Fukuzawa Yukichi*, translated by Eiichi Kiooka (Tokyo: Hokuseidō Press, 1960), p. 214.

their statesmen are successful in managing their national affairs, and
the businessmen in theirs, and the people generally ardent in their
patriotism and happy in their family circles.

I regret that in our country I have to acknowledge that people are
not formed on these two principles, though I believe no one can escape
the laws of number and reason, nor can anyone depend on anything
but the doctrine of independence as long as nations are to exist and
mankind is to thrive. Japan could not assert herself among the great
nations of the world without full recognition and practice of these two
principles. And so I reasoned that Chinese philosophy as the root of
education was responsible for our obvious shortcomings.[16]

Fukuzawa here played a role of the enlightener in several aspects, rationalist
like René Descartes, utilitarian like Francis Bacon, and finally, critic of the official
ideology of the *ancien régime* like Voltaire. What kind of place could the traditional
Japanese mathematics *wasan* occupy in this thought? I have already characterized
wasan as having almost no relationship with other sciences, with the exception
of calendrical calculations. Clearly Fukuzawa's "number and reason" is connected
with practical purposes as well as with a metaphysical basis.

Fukuzawa's main principle pf learning can be concisely summarized in the
word '*jitsugaku*'— real, practical and useful learning, which is contrasted with '*kyo-
gaku*' — empty, useless learning, actually Confucianism. It was this real learning,
namely Western learning, that the young Meiji leaders adopted.

Fukuzawa had been an enlightened official working for the Tokugawa government. Why could Japan so easily discard Confucianism and adopt Western
learning, in contrast with China? The excellent historian of modern Japan, Herbert Norman, has provided a social explanation in his classical book published in
1940 *Japan's Emergence as a Modern State*:

It is easy to see how the acceptance of Western learning and science
would spell the end of the monopoly of learning and office enjoyed by
the Confucian bureaucracy. Christianity, the natural sciences, even the
military sciences were all repugnant to it. It is clear from the fiasco of
the Hundred Days of Reform of 1898 that the mandarinate was inca-
pable of undertaking reforms — that China could only be modernized,
that is to say *industrialized*, by an uncompromising political revolution
which first and foremost would have to sweep away the self-sufficient,
conservative mandarinate. In Japan there was no similar dominant
class or caste with a vested interest in maintaining Confucian or even
Shintoist learning. In Japan, the class from which administrators and
councilors were drawn was the *bushi* or worrior class. This governing
class had often learned from the foreigner whether Chinese, Korean,
Portuguese, Dutch; now when they were given a practical illustration

---

[16] *Ibid.*, pp. 214–215.

of the superiority of Western military science, they hastened to adopt
it not only in order to protect national independence, but to maintain
their own prestige in a society which glorified the military virtues.[17]

With the Meiji Restoration, Western learning in Japan was no longer confined
to military science. Western learning was to be the official doctrine in the new
Japan. Fukuzawa's thought played an important role in this generalization. To
the Meiji leaders, educational reform ranked as one of the key measures in the
transformation of Japan from a feudal to a modern, unified national state. Aware
of the urgent need for a modern unified educational system in order to catch up
with the West, they came down firmly on the side of Western learning.

In 1871 the Ministry of Education was established, and in 1872 the Fundamental Code of Education, representing a complete victory of the Westernizers,
was promulgated. The first Minister of Education was Ōki Takatō (1832–1899),
who is known to have been very instrumental in the transfer of the capital from
Kyoto to Edo (now Tokyo), and who later became a Governor of Tokyo. Ōkuma
Shigenobu (1838–1922), who came from the same fief as Ōki, assisted the Minister.
Education, the preamble of the Code of Education declared in strongly utilitarian tones, is "the key to success, and no man can afford to neglect it. Learning
having been viewed as the exclusive privilege of the samurai and his superiors,
it was neglected by farmers, artisans, merchants, and women. Even those among
the samurai who did pursue learning indulged in poetry, empty reasoning, and
idle discussions, and their dissertations, while not lacking in elegance, were seldom
applicable to life. There shall, in the future, be no community with an illiterate
family, or a family with an illiterate person."[18]

As a result of the educational reform closely tied with the enlightenment
movement, Western mathematics, not *wasan*, was to be adopted as the basis of
Western learning, and was actually incorporated into the curriculum of the new
schools.

In both the curriculum and methods, American influence was very strong.
In 1873 David Murray, professor of mathematics at Rutgers College, was brought
to Japan as an adviser of the Japanese Ministry of Education. Murray remained
in Japan, either as advisor or educational inspector, until 1878. In 1876, when
Murray and the Vice-Minister of Education of Japan were in the United States
to attend the Philadelphia Centennial Exposition, they spent a considerable sum
of money in Boston buying a set of school furniture and teaching materials in
use in the schools there. Thus one of the models for the Japanese classroom was
the Boston schools of the 1870s. Western arithmetic became the official subject,
while Japanese mathematics, even the very polular *soroban* (Japanese abacus)

---

[17] *Origins of the Modern Japanese State: Selected Writings of E. H. Norman*, edited by John
Dower (New York: Pantheon Books, 1975), pp. 138–139.

[18] *"Gakusei"* (The System of Learning), in The Ministry of Education, *Gakusei Hyakunen-shi*
(100 Year History of the System of Learning), *Shiryō-hen* (Materials)(Tokyo: Monbushō, 1974),
p. 11.

was once expelled. But an important problem occurred. Who could teach Western mathematics? In the early Meiji period, however enthusiastic and dedicated, the hastily assembled teachers lacked enough training. The new texts were too difficult for them to handle. Certainly few of them knew anything of science or arithmetic, particularly since the new texts used unfamiliar Arabic numerals. Consequently in 1873 the curriculum was revised and the teaching of arithmetic by using the *soroban* abacus was permitted. This situation continues even in today's elementary education.

4   The University of Tokyo and the Tokyo Mathematical Society, 1877–1881

In 1877, the tenth year after the Restoration, two important events took place. In April the first national univeristy, the University of Tokyo, was established. Looking at the faculty of the Department of Mathematics-Physics-Astronomy, we are astonished to note that four professors out of five were foreigners. William E. Parson was American, and G. H. Berson, Stephane Mangeot and A. Dybouski were Frenchmen. Only one Japanese professor, Kikuchi Dairoku (1855–1917), can be found. Kikuchi is an extremely important scholar in the history of mathematics in Japan. In his childhood he studied at the *Bansho Shirabe-sho*, and was later sent to school at London by the order of the Tokugawa government. Because of his brilliance, upon his return to Japan the new government told him to go back to continue his education in London and then at the University of Cambridge. He was the 19th Wrangler at the Cambridge Mathematical Tripos of 1877 out of 110 candidates.[19] Soon after graduation in 1877, he was nominated as professor of mathematics at the University of Tokyo. The mathematical thought which influenced Kikuchi may be gathered by Joan Richards' book on geometrical thought in nineteenth century England.[20]

At first, as I stated above, important chairs of the university were mostly occupied by foreigners, and lectures, even those by Kikuchi, were conducted in English or French. However, it must be noted that the professors from foreign countries were dismissed or resigned of their own free will after a few years' teaching, when their first students graduated. They were replaced by these excellent students. The new leaders of the university preferred scholars of Japanese nationality to foreigners despite the immaturity of their knowledge, and the students could rapidly adopt the new learning.

The second and historically more interesting event of 1877 was the establishment of the Tokyo Mathematical Society (*Tokyo Sūgaku Kaisha*). This society

---

[19]Nakayama Shigeru, "Baron D. Kikuchi's Cambridge Days," *Kagakushi Kenkyū*, No. 65 (1963), pp. 36–37.

[20]Joan L. Richards, *Mathematical Visions: The Pursuit of Geometry in Victorian England* (Boston-San Diego-New York: Academic Press, 1988). Kikuchi seems to have been influenced especially by the mathematical thought of Isaac Todhunter and of William K. Clifford.

was private. So far we have treated the higher mathematics which was the object of research of professional mathematicians. Even Kikuchi should not be called a professional mathematician who obtained original results. As his later career shows, he was essentially a man of education. To discuss higher mathematics, we have to refer to *wasan* mathematicians. Professional mathematicians who were educated outside of the military institutes or *Bansho Shirabe-sho* were at that time *wasan* mathematicians. One of these *wasan* mathematicians, Kawakita Tomochika (1840–1919) stated in 1887:

> The establishment of the Tokyo Mathematical Society in the autumn of 1877 was the beginning of meeting of mathematicians in our country. Seventy percent of the members of the society were those who had once studied Japanese native mathematics. But after the society was renamed the Tokyo Mathematico-Physical Society, it consisted mostly of men of Western learning. In contrast, the members of *wasan* mathematicians were only twenty percent of them.[21]

In September 1877, the Tokyo Mathematical Society was founded with 117 members. From November, a monthly journal for the society was published. This society was the first of any learned societies in Japan. We know the object of the society from the introduction of the first issue of the journal. This introduction was written by one of the first two presidents, Kanda Takahira, the other president being Yanagi Narayoshi (1832–1891), a *wasan* mathematician. It might seem curious that the society had two presidents. But one came from mathematicians in Western style, and the other from *wasan* practitioners, the two being on the equal footing. The introduction reads as follows:

Introduction to the Journal of the Tokyo Mathematical Society
The object of the recent establishment of the Mathematical Society is to promote increasingly science. It means to enlighten people greatly by means of the principles of real learning (*jitsuri*). Because numbers are representations of the principles (*ri*) and without mathematical proofs the principles cannot be acquired, it is necessary to elucidate numbers if one wants to inquire into the principles closely. In this country, there have been not a few mathematicians from ancient times. On the occasion of the importation of Western learning, mathematics was greatly promoted and some eminent scholars came to unite the good of both the West and the East and to reform this science. Turning now to our past, the feudal era, *samurai* exclusively respected not the strength of their intelligence but that of their physical constitution. Confucian and Buddhist scholars all pursued the empty principles, not the real, useful principles. In particular, regarding arithmetic with contempt as a business of merchants, they set it aside. Although such an attitude

---

[21]Quoted from Ogura Kinnosuke, *Op. cit.* (n. 13), p. 168.

has gradually come to disappear in modern times, it still remains. We do not mention common people. Even those people who are in civil or military service and who teach quite often disregard mathematics. To make matters worse they do not feel ashamed of their attitude. The reason is that they do not know that they cannot acquire the principles without the pursuit of numbers. On the other hand, those people who intend to reform this science only gather among themselves and do not let the general public know its usefulness. This is the very object of the society. This society has the object of letting the general public know mathematics by all means. So we have established the society to pave the way for this object as far as we can. For this object the following are crucial:

That we collect mathematical books, ancient and modern, foreign and domestic.

That we necessarily answer questions whatever they may be.

That we openly ask the general public undecided things of the society.

That we translate mathematical books of the West.

That we publish those books that have been already translated.

That we fix Japanese mathematical terminology.

That we publish the reports of the meetings of our society.

These are rough proposals so that the details should be discussed. This time the first issue of the reports was accomplished. The title is the *Tokyo Sūgaku Kaisha Zasshi* (Journal of the Tokyo Mathematical Society), which is being put into print. This is the intention of the establishment of the society.

October, the tenth year of Meiji, 1877

Kanda Takahira.[22]

We notice that Kanda's introduction shares the same enlightened thought as Fukuzawa's. In fact Kanda was a close friend of Fukuzawa. Looking at the first issue of the journal, we can well understand the nature of the society. Despite Kanda's introduction, it is very doubtful that the members could understand what the academic society was. The journal was occupied with problems which the members proposed to each other, most of them written with Western symbolism, but usually educational and elementary. Occasionaly difficult and original problems were also submitted. They were written in the traditional Japanese way, both in symbolism and methods. The society could not function without *wasan* mathematicians, and mathematicians of Western style could not yet tackle such difficult problems in their own way.

In 1880, the committee of the society consisting of twelve members was selected by election. Three of them were working for the Ministry of Education or

---

[22]Quoted from The Japan Mathematical Society, *Op. cit.* (n. 14), pp. 83–84.

the university. Six members were related to the navy. Three members were selected from among the *wasan* mathematicians. We can understand from this that the main thrust of the Tokyo Mathematical Society was related to the navy, producing a continuity in the nature of Western mathematics before and after the Restoration. Throughout the early stage of the adoption of Western mathematics, military aspects colored its nature.

As for the nature of the society, it should be noted that the Tokyo Mathematical Society was different from its Western counterparts. In the first place, it was private, and was not so well organized as the Paris Academy of Sciences, which accomodated scientific experts by providing them with pensions. But, at the same time, it was not a society of semi-amateurs, like the Royal Society of London of the eighteenth century. Secondly, the establishment of the society gradually accelerated the dissolution of institutional basis of *wasan*, because *wasan* was mainly taught in its guilds and was more or less as esoteric as the other Japanese arts. Now members of the society could openly discuss solutions of mathematical problems and methods.

5   The Establishment of the Research Tradition of Western Mathematics: The German Model in Japan — The Tokyo Mathematico-Physical Society and the Imperial University of Tokyo, 1881–1903

So far I have described the process of institutionalization of Western mathematics as one of the components of the modernization of Japan as a whole. A decisive event related to this process took place in May of 1884 when the Tokyo Mathematical Society was renamed the Tokyo Mathematico-Physical Society. This event was not a simple change in name. A drastic reorganization was carried out. I mentioned that *wasan* was basically 'pure' mathematics and had almost no relationship with sciences other than calendrical calculations. The new society's name was the Tokyo *Mathematico-Physical* Society, which gathered not only mathematicians but also physicists. In this society *wasan* mathematicians must have felt powerless. The steering committee was consisted of mainly university-related people. Frustrated *wasan* mathematicians led by Kawakita had to organize a new society called *Sūgaku Kyōkai* (The Mathematical Association) three years later in 1887. But the latter society was just a club for those without university positions sharing a common interst. Some of them began to write their swan song, the history of Japanese mathematics. Endō Toshisada (1843–1915) was one of these scholars.[23] However, we have to admit in all fairness that *wasan* was defeated because of its

---

[23]In 1896 Endō Toshisada published the first critically written history of *wasan*: *Dai-Nippon Sūgakushi* (A History of Mathematics in Great Japan). See my review of a recent version of this book published by Kōseisha-Kōseikaku, Tokyo, 1981 in *Historia Scientiarum*, No. 21 (1981), pp. 123–124.

basic thought and the change of environment, not because of its supporters' lack of ability. The historian scarcely need mention Max Planck's well-known passage about physicists: "They gradually die off, and a rising generation is familiarized with the truth from the start."[24]

In any case, it can be said that the Tokyo Mathematico-Physical Society was such a full-fledged learned organization as we can see in nineteenth century Europe. It continued until 1945, the year of defeat of Japan in World War II.

This reorganization of the Tokyo Mathematical Society into the new Tokyo Mathematico-Physical Society was carried out under the leadership of Kikuchi Dairoku. This reorganization was approved with almost no strong opposition, but as Ogura Kinnosuke (1885-1962), a leading historian of modern Japanese mathematics, pointed out, many members of the old society judged it not as a natural or spontaneous event but as a kind of *coup d'état* attempted at by Kikuchi.[25] At all events, the establishment of the Tokyo Mathematico-Physical Society marked a crucial step in the institutionalization or professionaization of Western mathematics in Meiji Japan.

The next problem to be answered is as follows: When was the research tradition of Western mathematics established in Japan? When could Japanese mathematicians begin to produce original results? No one would deny that Takagi Teiji (1875-1960) was a first-class mathematician who was internationally admired. In 1903, he solved a difficult problem in the theory of numbers which consitituted a part of the conjecture called 'Kronecker's *Jugendtraum*'.[26] Also, in 1920, he solved the ninth problem which David Hilbert submitted to the second International Congress of Mathematicians in 1900.[27] That is, he established a very general theory of the class field theory, and succeeded in resolving generally Kronecker's conjecture. We should not ascribe the establishment of the research tradition only to Takagi's personal ability. Rather we have to answer a question: In what kind of intellectual milieu was he educated?

As I stated above, from 1884 onward the initiative of the Tokyo Mathematico-Physical Society was held by university men. Further, in 1886, the national university system was reorganized. In this reform, the University of Tokyo became the Imperial University of Tokyo. In the "Imperial University Order," German university systems and ideas were strongly reflected: "The purpose of the Imperial University is to teach and investigate those mysteries of science and learning, of arts and crafts, which are of practical service to State necessity."[28] Educational

---

[24] Max Planck, "Persönliche Erinnerungen aus alten Zeiten," in *Vorträge und Erinnerungen*, (Stuttgart: Hirzel, 1949), p. 13 (my translation).

[25] Ogura Kinnosuke, *Op. cit.* (n. 13), p. 171.

[26] Teiji Takagi, "Über die im Berichte der rationalen komplexen Zahlen Abel'schen Zahlkörper," *Journal of the College of Science, Imperial University of Tokyo*, **19** (1903), pp. 1–42; Collected Papers (Tokyo–New York–Berlin–Heidelberg: Springer-Verlag, ₂1990), pp. 13–39.

[27] Idem, "Über eine Theorie des relativ Abel'schen Zahlkörpers," *Journal of the College of Science, Imperial University of Tokyo*, **41** (1920), pp. 1–133; *Collected Papers*, pp. 73–167.

[28] *"Teikoku Daigaku Rei"* (Imperial University Order), in The Ministry of Education, *Op. cit.* (n. 18), p. 152.

principles of German universities were different from those of British universities; the latter were intended to form an ideal gentleman, while the former were marked by a nationalistic character and promoted specialized and detailed study in science and learning, the professors being government officials.

Until 1886 the Department of Mathematics at the University of Tokyo had been led by Kikuchi Dairoku, the only titular professor of mathematics. He graduated from the University of Cambridge, so his idea of mathematical education was strongly influenced by the British style. We may say that the mathematical education at Cambridge which Kikuchi Dairoku received was not so high in comparison with that of leading German universities.

But an important event took place in 1887. Fujisawa Rikitarō (1861-1933), a former disciple of Kikuchi, came back from Germany to the Imperial University of Tokyo and was made the second titular professor of mathematics there. He had been educated for about four years at the University of Berlin under Leopold Kronecker and Karl Weierstrass and then at the University of Strasbourg, then a German university. We know that at Berlin Ernst Edward Kummer lectured on the theory of numbers, Kronecker on the theory of algebraic equations, and Weierstrass on the theory of elliptic functions. Fujisawa returned to Japan in 1887 soon after having taken his Ph. D. degree with his thesis on Fourier series at Strasbourg under the guidance of Erwin Christoffel in 1886. He was mentioned in Wilhelm Lorey's book on the study of mathematics in German universities.[29]

Fujisawa, on becoming professor of mathematics, seems to have proposed a reform of the Department of Mathematics curriculum. He began to teach the theory of functions and the theory of elliptic functions, and opened research seminars in the German style. The research tradition in Japan, accordingly, began with the arrival of Fujisawa. Fujisawa published seminar notes whose standard was very high. One of these contained Takagi Teiji's article "On Abelian Equations." Takagi Teiji later studied at the Universities of Berlin and Göttingen and obtained the brilliant results stated above. Takagi was educated along the German style research tradition. Herewith the independent research tradition was established, in which the German *Wissenschaft* ideology, to use Roy Steven Turner's terminology, was dominant and persistent.[30]

We ask again: What caused the appearance of Takagi Teiji as an internationally renowned mathematician? One answer may be proposed: "Because of his brilliance." We approve of this statement, of course. But if this statement is interpreted in a strictly internal manner, we must make one reservation. He was educated through research seminars led by Fujisawa and later in Germany under

---

[29]Wilhelm Lorey, *Das Studium der Mathematik an den deutschen Universitäten seit Anfang des 19. Jahrhunderts* (Leipzig-Berlin: Teubner, 1916), p. 158.

[30]The expression "*Wissenschaft* ideology" is used by R. S. Turner in his "The Growth of Professional Research in Prussia, 1818-1848 — Causes and Context," *Historical Studies in the Physical Sciences*, **3** (1971), pp. 137–182. This ideology emphasizes in particular original research. See also his "The Prussian Universities and the Research Imperative, 1806 to 1848" (Princeton University unpublished Ph. D. dissertation, 1973).

the guidance of Felix Klein and David Hilbert. That is say, he was a product of the German mathematical culture, in which the research imperative was pervasive.

How and why did the German style of scholarship begin to be dominant in Japan? Up until 1881, the leaders of the Japanese government, in general, borrowed and discarded as seemed fit for their nation, sometimes from England or the United States, sometimes from France, and sometimes from Germany. But with the turning point termed the *seihen* (literally, political change, i. e. *coup d'état*) of the fourteenth year of Meiji, i. e. 1881, German learning began to be the most powerful, when Itō Hirobumi (1841–1909), later Japan's first prime minister, achieved political hegemony in the government by expelling Ōkuma Shigenobu, a leader of the movement for 'liberty and people's rights' (*jiyū minken*) strongly influenced by British and French political thought, particularly the former. Itō was such a politician who thought that the Japanese centralized political system should be formed around the Emperor, following Prussia as a model. Itō is known to have been assisted at the time by his ideologue Inoue Kowashi (1844–1895). During the *coup d'état* of 1881, Inoue wrote several memoranda to be proposed to Itō in which he insisted that German learning must become central, while the systems of learning of both Britain and France be suppressed.[31] In Inoue's opinion the latter two tended to promote political radicalism.

Significantly in 1881, the Departments of Mathematics, of Physics, and of Astronomy of the University of Tokyo began to require their students to study German instead of French. Of course, the establishment of the Imperial University of Tokyo was accomplished along the intellectual and political line of Itō and Inoue. It does not seem to be simply accidental that Fujisawa Rikitarō, who was nationalist mandarin-mathematician, instead of Kikuchi, displayed his academic power sufficiently within such a regime. This was one condition under which the talented mathematician Takagi emerged as a symbol of Japan's successful adoption of Western mathematics.

Conclusion

In concluding this report, I will briefly reflect on why Japan could so rapidly adopt Western mathematics in comparison with China.

Before the Meiji Restoration, the Japanese studied Western mathematics directly through Dutch instructors and books mainly in military institutes. It should be emphasized that the Japanese began to study on their own initiative and that the books were translated by the Japanese themselves. On the other hand, in China mathematical texts were translated mainly by foreigners. It can be said that the principal contributors of the adoption of Western mathematics in China were

---

[31] Inoue Kowashi, *"Jinshin Kyōdō Iken-an"* (A Draft of Opinions for Directing People's Minds), in *Inoue Kowashi Den* (A Biography of Inoue Kowashi), Vol. 1, *Shiryō-hen* (Historical Material) (Tokyo: Kokugakuin Daigaku Toshokan, 1966), pp. 248–251; *"Doitsu Sho-seki Honyaku Iken"* (An Opinion on the Translation of German Books), *Ibid.*, pp. 254–255.

missionaries from the West. For example, Euclid's *Elements* was translated into Chinese by Matteo Ricci, a Jesuit scholar. Moreover, the Japanese totally accepted Western numerals and symbolism, while the Chinese continued for a long time to use their own Chinese-style symbolism. Most importantly, in Japan the adoption of Western mathematics was carried out through rapid educational reform. But this was not the case in China. These differences were caused by the traditional thought that was so deeply rooted among the Chinese people. Fortunately or unfortunately Japan did not have such a tradition. Or I would rather say that Japan had a traditional mathematical thought related to *wasan*, and that, although *wasan* was totally discarded, the spirit of loving pure mathematics was tenaciously preserved.

Importantly, however, I do not necessarily want to present herein a simple success story. As I stated in Section 5, it was a modified German model of pure mathematics that Japanese mathematicians tried to accept after 1887. This process of institutionalization, as can be easily imagined, was keeping step with the intellectual and political consolidation of Imperial Japan.

How did Japanese contemporary intellectuals look at the unexampled rapid modernization in Meiji Japan? The representative novelist of modern Japan Natsume Sōseki (1867–1916) stated the following observation in a lecture entitled "*Gendai Nippon no Kaika*" (The Enlightenment of Modern Japan) in 1911:

> The Western enlightenment (i. e. ordinary enlightenment) was spon-
> taneous, while today's enlightenment in Japan is externally deter-
> mined. [...] Today's enlightenment in Japan can be briefly characterized
> as a superficial enlightenment. [...] We have to slide down the surface,
> or to suffer a nervous breakdown if we resist sliding down so. Accord-
> ingly, we Japanese have fallen into a difficult circumstance which can
> be said to be unfortunate and miserable. My conclusion is only this. I
> do not suggest you do this or do that. We cannot act in another way.
> My conclusion is extremely pessimistic: We can do nothing and we say
> with a sigh that we are having a hard time.[32]

Herbert Norman, significantly, arrived at a similar conclusion in 1940 in a section titled "How the Struggle for National Independence Inevitably Led to Expansion" of his *Japan's Emergence of a Modern State*:

> History is a relentless task-master, and all its lessons warned the
> Meiji statesmen that there was to be no half-way house between the
> status of a subject nation and that of a growing, victorious empire
> whose glory, to paraphrase that gloomy realist Clemenceau, is not un-
> mixed with misery.[33]

The two statements quoted above do not necessarily imply a determinist view of history but rather indicate that we can understand the history of modern Japan

---

[32]Sōseki, *Zenshū* (Collected Works), Vol. 11 (Tokyo: Iwanami Shoten, 1966), pp. 333–342.
[33]E. H. Norman, *Op. cit.* (n. 17), p. 305.

well only when we analyze its destiny taking into account the surrounding international circumstances.

The history of mathematics in modern Japan is not a simple success story
and is a typical history in which internal and external factors were inseparably
interwoven.

The Philosophical Views of Klein and Hilbert

David E. Rowe

There is always a question in writing the history of mathematics about how to bring the purely biographical elements and the mathematics itself together. The late Otto Neugebauer was a strong exponent of a tradition in which all but the most essential information about a scientist's life was consciously suppressed. Before his death, Neugebauer evidently destroyed all his correspondence as editor of the *Zentralblatt* because he did not think the information it contained was anybody's business. This action may appear rather extreme, but it actually accords very well with a tradition that essentially views the history of mathematics in Platonic terms as a collection of disembodied ideas. Until fairly recently, this was also the mainstream approach to the history of science, and in defense of the old history I think it needs to be said that studies, for example, of Newton's personality that try to link his creative drive to the rage he felt for the mother who abandoned him, whatever the merits of the argument may be, are not going to be very enlightening for someone who is primarily interested in knowing about what was whirling through Newton's mind during that year and a half that he spent writing his *Principia*.[1]

In my opinion, biographical details and information about the cultural and intellectual milieu in which a mathematician lived are often of decisive significance in shaping the larger context in which his work develops. In such cases, ignoring this larger framework will produce, at best, an incomplete picture of a mathematician's work, and is likely to obscure and misinterpret the motivations behind it. I believe the cases of Hilbert and Klein illustrate this problem very well, for much that has been written about them gives an inaccurate picture of what they thought mathematics was all about.

Hilbert and Klein have generally been portrayed as very different types of mathematicians, and some of these distinctions are obvious and important. Herbert Mehrtens, for example, has recently written a provocative work on the "mathematical modernists" and their opponents around the turn of the century, in which Hilbert and Klein appear as leading representatives of these two opposing camps.[2] There is no question that beginning around the turn of the century, Klein, like

---

[1]This psychoanalytic interpretation is argued in Frank E. Manuel, *A Portrait of Isaac Newton*. Cambridge, Mass.: Harvard Univ. Press, 1968. Newton's scientific achievements are discussed in detail in Richard S. Westfall, *Never at Rest. A Biography of Isaac Newton*. Cambridge: Cambridge Univ. Press, 1980.

[2]Herbert Mehrtens, *Moderne-Sprache-Mathematik*. Frankfurt am Main: Suhrkamp, 1990.

Poincaré, saw the burgeoning interest in abstract structures and axiomatics as a potential threat to the lifeblood of mathematics, whereas Hilbert appears to have had few of these reservations. Nevertheless, it would be very mistaken to think that Klein had no appreciation for axiomatic thinking, or that Hilbert was a reductionist who valued nothing else. Mehrtens' study thus presents what I might call a fair first-order approximation of Hilbert's and Klein's views. But as I will try to indicate here, its verisimilitude breaks down as soon as one begins to examine the higher-order terms.

Whereas Klein had both feet firmly rooted in the nineteenth century and consciously looked backward, Hilbert cast his gaze toward the future, and he profoundly influenced the course of twentieth century mathematics, as everyone knows. After his death in 1943, Hilbert's leading disciple, Hermann Weyl, wrote that "the era of mathematics upon which he impressed the seal of his spirit... achieved a more perfect balance than prevailed before and after, between the mastering of concrete problems and the formation of general abstract concepts."[3] Weyl attributed this "happy equilibrium" in large part to Hilbert's work and influence, adding that "no mathematician of equal stature has risen from our generation."[4] During the early 1920s, of course, Weyl had found himself in profound disagreement with his mentor over the issue of constructive versus axiomatic foundations. Summing up Hilbert's position some twenty years after this "Grundlagenkrise" had receded in the past, Weyl wrote:

> Hilbert is the champion of axiomatics. The axiomatic attitude seemed to him one of universal significance, not only for mathematics, but for all sciences. His investigations in the field of physics are conceived in the axiomatic spirit. In his lectures he liked to illustrate the method by examples taken from biology, economics, and so on. The modern epistemological interpretation of science has been profoundly influenced by him. Sometimes when he praised the axiomatic method he seemed to imply that it was destined to obliterate completely the constructive or genetic method. I am certain that, at least in later life, this was not his true opinion.[5]

Considering his deep roots in Göttingen mathematics and subtle grasp of philosophical issues, Weyl was perhaps uniquely qualified to pass judgement on the philosophical views of both Hilbert and Klein. His principal contact with Klein came during the years just prior to the outbreak of World War I when Weyl was a *Privatdozent* in Göttingen. During the winter semester of 1911–12 he lectured on Riemann surfaces, one of Klein's pet subjects, and these led to the publication of *Die Idee der Riemannschen Fläche*, the book that effectively put Klein's

---

[3]Hermann Weyl, "David Hilbert and his Mathematical Work," *Bulletin of the American Mathematical Society*, **50**(1944), 612–654, on p. 612.

[4]*Ibid.*

[5]*Ibid.*, p. 630.

earlier treatment of the subject out of circulation. Nevertheless, in the forward, Weyl stressed his indebtedness to Klein and his booklet of 1882, *Ueber Riemann's Theorie der algebraischen Funktionen und ihrer Integrale*. According to Weyl, it was Klein's more general conception of a Riemann surface, in which the complex plane is replaced by an arbitrary closed surface, that gave the modern theory its decisive vitality and power.[6]

But whereas Weyl said many insightful things about Hilbert's views on the foundations of mathematics, when it came to Klein's opinions he wrote these off as nothing more than a spin-off of a kind of naive positivism which he associated with Ernst Mach. Weyl had no sympathy for the distinction Klein made between pure mathematics and so-called *Approximationsmathematik*; nor could he abide Klein's attitude toward modern axiomatics. He once characterized Klein's views on this subject by recalling a conversation in which Klein had belittled the work of certain people by saying that after **he** (Klein) had shown how to jump over a ditch, these axiomatists would come along and show how one can still clear it with a table tied to one leg.[7]

As Weyl emphasized, Klein's work sparkles with an intuitive, aesthetic feel for mathematical ideas. He was the geometer *par excellence*, a grand synthesizer, forever seeking out the visual elements that bring a theory to life, leaping over the conventional boundaries that separate algebra, number theory, and analysis from geometry proper. His style was highly eclectic, based on a genetic approach to mathematical ideas, and the breadth of his knowledge truly astounding, as may be judged by the mammoth *Encyklopädie der mathematischen Wissenschaften*, a project he managed for more than twenty years. In this respect he even surpassed Hilbert, but of course he lacked the latter's penetrating depth. Hilbert's work, especially his most profound work—the papers on invariant theory, his *Zahlbericht*, and his booklet on the foundations of geometry—is characterized by streamlined,

---

[6]Hermann Weyl, *Die Idee der Riemannschen Fläche*. Leipzig: Teubner, 1913, pp. iv-v: "Die Riemannsche Fläche ist ein unentbehrlicher *sachlicher* Bestandteil der Theorie, sie ist geradezu deren Fundament. Sie ist auch nicht etwas, was a posteriori mehr oder minder künstlich aus den analytischen Funktionen herausdestilliert wird, sondern muß durchaus als das prius betrachtet werden, als der Mutterboden, auf dem die Funktionen allererst wachsen und gedeihen können. Es ist freilich zuzugeben, daß Riemann selbst die wahre Verhältnis der Funktionen zur Riemannschen Fläche durch die Form seiner Darstellung etwas verschleiert hat—vielleicht nur, weil er seinen Zeitgenossen allzu fremdartige Vorstellungen nicht zumuten wollte; dies Verhältnis auch dadurch verschleiert hat, daß er nur von jenen mehrblättrigen, mit einzelnen Windungspunkten über der Ebene sich ausbreitenden Überlagerungsflächen spricht, an welche man noch heute in erster Linie denkt, wenn von Riemannschen Flächen die Rede ist, und sich nicht der (erst später von Klein zu durchsichtiger Klarheit entwickelten) allgemeineren Vorstellung bediente, als deren Charakteristikum man dieses nennen kann: daß in ihr die Beziehung zu der Ebene einer unabhängigen komplexen Veränderlichen, sowie überhaupt die Beziehung zum dreidimensionalen Punktraum grundsätzlich gelöst ist. Und doch ist darüber kein Zweifel möglich, daß erst in der Kleinschen Auffassung die Grundgedanken Riemanns in ihrer natürlichen Einfachheit, ihrer lebendigen und durchschlagenden Kraft voll zur Geltung kommen. Auf dieser Überzeugung basiert die vorleigende Schrift."

[7]H. Weyl, "Axiomatic versus Constructive Procedures in Mathematics," edited by Tito Tonietti, *Mathematical Intelligencer*, **7**(4)(1985), 10–17, on p. 16.

axiomatic rigor. An even more suggestive description of the character of Hilbert's work comes from his first doctoral student, Otto Blumenthal:

> In the analysis of mathematical talent one has to differentiate between the ability to create new concepts that generate new types of thought structures and the gift for sensing deeper connections and underlying unity. In Hilbert's case, his greatness lies in an immensely powerful insight that penetrates into the depths of a question.... Insofar as the creation of new ideas is concerned, I would place Minkowski higher, and of the classical great ones, Gauss, Galois, and Riemann. But when it comes to penetrating insight, only a few of the very greatest were the equal of Hilbert.[8]

Felix Klein certainly would have mentioned Sophus Lie in this context, for he regarded Lie as the most impressive creative talent he had ever known.[9] But the point I wish to emphasize with this passage is that neither Hilbert nor Klein was principally a productive talent. Their ideas were nearly always closely tied to the work of important predecessors, and the most distinctive difference between them, to my mind, lies not in their attitudes toward "modernism" but in their orientation toward the achievements of the preceding generation. This issue not only reflects Hilbert's depth, as opposed to Klein's relative superficiality, it also relates to an issue of key importance that deserves considerably more attention than it has thus far received—the issue of modes of communication and presentation within the social structure of mathematics.

The tension between a mathematician's path to discovery and the mode of formal presentation he employs in conveying his results can be traced all the way back to the ancient Greeks. This theme forms the background against which the whole debate over the status of so-called Greek "geometrical algebra" has been played out. And whereas the recovery of Archimedes lost text *The Method* gave us a wonderful glimpse into the mental workshop of that era's greatest mathematician, the classical position was aptly stated by Gauss, who, when asked why he chose to leave his readers in the dark as to how he discovered the prinicipal propositions in his *Disquisitiones Arithmeticae*, responded that to do so would be like erecting a cathedral but leaving all the scaffolding behind.[10] Klein stood at the far end of the other side of the spectrum in a lineage that includes Descartes and Kepler. Unlike Gauss, he was deeply interested in pedagogy, and his works very often have an almost conversational tone about them. The message, however, is often the same: mathematics is guided in the first instance by inspired ideas; the formal development of a theory will then lead to a refinement of these basic insights, but

---

[8]O. Blumenthal, "Lebensgeschichte," in *David Hilbert Gesammelte Abhandlungen*, Bd. 3. Berlin: Springer, 1935, p. 429.

[9]D. E. Rowe, "Der Briefwechsel Sophus Lie – Felix Klein, eine Einsicht in ihre persönlichen und wissenschaftlichen Beziehungen," *NTM. Schriftenreihe für Geschichte der Naturwissenschaften, Technik und Medizin*, 25(1)(1988), 37–47.

[10]Wolfgang Sartorius von Walterhausen, *Gauss zum Gedächtnis*. Leipzig: Hirzel, 1856.

one should not for that reason abandon the latter nor overlook the indispensible role they play.

A watchword for much of Klein's mathematics is the role of *Anschauung*, that is an appeal to structures and relationships that can be visualized or imagined. The roots of this can be found in the geometrical legacy of his teacher, Julius Plücker, one of the creators of modern projective geometry. In his obituary to Plücker, Alfred Clebsch wrote: "Es ist die Freude an der Gestalt in einem höheren Sinne, die den Geometer ausmacht,"[11] and this delight in studying concrete, visualizable form proved to be a lasting inspiration for Klein as well.[12] The novelty in Klein's work came in the unexpected way in which he brought this *anschauliche* approach to bear on other branches of mathematics—algebra, function theory, potential theory, etc.—bearing little immediate connection to geometry.

As a student of Plücker, Klein adopted algebraic techniques right from the beginning. It was Clebsch, however, who introduced him to the advantages of the projective point of view, in particular the possibility of exploiting the formal apparatus of algebraic invariant theory. But Klein was never dogmatic in his preferences for one approach over any other. His principal criticism of Jacob Steiner and other synthetic geometers was that in their efforts to preserve the *anschauliche* elements they threw out one of its richest resources, namely the interplay between algebraic formalism and geometric structures.[13]

During his early career, Klein was constantly on the alert for new ideas which he assimilated with uncanny ease, and his sources—Plücker, Clebsch, Lie, Weierstrass, and Riemann—were fertile indeed. Riemann enjoyed a special place in his pantheon of heroes, and for reasons that should be clear from these remarks taken from his well-known lectures on the history of 19th-century mathematics:

> I must explain so many things here so that one can realize at least to some degree the extent to which Riemann's incomparable genius was ahead of its time and thus influenced considerably further production. Certainly the capstone to every mathematical theory is provided by compelling proofs for all its assertions.... The secret of brilliant productivity, however, will forever remain the posing of new questions and the sensing of new theorems that contain worthwhile results and connections. Without the disclosure of new viewpoints and without providing new goals, mathematics would soon become exhausted through the rigor of its logical proofs and begin to stagnate for lack of substance. Thus, in a certain sense, mathematics is promoted most by those who distinguish themselves through their intuition rather than through the

---

[11] A. Clebsch, "J. Plücker, in Memorium," in A. Schoenflies, F. Pockels, eds., J. Plücker, *Gesammelte Mathematische Abhandlungen*, Bd. I. Leipzig: Teubner, 1895.

[12] See, for example, the section on "Anschauliche Geometrie" in F. Klein, *Gesammelte Mathematische Abhandlungen*, Bd. 2. Berlin: Springer, 1922, pp. 3–251.

[13] Typical remarks on the Steiner school can be found in F. Klein, *Vorlesungen über höhere Geometrie*, 3rd ed. edited by Wilhelm Blachke. Berlin: Springer, 1926, pp. 116–117.

strict logic of their proofs. There is no doubt that among the mathematicians of recent decades, Riemann is the one whose impact is felt the strongest today.[14]

Perhaps the sharpest and most meaningful way to sum up the differences between Klein and Hilbert is by saying that Hilbert had no need for heroes. In contrast with Klein, he was an utterly ahistorical thinker who measured the quality of a mathematician's work by the number of earlier investigations it rendered obsolete. The achievements of Kummer and Kronecker, for example, in the theory of algebraic number fields are practically unrecognizable in Hilbert's *Zahlbericht*. This uncharitable attitude toward ones forerunners apparently had nothing malicious about it; Hilbert just was not capable of nor interested in seeing things from anything other than his own point of view. Thus when he and Hurwitz had earlier discussed the two approaches of Kronecker and Dedekind to factorization into prime ideals in an algebraic number field, they concluded that "both were abominable."[15] One detects the same sort of arrogance in other pronouncements of Hilbert, including his certainly jestful remark that "physics had become too difficult for the physicists."[16]

With this general background in mind, I would like to take a closer look at what Klein and Hilbert had to say about the role of the axiomatic method in mathematics. First, a typical pronouncement by Klein concerning the axioms for a group:

> Here the appeal to fantasy recedes fully in the background, while the logical skeleton is carefully exhumed. This abstract formulation is splendid for working out the proofs, but is entirely unsuitable when it comes to finding new ideas and methods; rather it represents the culmination of preceding developments. Thus it eases instruction insofar as with its help one can easily provide complete proofs for known propositions. On the other hand, the matter is made more difficult for the student, who is placed before something that is already complete (*etwas Abgeschlossenes*) and who cannot imagine how one could ever arrive at these definitions about which he is unable to form any picture whatsoever. Indeed the method has the general disadvantage that it does not stimulate thinking; one must only be careful (*aufpassen*) not to collide with any of the four commandments [the axioms].[17]

---

[14]F. Klein, *Vorlesungen über die Entwicklung der Mathematik im 19. Jahrhundert*, Bd. I. Berlin: Springer, 1926, pp. 271–272.

[15]Constance Reid, *Hilbert*. Berlin/Heidelberg/New York: Springer, 1970, p. 42.

[16]An account of Hilbert's (and Klein's) contributions to general relativity theory is given in Jagdish Mehra, *Einstein, Hilbert, and the Theory of Gravitation*. Dordrecht/Boston: Reidel, 1974.

[17]F. Klein, *Vorlesungen über die Entwicklung der Mathematik im 19. Jahrhundert*, Bd. I, pp. 335–336.

Klein's concerns with pedagogy came clearly to the fore here; and his remarks reveal that he would have had no sympathy with approaches like the Moore method introduced by the influential topologist R. L. Moore at the University of Texas. He felt that it was indispensible for mathematics teachers to know something about the historical development of mathematical ideas. That way they could appreciate the long and torturous journey that even the most elementary notions had travelled before they could emerge as the crystallized and systematic forms we encounter today.

But to return to the passage on the group axioms, it is important to notice that Klein does not just criticize this approach. Indeed, he always recognized the merits of the axiomatic method when—as in the case of group theory—the axioms organically emerge as the logical core underlying a significant body of already developed mathematics. What he objected to was "the opinion *that the axioms are only arbitrary statements which we set up at pleasure and the fundamental concepts, likewise, are only arbitrary symbols for things with which we wish to operate* (his italics)."[18] Klein identified this view with the sterile philosophy of nominalism, and regarded it as "the death of all science."[19] As regards geometry proper, he emphatically stated that "*the axioms of geometry are—according to my way of thinking—not arbitrary, but sensible statements, which are, in general, induced by space perception and are determined as to their precise content by expediency* (his italics)."[20]

So how did Klein react to Hilbert's famous *Grundlagen der Geometrie* of 1899, the latter's first serious foray into formal axiomatics? One might guess the answer to this by bearing in mind that it was Klein who prompted him to edit his lectures on the foundations of geometry in the first place.[21] Although Hilbert had published almost nothing on geometry before this, Klein had been following his work in this field for some time, and with his usual flair for the dramatic he found the perfect occasion for Hilbert to present the fruits of his work: as part of the celebration unveiling the Gauss-Weber monument in Göttingen.

Klein fully appreciated the significance of Hilbert's axiomatic treatment of the foundations of geometry, which he referred to as "the most important work" of its kind.[22] But then Hilbert was not just fiddling around with any old system of axioms; the axioms he replaced were those of Euclid. That the Euclidean edifice stood in need of large-scale renovation had become clear to geometers long before Hilbert published his booklet—one detects cracks in its *stocheia* already in the very first proposition of the *Elements* ("To construct an equilateral triangle on a given

---

[18] F. Klein, *Elementarmathematik vom höheren Standpunkte aus*, Bd. 2, 3rd. ed. Berlin: Springer, 1925, p. 202.

[19] *Ibid.*

[20] *Ibid.*

[21] Reid, *Hilbert*, p. 61. For a detailed analysis of the evolution of Hilbert's thought on the foundations of geometry, see Michael M. Toepell, *Über die Entstehung von Hilberts "Grundlagen der Geometrie"*, Studien zur Wissenschafts-, Sozial- und Bildungsgeschichte der Mathematik, Bd. 2, (Göttingen: Vandenhoeck & Ruprecht, 1986.

[22] F. Klein, *Elementarmathematik vom höheren Standpunkte aus*, Bd. 2, p. 200.

line segment"). Hilbert was continuing the work of Moritz Pasch, who was the first modern geometer to fill one of the significant gaps in the axiomatic system set forth by Euclid. Pasch's axioms of order established a betweenness relation for the points on a line as well as the property that a line intersecting one side of a triangle must necessarily intersect another. By building on the work of his predecessors—Pasch, Schur, Peano, and Veronese—Hilbert succeeded in erecting a new foundation that was sufficiently secure to withstand the weight of the theory it was built to support. True, certain details had to be altered after redundancies in some of the axioms were found, but this was only so much *Flickwerk*, and the five groups of axioms Hilbert laid down for incidence, order, congruence, parallelism, and continuity remained essentially intact.

Moreover, Hilbert went decidedly beyond his predecessors by systematically demonstrating the independence of the axioms in his system. He did this by showing that an alternative postulate could be embodied in a "non-Euclidean" model for geometry. Here I am using the term non-Euclidean geometry in the broad sense to connote any system of geometry distinct from the classical Euclidean case; e.g. systems like the non-Desarguesian geometries Hilbert obtains by dropping the congruence axiom III.5 and retaining the other axioms in the first three groups. Axiom III.5 is essentially the SAS congruence theorem for triangles; it forms another glaring gap in the *Elements* where Euclid attempts to prove it in I.4 by the method of "applying" one figure to another. (It is indeed quite curious to see how Euclid goes to considerable trouble to prove the proposition that a given line segment can be "applied" to a point, i. e. it can be transported in the plane so as to make a given point occupy an endpoint of the segment, and then without batting an eye he simply transports a whole triangle onto another triangle so that its two given sides and their included angle coincide with those of the other given triangle.)

One of the earliest uses of models to demonstrate the relative consistency of a non-Euclidean geometry was the one introduced by Klein back in 1871. It utilized Cayley's theory in which the metric relations of Euclidean geometry are derived by referring to an invariant expression associated with a particular conic section or quadric surface in the projective plane or three-space. Cayley had introduced this invariant expression for a general quadratic form, but he only interpreted it geometrically in the case where the figure degenerates into the circular points at infinity in the plane. Klein conjectured almost immediately that the so-called Cayley metric had a more general significance when he first encountered Cayley's "Sixth Memoir on Quantics" during his sojourn in Berlin in early 1870. But it took him nearly a year and one-half to convince himself that he had a proof. In fact, Klein's subsequent paper "Ueber die sogenannte Nicht-Euklidische Geometrie" failed to persuade a number of authorities, including Cayley himself.[23]

Klein's formulas for the distances between points and between lines were almost identical with Cayley's except that he used cross-ratios, which are projective

---

[23]F. Klein, "Ueber die sogenannte Nicht-Euklidische Geometrie," (Erster Aufsatz) *Gesammelte Mathematische Abhandlungen*, Bd. I, pp. 254–305. Regarding Cayley's doubts, see Klein's remarks on p. 242.

invariants, in place of Euclidean distance measures. He then merely alluded to the fact that von Staudt had shown how one could coordinatize projective geometry without any recourse to metric notions and without employing the parallel postulate. This meant that there was a purely projective way to introduce cross-ratios, and therefore Klein's formulas could be interpreted without borrowing anything from other metric geometries. Klein tried to spell out how one could carry out this coordinatization in a second article with the same title, but apparently with little success. It was one of his few forrays into the foundations of geometry, and since logic was not Klein's strongest suit these waters remained as muddy as ever. In 1874 he published a third short article on this subject after Zeuthen and Lüroth had informed him of an error in his argumentation. But Klein still could not put the matter right, and with noticeable embarassment he published one final note explaining how Darboux had pointed out another error, and that the matter was really much simpler than Klein had portrayed it the third time around.[24]

Klein's most glaring weakness was, of course, one of Hilbert's greatest strengths. His *Grundlagen der Geometrie* was a showcase for all kinds of geometries, but its principal goal was to characterize Euclidean geometry by slowly building up its axiom system and deriving various "pseudo-geometries" along the way, thereby illustrating the necessity of additional axioms. With his last group of axioms, Hilbert took up essentially the same question Klein had struggled with earlier, namely what must one assume in order to establish a one-to-one correspondence between the points on a line and the real numbers. Hilbert set down two axioms: 1) the so-called axiom of Archimedes restricted to line segments (actually this axiom is Definition 4 in Book V of Euclid's *Elements*, a key ingredient in Eudoxus's theory of proportion) ; and 2) what Hilbert called the axiom of line completeness, which says that it is impossible to extend the set of points on a line and still preserve the axioms for incidence, order, and congruence, as well as Archimedes' axiom.

The possibility of employing a non-Archimedean system had been hotly debated ever since the publication of Veronese's *Fondamenta di geometria* in 1891. Others who had introduced infinitesimals around this time included Otto Stolz, Paul Dubois-Reymond, and Karl Johannes Thomae. Interestingly enough, Georg Cantor was one of the harshest critics of this approach, perhaps in part because it offended certain theological preconceptions that were dear to him. He once called infinitesimals the "cholera bacillus of mathematics" and on another occasion referred to them as "paper magnitudes that have no other existence other than to be on the paper of their discoverers and disciples."[25]

Hilbert and Klein, on the other hand, both recognized the validity of non-

---

[24]See F. Klein, "Ueber die sogenannte Nicht-Euklidische Geometrie," (Zweiter Aufsatz) *Gesammelte Mathematische Abhandlungen*, Bd. I, pp. 311–343; "Nachtrag zu dem 'zweiten Aufsatz über nicht-Euklidische Geometrie," *Ibid.*, pp. 344–350; "Über die geometrische Definition der Projektivität auf den Grundgebilden erster Stufe," *Ibid.*, pp. 351–352.

[25]See Joseph W. Dauben, *Georg Cantor. His Mathematics and Philosophy of the Infinite.* Cambridge, Mass.: Harvard University Press, 1979, pp. 128–132.

Archimedean systems in geometry, a concept already implicit in the ancient Greek notion of horned angles. In Hilbert's *Grundlagen* they play the role of a non-Euclidean system in which all the usual axioms hold with the exception of the axioms of continuity, which are therefore necessary. After introducing this type of system, he went on to show that his axioms for Euclidean geometry are consistent in the sense that they lead back to the usual axiom system for the real numbers. Thus a contradiction in Euclidean geometry would necessarily lead to a contradiction in the ordinary arithmetic of the real numbers, which are characterized by the axioms for a complete ordered field.

In the course of his argument, Hilbert laid down some of the basic principles that later formed the heart of the formalist program. An axiomatic system worthy of its salt must be consistent—otherwise it will ultimately lead to a contradiction—and independent (i.e. free from redundancies). Later he would pose the even more challenging demand that the system be complete, that is capable of providing the ammunition necessary in order to prove any "true" proposition within the theory. This approach ultimately led him to erect a proof theory with which he hoped to show that one could always decide the truth or falsity of any proposition that could meaningfully be expressed within the formal language of the system.

But what came to be known as Hilbert's formalist program was something quite different from his overall philosophy of mathematics. Hilbert's program arose only later as a reaction to what Hilbert viewed as a frontal assault on the temple of mathematics. Back in 1900, when he delivered his famous speech at the International Congress in Paris, he cited the resolution of the status of the parallel postulate and the impossibility of squaring the circle in flatly proclaiming that **all** mathematical problems are capable of being answered with finality. Indeed, he went so far as to call this a "conviction that every mathematician certainly shares." The full passage reads as follows:

> This remarkable circumstance [namely the resolution of these ancient problems] along with other philosophical reasons probably accounts for the conviction that every mathematician certainly shares [die jeder Mathematiker gewiß teilt], but which, up to now at least, no one has supported by proofs—I mean the conviction that every mathematical problem is necessarily capable of strict resolution, be it by way of a successful answer to the question raised, or by proving the impossibility of its solution and thereby the inevitable miscarriage of all such attempts.... This conviction in the solvability of every mathematical problem provides powerful encouragement [Ansporn] to us during our work; we hear within ourselves the constant cry: *There is the problem, seek the solution. You can find it through pure thought, for there is no Ignorabimus in mathematics.*[26]

---

[26]D. Hilbert, "Mathematische Probleme," in *David Hilbert Gesammelte Abhandlungen*, Bd. 3, pp. 290–329, on p. 297.

This remarkable pronouncement reveals much about Hilbert's philosophical views; in particular, it shows that in 1900 at least there was nothing to his mind at all problematic about mathematical knowledge. The ultimate solvability of every meaningful mathematical proposition was simply a truism. As we shall see, Hilbert sang a very different tune twenty years later.

As indicated above, Klein showed that the consistency of non-Euclidean geometry can be reduced to proving that the axioms for Euclidean geometry are consistent, and Hilbert, in his *Grundlagen der Geometrie*, succeeded in pushing this argument one step further by showing that a contradiction in Euclidean geometry would necessarily lead to a contradiction within the axiom system for the real numbers. Of course the fact that their research results had something in common reveals very little about their perspectives on the foundations of geometry. Yet for all their apparent differences, it must be emphasized that Hilbert, like Klein, pursued his investigations as part of what both saw as the natural organic development of geometrical knowledge. Indeed, Hilbert's book commences with a motto taken from Kant's *Critique of Pure Reason*: "So fängt denn alle menschlichlichen Erkenntnis von Anschauungen an, geht von da zu Begriffen und endigt mit Ideen."[27] The use of the term "Anschauung" recurs over and again in discussions of the foundations of geometry in the 19th century, and it is always problematic trying to determine the degree to which it is colored by Kantian conceptions. The striking feature here, however, is how Hilbert takes up this typically Kleinian leitmotive and invokes it here as a guiding principle. As Michael Toepell has convincingly shown, *Anschauung* held an important place in Hilbert's geometrical work, and he was by no means convinced that one could ultimately dispense with it altogether.[28]

Another example of how Hilbert conceived the organic development of mathematics comes from his earlier work on invariant theory. In an expository article presented by Klein at the 1893 Chicago Mathematics Congress, Hilbert summarized recent work on algebraic invariants in the following terms: "In the history of a mathematical theory one can usually distinguish three developmental phases without difficulty: the naive, the formal, and the critical. With regard to the theory of algebraic invariants, its founders, Cayley and Sylvester, may be regarded as representatives of the naive period. It was they who had the immediate joy of exhibiting the simplest invariant structures and their application to the solution of equations of the first four degrees. The inventors and perfecters of the symbolic method, Clebsch and Gordan, are the champions of the second period. The critical period finds its expression in the theorems listed above [Hilbert's own]...."[29] Hilbert and Klein were both universalists who shared a vision of mathematics as

[27]D. Hilbert, *Grundlagen der Geometrie*, 6th ed. Leipzig/ Berlin: Teubner, 1923, p. 1.

[28]See the discussion on "Axiomatische Methode, Anschauung, Intuition" in M.-M. Toepell, *Über die Entstehung von Hilberts "Grundlagen der Geometrie"*, pp. 257–261.

[29]D. Hilbert, "Ueber die Theorie der algebraischen Invarianten," in E. H. Moore, et al., eds., *Mathematical Papers Read at the International Mathematical Congress held in Connection with the World's Columbian Exposition Chicago 1893*. New York: Macmillan, 1896, pp. 116–124, on p. 124.

one of most sublime ornaments of human culture. Indeed, they often played the role of emissaries who spoke on behalf of the larger cultural mission of mathematics, although each had his own special axe to grind on such occasions. But even their mathematical work had much in common that is not generally appreciated. Their correspondence, for example, reveals how closely Klein followed Hilbert's work on invariant theory during the latter's formative years.[30] Although no great expert in this field, Klein utilized algebraic invariants heavily in his work on geometry and higher-order equations and the related theory of elliptic modular functions. As a product of the Clebsch school, he learned the basic techniques of the theory early on and fully appreciated the importance of the theory for geometry. When he needed the advice of an expert, he generally turned to Paul Gordan, known to many as the "King of Invariants" after he solved the difficult problem of determining a finite system of basis invariants for binary forms. Klein and Gordan had been colleagues for a brief time at Erlangen until the former accepted a position at the Technische Hochschule in Munich. After that they used to meet fairly regularly in Eichstätt, roughly halfway between Erlangen and Munich.

Klein was placed in an awkward position when Hilbert began attacking the famous finite basis problem using new methods that broke with the computational techniques utilized by Gordan and others. Hilbert's *Doktorvater*, Ferdinand Lindemann, called his methods "unheimlich," and Gordan himself went so far as to accuse the young upstart of writing "theology" rather than mathematics.[31] Hilbert remained adamant in defending his existence proofs, but it should not be overlooked that he also apparently took these criticisms to heart. In Herman Weyl's opinion, it was Hilbert's desire to frame a proof that was, at least in principle, constructive that led him to prove his Nullstellensatz.[32] By combining this with Cayley's Ω-differentiation process he was able to tie his approach in with Kronecker's constructive methods. This apparently quelled Gordan's objections; it also prompted Minkowski to congratulate his friend for discovering what he called "smokeless gunpowder." He wrote to Hilbert that "it really is high time to decimate the fortresses of the robber-knights—Stroh, Gordan, Stephanos, and whoever they all are—who held up the individual traveling invariants and locked them up in their dungeons, considering the danger that new life may never spring from these ruins again."[33] Klein called Hilbert's work simply "the most important work on general algebra ever published in *Mathematishe Annalen*," the journal founded by Clebsch in 1868 and which had since become under Klein's editorship the leading mathematics periodical of the day.[34]

---

[30] *Der Briefwechsel David Hilbert - Felix Klein (1886–1918)*, ed. Günther Frei (Arbeiten aus der Niedersächsischen Staats- und Universitätsbibliothek Göttingen, **19**). Göttingen: Vandenhoeck & Ruprecht, 1985, pp. 3–55.

[31] O. Blumenthal, "Lebensgeschichte," p. 394.

[32] H. Weyl, "David Hilbert and his Mathematical Work," p. 617.

[33] H. Minkowski to D. Hilbert, 9 Feb. 1892, in Hermann Minkowski, *Briefe an David Hilbert*, ed. L. Rüdenberg and H. Zassenhaus. Berlin/ Heidelberg/ New York: Springer, 1973, p. 45.

[34] F. Klein to D. Hilbert, 18 Feb. 1890, in *Der Briefwechsel David Hilbert - Felix Klein (1886–1918)*. On the emergence of *Die Mathematische Annalen*, see the *Einleitung* to R. Tobies and

Hilbert's conflict with Gordan foreshadowed a mathematical debate concerning the ontological status of mathematical constructs and the unrestricted use of the law of the excluded middle, a debate that took on growing significance for Hilbert in the years ahead when he championed Georg Cantor's theory of transfinite numbers while attacking the constructivist views of Cantor's archrival, Leopold Kronecker. In his famous Paris lecture of 1900, Hilbert chose to put two conjectures of Cantor at the top of his list of 23 unsolved problems: the continuum hypothesis and the assertion that the real numbers can be exhibited as a well-ordered set. The problem that followed (number two) stemmed directly from Hilbert's own recent work on the foundations of geometry. Since he had proven that the axioms for Euclidian geometry were consistent provided that the same was true for the axioms of arithmetic for the real numbers, proving the latter was clearly a matter of special concern to him, especially since this would "simultaneously provide a proof for the mathematical existence of the real numbers, i.e., the continuum. In fact, if the proof of consistency fully succeeds, then there will no longer be any justification for the objections that have formerly been raised against the existence of the real numbers."[35] These remarks were clearly directed against the views of Leopold Kronecker, although it is quite clear that Kronecker would have found Hilbert's approach neither congenial nor persuasive. But Hilbert went still further, proclaiming his conviction that by means of a self-consistent system of axioms one could prove not only the existence of the continuum but also the validity of the higher Cantorian number classes of ordinals and cardinals.

Still, none of these foundations problems held an overriding importance for Hilbert, and aside from a single speech delivered at the International Congress held in Heidelberg in 1904, he remained mute on this topic until 1917. By then he could no longer avoid the "foundations crisis" that was beginning to reach the boiling point in mathematical circles, and a confrontation with Brouwer was no longer avoidable. As is well known, during this later period Hilbert paid considerably more attention to foundations questions. Nevertheless, the full spectrum of his mature views seems not to have received the attention it deserves. In order to achieve a more balanced perspective as to what these were, I would like to present some extensive passages from Hilbert's recently published lectures on "Natur und mathematisches Erkennen."[36] In these lectures, delivered in 1919 and originally compiled and edited by Paul Bernays, Hilbert sought to counter some commonly held beliefs about the nature of mathematical knowledge:

> I begin by repeating the conventional popular view of mathematics and mathematical thinking, which may be expressed as follows: "Mathematical truths are absolutely certain, for they are proven by means of infallible reasoning from definitions. They must therefore accord every-

D. E. Rowe, eds., *Korrespondenz Felix Klein - Adolf Mayer* (Teubner-Archiv zur Mathematik, 14). Leipzig: Teubner, 1990.

[35]D. Hilbert, "Mathematische Probleme," p. 301.

[36]D. Hilbert, *Natur und mathematisches Erkennen*, ed. D. E. Rowe. Basel: Birkhäuser, 1992.

where with reality, and this they do." According to this view, mathematics would be, strictly speaking, nothing but a monstrous tautology.
All propositions, as Poincaré formulated it, would express nothing more
than a circuitous way of saying that a=a. In philosophic language we
can therefore say that the propositions of mathematics are analytic
propositions.[37]

Hilbert then illustrated this with the theorem: $2 + 2 = 4$ by defining $2, 3, 4$
recursively through addition of 1 and by defining $+2$ as the operation:

$$x + 2 = (x + 1) + 1,$$

thus

$$2 + 2 = (2 + 1) + 1 = 3 + 1 = 4$$

He then criticizes this approach by saying:

> Were this viewpoint correct, then mathematics should appear as
> nothing more than a series of logical arguments heaped one upon the
> other. One would find nothing but an arbitrary series of conclusions
> driven by the power of logic alone. But in reality nothing of the kind
> exists; indeed, the conceptual structure of mathematics is constantly
> led by intuition (*Anschauung*) and experience so that mathematics for
> the most part represents a closed structure free of arbitrariness.[38]

Hilbert next gives a lengthy illustration of this last assertion by considering
the classical construction of the continuum, beginning first with an intuitive argument to show that a line segment must contain infinitely many points. He then
proceeds to show why a countably infinite collection still leaves holes—a realization known to the ancient Pythagoreans. Hilbert's answer to this was to introduce
Dedekind cuts, although he could just as well have referred his listeners to the
Eudoxian theory of proportion set out in Book V of Euclid's Elements. This leads
to what he calls a continuum of the second type.

Following the same line of reasoning he adopted in his *Grundlagen der Geometrie*, Hilbert next examines the possibility of introducing infinitesimals. He
then shows that such a system is logically tenable and leads to a continuum of the
3rd kind. One need only drop the condition that all magnitudes bear a finite ratio
with one another i.e., one adopts a non-Archimedean number system. But then
Hilbert makes the following highly significant remarks:

> Now we would not wish to give up this demand [of the Archimedean
> postulate] without compelling reason. And such reason certainly does
> not lie within our experience. On the contrary, it is a noteworthy fact

---

[37] *Ibid.*, p. 4.
[38] *Ibid.*, p. 5.

that all distances that occur in nature can be measured against one another. From the largest astronomical expanses to the smallest distances
as we find them in atoms and electrons, all can be expressed as mul-
tiples or fractions of a centimeter. Therefore the continuum of the 3rd
kind, which naturally has its importance as a number system within
mathematics, is to be rejected for purposes of mathematically concep-
tualizing the continuous, and thus we remain with the continuum of
the 2nd kind. And thus we see through this example of the continuum
how mathematical conceptualization is stimulated by intuition and led
by experience. One could give numerous other examples of this. In particular, it should be mentioned that the impulse to develop differential
calculus came from geometry and mechanics, and that its further development was decisively influenced by scientific and technical problems,
as it will be in the future.[39]

These remarks clearly reveal how far removed Hilbert was from the logicist
views of Bertrand Russell as expressed in the famous dictum that "mathematics
is the subject in which we do not know what we are talking about nor whether
or not what we are saying is true." Hilbert also discussed other commonly held
misconceptions of mathematics, for example, that its inferences are based on self-
evident axioms and that they are drawn by employing a logical system that is
beyond any possible reproach. After a lengthy discussion of these matters, he
summarized his conclusions as follows:

> Looking over the course of our observations, we can say that the
> three viewpoints along which we have modified the primitive concep-
> tion of mathematics follow the pattern of the philosophical scheme:
> "Begriff, Urteil, Schluss." We determined that the Begriffsbildungen of
> mathematics are not arbitrary but rather are built up systematically
> for reasons that are both internal and external. Then we examined
> the Grundurteile of mathematical disciplines and found that absolute
> certainty regarding the initial propositions was not a necessary require-
> ment for the successful application of the mathematical method. And
> regarding mathematical Schliessen, we came to recognize that it is not
> confined to the usual methods of logic and that it is not free from errors
> and problems.[40]

These lectures of 1919 contain many similar pronouncements that I believe
help to put Hilbert's mature philosophical views in their proper perspective. Addressed to a lay audience, they reveal that Hilbert, like Klein, was a spokesman for
mathematical culture in the broadest sense of the word, and to me the parallels
between their thought and their careers are much more striking than the qualities

---

[39] *Ibid.*, p. 11.
[40] *Ibid.*, p. 35.

that distinguished them. They were both elitists who were not afraid to talk to the educated layman in simple terms. One finds this in Poincaré, as well, but certainly it was uncommon in Germany, at least outside of Göttingen. It just was not fashionable for German scholars to "stoop to the masses"—indeed it was regarded as a betrayal of the ideals of *Wissenschaft*. Klein paid a heavy price for doing so, having to suffer the scourn of the famous mathematicians of the Berlin school—Weierstrass, Kronecker, and Frobenius—who regarded him as a charlatan.[41] Of course, no one would ever have suggested this of Hilbert.

---

[41]See D. E. Rowe, " 'Jewish Mathematics' at Göttingen in the Era of Felix Klein," *Isis*, **77**(1986), 422–449, pp. 432–434.

Hermann Weyl's Contribution to Geometry, 1917–1923

Erhard Scholz*

1 Introductory Remarks

HERMANN WEYL (1885 – 1955) was formed during his time as a student in Göttingen (1903 to 1908, interrupted by two semesters in Munich) by Klein, Hilbert and the Göttingen tradition, before he himself later grew into the role of one of the most important representatives of this tradition. He tried to combine Klein's intuitive approach to mathematics with the conceptual and technical achievements of Hilbert (in particular with respect to analysis). In different contexts he took up some of Riemann's essential ideas and invigorated them in the context of 20th century mathematics. Perceived in a wider scope, Weyl brought his wide philosophical and literary interests to bear upon the style and orientation of his mathematical research and the expression of its results. These influences from diverging sources gave his work a particular personal profile and makes it interesting from a number of different aspects [Chevalley/Weil 1957].

Much of the historical literature on H. Weyl deals with his involvement in the debate on the foundations of mathematics, where he sided for some years (about 1919 to 1923) with Brouwer against Hilbert, after his initial more cautious encounter with foundational questions around 1910 at the time of his *Habilitation* and before he turned to a more reflective and open position with respect to foundational questions in the late 1920s. During the last few years more aspects of his work have started to attract attention from authors interested in Weyl's contributions to mathematics, mathematical physics, and philosophy of science [Chandrasekharan 1986; Deppert 1988; Sigurdsson 1991]. Still, large areas of Weyl's work, its position in the development of 20th century mathematics and its influences have not been investigated in satisfactory historiographical detail. This article intends to contribute a little to this immense task, concentrating on one phase and subject

*I want to thank S. Sigurdsson, U. Wernick, and D. Witte for their contributions to a better understanding of H. Weyl and his *Raum – Zeit – Materie*, as well as the Mittag-Leffler Institut where large parts of this paper have been prepared under splendid working conditions during the program on History of Mathematics in early 1991.

of Weyl's work, the years of his most intense involvement in relativity theory and differential geometry.

This involvement started in 1917, shortly after Einstein's breakthrough to the fundamental principles of general relativity theory late in 1915. It led Weyl to a number of contributions to the subject, in particular to the publication of his book on *Space – Time – Matter* in 1918, which was revised and extended in five consecutive editions up to 1923 (Weyl RZM). The year 1923 demarcates the end of Weyl's most intense work in relativity and differential geometry and his move towards the next field of research, the representation theory of Lie groups and its links to quantum theory [Borel 1986; Speiser 1988]. All this happened during his time as a professor at the *Eidgenössische Technische Hochschule* (ETH) in Zürich where he worked between 1913 and 1930 (Frei 1989). In 1930 he returned to Göttingen as Hilbert's successor, but only for a short interlude, because the rapidly deteriorating cultural and working conditions after the Nazis seized power induced him to accept a call to the recently founded *Institute for Advanced Studies* at Princeton in summer 1933. There he worked until his retirement in 1950.

The goal of this paper is to follow Weyl's geometric thought during the years of his most intense involvement in relativity theory. After a discussion of Weyl's road into relativity (section 2) and some global features of *Space – Time – Matter* (section 3) I shall discuss three central topics of Weyl's geometric research in the period, affine connections (section 4), a generalization of Riemannian geometry, which he called "purely infinitesimal geometry" (in today's terminology the geometry of Weylian manifolds – section 5), and Weyl's formulation and solution of the generalization of the Helmholtz-Lie space problem to differential geometry (section 8). This discussion is extended by a short discussion of Weyl's unified field theory which was based on his "purely infinitesimal geometry" and its reception by the physicists (sections 6 and 7).

2   Weyl's Road into Relativity

When Weyl came to Zürich in the fall of 1913, EINSTEIN was still working at the ETH but had already decided to accept the combined call to the University and the Academy of Science at Berlin. Einstein had just published his joint work with Marcel Grossmann (Einstein/Grossmann 1913), which contained a discussion of the relativistic principle of general covariance from the point of view of Riemannian geometry. A misinterpretation of two facts of differential geometry (by Einstein) and a mistaken evaluation of a limit discussion (by Grossmann) led Einstein to the impression that general covariance was not the right thing to look for and to a two-year-detour searching for a principle of "restricted covariance" [Stachel 1989]. Einstein was just heading for this detour when Weyl arrived at Zürich. He did not leave it until July/August 1915, and already three months later he arrived at his great breakthrough to general relativity [Pais 1986, 252ff.]. That included in particular the derivation of his field equations of gravitation (Einstein 1915b), the explanation of the precession of the perihelion of Mercury, and the (general relativistically) correct derivation of light deflection in a central symmetric field (Einstein 1915a). In the autumn of 1913 there was no particular reason for Einstein to seek for cooperation with Weyl, nor for Weyl to approach Einstein.

Leaving Göttingen behind, Weyl had just ended his first phase of most active involvement in real and complex analysis. Early in 1914 he had worked on questions of equidistribution in number theory (with application to celestial mechanics) [Binder/Hlawka 1986]. On May, 15th, 1915 he was drafted into the German army, where he was stationed near Saarbrücken. At the request of the Swiss government he was released in spring 1916. The year in the army had interrupted his work completely. In retrospect Weyl described his situation as follows:

> "My mathematical mind was as blank as any veteran's and I did not know what to do. I began to study algebraic surfaces; but before I had gotten far, Einstein's memoir [(Einstein 1916)] came into may hand and set me afire." (Weyl at the Bicentennial Conference, Princeton 1946, cited from [Sigurdsson 1991, 62]).

In the course of one year he worked through the most essential contributions to general relativity then available: Einstein's publications of 1915 to 1917, Hilbert's contributions to the "foundations of physics"(Hilbert 1915, 1917), Schwarzschild (1916), de Sitter (1916/1917; 1917), and moreover Levi-Civita on parallel transport (1917a) and on relativity (1917b). In the summer of 1917 Weyl gave a course on general relativity at the ETH and published his first contribution on the subject (Weyl 1917), containing, beside other points, a new exact (axial symmetric) solution of the Einstein equations and a correct interpretation of invariance principles of relativity, which had been hinted at by Hilbert (1915) but given a rather misleading physical interpretation. Weyl's famous book *Raum – Zeit – Materie* (RZM [1]1918) was an elaboration of this lecture course.

3   Some Characteristics of R-Z-M

Weyl's *Raum - Zeit - Materie* was the first high level textbook on general relativity and moreover one breathing a philosophically inspired balance of mathematical and physical reasoning. It was well accepted and went through four reeditions in the next five years (RZM [2]1919, [3]1919, [4]1921, [5]1923), during which it was amended and changed step by step.[1] Weyl included his new insights or positions gained in differential geometry, field theory, and relativity. Thus the successive editions of RZM provide an easily accessible, paradigmatic documentation of the evolution of Weyl's thoughts on geometry and relativity during the phase of his most intense involvement in the field. Several general remarks on this classical work might be

---

[1]With the exception of the second edition which was an unaltered reedition of the first, RZM [2]1919 =[1] 1918.

appropriate before we proceed to a more detailed discussion of some more technical
aspects of wider importance:

1. The first two parts of RZM consist of a self-contained introduction to linear
   geometry (affine, vectorial, and metrical geometry, with an introduction to
   tensor algebra – part I) and to (semi-)Riemannian geometry, which in itself can be considered as a first rate introductory textbook on differential
   geometry (part II). Already the first edition (RZM [1]1918) contains a short
   introduction to the concept of parallel displacement in Riemannian manifolds, and later editions (from [3]1919 onwards) have included the concept of
   general affine connections and a short discussion of Weylian geometry.

2. Part III contains a lucid introduction to the geometry of special relativity, to
   relativistic mechanics, and electromagnetism, supplemented by a discussion
   of G. Mie's theory of matter.

3. Building upon Mie's theory and generalizing it, in the first three editions
   of RZM Weyl expounded a purely, or at least predominantly, dynamistic
   conception of matter. This view was increasingly counterbalanced in the last
   two editions by remarks and considerations tending towards a (stochastically
   determined) substrate concept of matter, suggested by the early quantum
   theory.

4. New ideas of mathematical physics, already in the first edition, were:
   - a strong emphasis on what Weyl later called the *guiding field* of
     spacetime as the physical interpretation of parallel displacement or
     affine connection in the underlying manifold,
   - a very principled (and correct[2]) discussion of causal structure in
     relativity,
   - a presentation of his new axially symmetric solution of the Einstein
     equations,
   - a first discussion of topological effects on space-time relations (in
     [1]1918 appearing as part of the discussion of the Schwarzschild so-
     lution).

Already in the months immediately following the preparation of the first
edition of *Raum-Zeit-Materie* Weyl radicalized his conception of parallel displacement. This led him to a new approach to differential geometry ("purely infinitesimal geometry"), consisting of the introduction of the concept of an affine connection, independent of any metrical consideration, and the generalization of a
(semi-)Riemannian metric to a particular type of gauge metric which later was
called a *Weylian metric* (Weyl 1918a).

These geometrical concepts became the basis for Weyl's idea of a unification
of gravitation and electromagnetism (Weyl 1918b). They led him in particular to

---

[2]In contrast to e.g. early considerations of Einstein and even Hilbert (1917).

propose the first *gauge field theory* in the history of theoretical physics. Both, the geometry of his gauge metric and the unified field theory approach, were included in the third edition (RZM [3]1919).

The reaction of the physicists was divided. Einstein immediately gave striking counterarguments against Weyl's theory as an acceptable physical approach, others were more receptive for at least one or two years (see section 7). Taking into account the growing scepticism among the physicists with respect to his approach, Weyl went back to a more conceptual, and even philosophical, argumentation in favor of his new geometry. His arguments, based upon what he called the *mathematical analysis of the space problem*, were first stated in (Weyl 1921) and RZM [4]1921, where he adapted the Helmholtz-Lie idea to characterize a physically relevant geometrical structure by congruence postulates to differential geometry. In RZM [5]1923 Weyl could already refer to a first, although not yet convincing, solution to this problem (Weyl 1922a) and also to a more refined version presented in lectures at Barcelona and Madrid in spring 1922. These were elaborated and published after the revision of (RZM [5]1923) in (Weyl 1923) (section 8).

Already this rough outline shows that H. Weyl's geometrical thoughts in the period between 1917 and 1923 are deeply intertwined with physical and philosophical considerations on spacetime and field structures. A discussion of his most important geometrical contributions of this period must take these contexts into account.

4 Affine Connections

Already in the first edition of *Raum – Zeit – Materie* Weyl took up Levi-Civita's idea of interpreting covariant differentiation and the meaning of the Christoffel symbols of Riemannian geometry as *parallel displacement* of vectors in a manifold. In the following years he generalized the concept and initiated a turn towards a more principled understanding of this concept and the use of it under the heading of *affine* or even more general *connections*.

Levi-Civita had studied in (1917a) a Riemannian manifold M embedded in a euclidean space[3] and defined *parallel transport* of a tangent vector $\xi(s)$ along a curve in M by parallel displacement in euclidean space and projection into the tangent space of M. For a one-parameter family of vectors displaced parallelly

---

[3]In fact Levi-Civita claimed as "well known (... come è ben noto...)" that a Riemannian manifold of dimension n can always be isometrically embedded in a euclidean space of dimension $\leq \frac{n}{2}(n+1)$ (1917, 176). That is obviously wrong for $n = 2$. Hilbert had shown in 1901 that a complete isometric embedding of the hyperbolic plane in 3-space is impossible. For higher dimensions non-isometric differentiable embeddings into $\mathbf{R}^{2n+1}$ were studied by H. Whitney in 1936 [Dieudonneé 1989, 60f.].

along a curve $x(s)$ he thus derived the differential equation (Levi-Civita 1917a, 179)

$$\frac{d\xi^i}{ds} + \sum_i \Gamma^i{}_{jk} \frac{dx^j}{ds} \xi^k = 0 \ .$$

In this way the Christoffel symbols $\Gamma^i{}_{jk}$ which up to then used to play only a purely formal, although important, role in Riemannian differential geometry were endowed with an intriguing geometrical meaning. In consequence covariant differentiation introduced by Levi-Civita's teacher Ricci could be understood more geometrically, as a differential procedure derived from the infinitesimal changes of vector or tensor fields in comparison with the parallelly transported vector/tensor at the point in question.

For Weyl this geometrization of the Christoffel symbols seemed of particular importance, because Einstein had included in his memoir of 1916 a passage where he gave a fundamental physical interpretation to the Γ's, although only in passing and without drawing full consequences. In order to explain the conceptual link between metric and gravitation Einstein had discussed the equation of the geodetical line

$$\frac{d^2 x^i}{ds^2} = -\Gamma^i{}_{jk} \frac{dx^j}{ds} \frac{dx^k}{ds} \ .$$

He put the right hand side in a close analogy to gravitational acceleration of a mass point and stated that

> "... these quantities determine the deviation of the motion from uniformity. *They are the components of the gravitational field.*" (1916, 802; emphasis E. S.).

Einstein considered throughout this article $\Gamma^i{}_{jk}$ consistently as coefficients of the gravitational field and the g_{ij} as gravitational potentials. He even went so far as to write the Ricci tensor term of the field equations with the appropriate expression in the Γ's (ibid. 808). Weyl accepted this point of view fully. He not only took up the viewpoint of the $\Gamma^i{}_{jk}$'s as "gravitational field", but even went so far to call them, mathematically considered, the "components of the metrical field" (RZM [1]1918, 189, 193 etc.) as a first attempt to make the underlying mathematics conceptually more explicit. From our point of view that sounds strange and even misleading, and Weyl indeed changed his terminology in the later editions.[4] But this first, even if not yet really convincing, approach to the problematics shows clearly the reasons why Weyl wanted to lay open the conceptual meaning of the Γ's.

---

[4]Weyl later changed the terminology to *(affine) connection* ([3]1919) and *guiding field* ([4]1921); see below. Einstein, in contrast, changed his terminology still in 1916 to a mathematically more conventional standpoint, So he introduced the *metrical tensor* as the proper mathematizatical expression for the gravitational field already in (1916b) *as if he had never done anything else before*: "Let us describe *as usually (wie üblich)* the gravitational field by the tensor of the $g_{\mu\nu}$..." (Einstein 1916b, 1111, emphasis E.S.).

In the first edition of RZM Weyl took up Levi-Civita's interpretation of the Christoffel symbols as defining parallel displacement and went only a small step further *by avoiding any reference to an embedding in euclidean space*. That was, of course, quite important for the context of semi-Riemannian structures, even if one did believe in Levi-Civita's argument of the isometric embeddability of any Riemannian manifold in euclidean space of sufficiently high dimension.

Weyl's procedure was as follows: First he defined *geodetical coordinate systems* in a point P by the condition $\frac{\partial g_{ik}}{\partial x^r} = 0$ in P. Then the *covariant gradient* λ of a *vector field* ξ was introduced by a formal manipulation showing how the elementary gradient of $g(\xi, \xi)$ can be modified to be covariant:

$$\lambda^i := \frac{\partial \xi^i}{\partial x^k} \xi^k + \Gamma^i{}_{jk} \xi^j \xi^k.$$

Thus Weyl could introduce *geodetical vector fields* ξ *in* P by the condition that the covariant gradient vanishes in P. Taking now a vector ξ attached to a point P of the manifold, and an infinitesimally close point P' (i.e. in slightly modernized language $P' - P = \vec{PP'} =: \delta x \in T_P M$) ξ undergoes a variation $\delta \xi$ from P to P', satisfying the conditions of a geodetical vector field, if

$$\delta \xi^i = -\Gamma^i{}_{jk} \xi^j \delta x^k.$$

Weyl explained that by this formula

> "... the original vector is, as one ought to express in natural terms, transplanted (verpflanzt) *unchanged* from P to P', it is displaced (verschoben) *parallely with itself* from P to P'" (RZM [1]1918, 100f.).

Only a little later Weyl radicalized his approach conceptually. He realized that the concept of the parallel transport of vectors is much more fundamental than Riemannian metrics for two reasons: It defines the differential geometric analogue for ordinary affine geometry and is therefore more fundamental than metrics; moreover it allows comparison of quantities in infinitesimally close points of the manifold only, whereas in Riemannian geometry, quantities (lengths etc.) can be compared even if attached to points in finite, possibly large, distance. The second aspect appeared to Weyl to be of utmost importance, because he looked for a geometry with only "purely infinitesimal" structures. Considered from this point of view there was a deficiency in Riemannian geometry. So Weyl proposed that a program of a more radical metrical geometry on manifolds ("purely infinitesimal geometry") be developed (see section 5). As the first step of this program Weyl refined Riemann's twofold distinction between analyis situs and metrical geometry in manifolds and introduced the concept of an "affinely connected manifold" independent of metrical considerations as an intermediate conceptual level (Weyl 1918a, 6ff.)[5]

---

[5]From the above it should be clear what Weyl alluded to, when in (1918a, 6) he added the

Weyl declared a point P in a manifold M as *affinely connected to its (infinitesimal) neighbourhood*, if for each vector ξ at P and for each point P' infinitely close to P, it is well-determined how ξ has to be *transported parallelly* from P to P', changing to ξ'. The whole manifold is affinely connected, if each point $P \in M$ is affinely connected to its infinitesimal neighbourhood (1918a, 6; RZM [3]1919, 113). So everything depends on how "parallel transport" is explained in a differentiable manifold.

Definition 1 (Weyl 1918a, 6f.): *Parallel transport* of a vector ξ at $P \in M$ to ξ' at an infinitesimal close point $P' \in M$ with $\vec{PP'} = dx$ is defined by a function $\Gamma(\xi, dx)$ with values in the tangent plane $T_{P'} M = T_P M$ such that:

  i. The change of the vector is given by

$$\xi' = \xi + d\xi = \xi + \Gamma(\xi, dx), \tag{1}$$

  ii. Γ is linear in ξ;

  iii. Γ is linear in dx, therefore in coordinates $\Gamma^i{}_{jk} \xi^j dx^k$;

  iv. Γ is symmetric, $\Gamma^i{}_{jk} = \Gamma^i{}_{kj}$.[6]

The $\Gamma^i{}_{jk}$ are called the *coefficients of the affine connection* (belonging to the parallel transport).

From this Weyl easily got

Lemma 1 (Weyl 1918a, 9): *Given any affinely connected manifold M and a point $P \in M$; then one can always choose coordinates x^i $(1 \leq i \leq n)$ near P such that $\Gamma^i{}_{jk}(P) = 0$, i.e. the coefficients of the affine connection vanish in P.*

Such a system is called *geodetical coordinate system at P*. In geodetical coordinate systems the parallel transport from P to P' obviously leaves vector coordinates unchanged. Thus they are particularly well adapted (not only technically but also conceptually) to express the nature of parallel transport as leaving vectors unchanged according to a specified criterion. Indeed Weyl introduced the concept of parallel transport in RZM ([3]1919, 113) by the conditions (i), (ii) as above and the existence of a geodetical coordinate system to each point $P \in M$. From this assumption he derived the conditions (iii), (iv) and showed the other way round that specification of symmetric $\Gamma^i{}_{jk}$ does the same as the determination of geodetical coordinate systems.

---

specification "world with gravitational field (Welt mit Gravitationsfeld)" to the property of a manifold to be "affinely connected". Superficial reading may take this as indication that Weyl considered affine connections still bound conceptually to a metrical structure. That is, however, *expressedly not the case*.

[6]Weyl expressed the symmetry conditions geometrically, as a closing condition for an "infinitesimal parallelogramm" generated by parallel transport of an infinitesimal vector δx along dx and then of dx along δx (1918a, 7).

After the discussion of *stationary vector fields*[7] and of *geodetical lines* $x^i(s)$,

$$\ddot{x}^i + \Gamma^i{}_{jk}\dot{x}^j\dot{x}^k = 0 \ ,$$

Weyl introduced the *curvature of an affine connection* by a beautiful and intriguingly geometric idea. He looked what happens to a vector ξ at P if parallelly transported about an infinitesimal parallelogram spanned by dx (let us express dx by a vector $v \in T_P M$) and δx (expressed by $w \in T_P M$) back to P, where it may arrive as ξ'. Then, so Weyl realized, $d\xi := \xi' - \xi$ *depends linearly* on ξ and on the parallelogram $v \wedge w$. In this sense parallel transport leads to an infinitesimal monodromy operation on $T_P M$, given by a $(1,3)$-tensor R called the *curvature tensor of the connection* Γ, such that $d\xi = R(\xi, v, w)$. In coordinates this means

$$d\xi^i = R^i{}_{jkl}\xi^j v^k w^l \quad \text{with} \ \ R^i{}_{jkl} = -R^i{}_{jlk} \tag{2}$$

As in Riemannian geometry, vanishing of the curvature tensor is a necessary and sufficient criterion for an affinely connected manifold to admit a finite path independent parallel transport and thus to be (at least locally) a linear affine space (Weyl 1918a, 10–12; RZM [3]1919, 117–121). Moreover Weyl showed that in Riemannian manifolds the Levi-Civita parallel transport is the uniquely determined connection which is compatible with the metrics. But for him that was a by-product of a more general metrical theory of manifolds discussed in our next section.

The concept of affine connection became of primordial importance for Weyl's understanding of relativity. Einstein had given up the characterization of the "gravitational field" by the $\Gamma^i{}_{jk}$ already in 1916 and preferred the characterization of the metric as zhe mathematical representative of the gravitional field.[8] Weyl, after having gained conceptual clarity on the nature of affine connections, could not easily be impressed by a methodological dogmatism maintaining that that physically relevant quantities have to be tensor fields: He had given a clear coordinate invariant description of the connections (although of course not covariant).[9] To avoid tiring discussions about "fictitious gravitational effects" by choice of coordinates and at the same time to point to the physical importance of the affine connection in relativistic spacetime, Weyl introduced the terminology of *guiding field (Führungsfeld)* in the fourth edition of RZM for the physical interpretation of the affine connection ([4]1921, 200, 257 etc.).[10] Moreover his concept allowed him

---

[7]The vector field ξ is *stationary* in P if its covariant derivative is 0 or, in Weyl's language, infinitesimal change of ξ coincides with parallel transport, i.e. $\frac{\partial \xi^i}{\partial x^j} + \Gamma^i{}_{\mu j}\xi^\mu = 0$.

[8]Cf. footnote 4. Einstein, however, returned for a short while in the early 1920s to the concept of an affine field theory under the influence of Eddington (and thus indirectly by Weyl; cf. section 7).

[9]Weyl mentioned that the components of an affine connection are obviously not covariant without making any fuss about it (RZM [3]1919, 114).

[10]This next step in the conceptual clarification of Weyl's standpoint seems to stand under the influence of the heated discussions at the Bad Nauheim meeting of the *Gesellschaft deutscher Naturforscher und Ärzte* (19.– 25. 9. 1920), at which Einstein had been attacked severely by Lenard [Pais 1986, 320]. Final preparation of the manuscript for the fourth edition of RZM took place in November 1920. See also Weyl's own report on the meeting (1922b).

to argue very distinctly about the involved interrelation of a priori (conceptual) and a posteriori (empirical) elements in the determination of the connection in spacetime.[11]

Also in Weyl's broader geometrical work the concept of connection appears time and again in changing contexts. It plays a central role in the program of "purely infinitesimal geometry"; a new type of connection is introduced there (*length connection*, a connection in a line bundle). In his analysis of the space problem Weyl finally worked essentially with certain types of connections in principal bundles and, in passing, with general linear connections in the tangent bundle (without, of course, using the bundle terminology; cf. section 8). So we have to keep in mind that the importance of affine connections in Weyl's geometric work must not be judged solely by their internal technical results in affine differential geometry, but even more from their meaning in his attempts to gain a deeper understanding of differential geometry in its entirety. And in fact, it was taken up by other mathematicians much more for this general importance than in the context of affine differential geometry.

5 "Purely Infinitesimal Geometry"

Mathematicians and physicists of the 19th and the early 20th century generally agreed that the growing understanding of causal relationships in the natural sciences depended to a high degree on the elaboration of infinitesimal structures within mathematics and their use for the formulation of natural laws (Riemann 1854, 285); Planck 1899, 603; Hilbert 1917, 64; 1924, 278). From this point of view, as Weyl remarked repeatedly (1918a, 14; 1919, 56), something strange can be observed in Riemannian differential geometry.

For vectors v and w (in local coordinates represented by v^i and w^i, $1 \leq i \leq n$), attached to points P, P' of finite distance in a Riemannian manifold, directions can be compared only after parallel transport of one of the vectors along a path connecting P and P', thus only by integration of an infinitesimally defined comparison procedure. That is compatible with the aforementioned principle of infinitesimal

---

[11]The a priori aspect is most fundamentally discussed in the analysis of the space problem (cf. section 8) but appears also in Weyl's way of handling cosmological models and the question of a matter-free universe. For Weyl's changing approach to the a posteriori determination of the connection compare the different methods he used to characterize measuring procedures for the spacetime metric in RZM

- by observation of zero geodesics on the light cone only ([1]1918, 182ff.),
- conformal structure by measuring the light cones and Riemannian calibration by measuring trajectories of mass points ([3]1919, 194ff.),
- by measuring the parallel transport of timelike vectors along timelike geodesics in addition to Einstein's hypothesis of Riemannian geometry ([5]1923, 224).

This very interesting question cannot be covered in this article. Compare from a more theoretical point of view [Ehlers 1988, 87f.].

comparison of quantities in differential geometry. Something completely different happens, however, for the length of vectors. They are obviously comparable without any integration:

$$\frac{|v|}{|w|} = \sqrt{\frac{g(v,v)}{g(w,w)}}$$

In this sense lengths are "absolutely comparable".

Weyl argued that this is a deeply rooted methodological insufficiency of Riemannian geometry and that one should look for a generalization of Riemannian metrics, which allows *direct comparison of lengths of vectors detached to infinitesimally close points only.* He therefore proposed that one start from a conformal structure on a manifold M, enriched by what he called a *length connection*, represented by a differential 1-form ω, allowing comparison of lengths of vectors ξ and η in infinitely close points P, P' $(P' - P = \vec{PP'} \in T_P M)$. To compare the length of ξ with the length of η one has to change $|\xi|_P$ to

$$|\xi|_{P'} := |\xi|_P (1 - \omega(P' - P)) \tag{3}$$

or, as Weyl wrote with $l(\xi)$ for the "length" of ξ

$$l'(\xi) = l(\xi)(1 - \omega(dx)), \quad \text{or } dl = -l\omega.$$

Recognizing that regauging the metric by multiplication with λ, thus the length function by $\tilde{l} = \lambda l$, corresponds to a transformation of the length connection ω according to

$$\begin{aligned} d(\tilde{l}) = d(\lambda l) \quad &= \quad l d\lambda + \lambda dl = l d\lambda - \lambda l \omega \\ &= \quad -\lambda l (\omega - \frac{d\lambda}{\lambda}) = -\lambda l(\omega - dlog\lambda) = -\lambda l \tilde{\omega}, \end{aligned}$$

Weyl was led to the consideration of "gauge transformations" for ω by $\tilde{\omega} = \omega - dlog\lambda$. Taken together he used

Definition 2 (Weyl 1918a, 15; RZM [3]1919, 110): A *purely infinitesimal metric* (or a *Weylian metric*) on a differentiable manifold M is given by

    (a) a *conformal* (semi-)Riemannian structure on M, represented by a quadratic differential form $g(-, -)$ (nondegenerate and of constant signature on M), locally $g_{ij} dx^i dx^j$, up to multiplication by a positive varying factor (the *gauge function* λ, $\tilde{g}_{ij} = \lambda g_{ij}$),

    (b) and a *length connection*, i.e. a class of differential 1-forms given by representatives $\omega, \tilde{\omega}$ with *gauge transformation* $\tilde{\omega} = \omega - dlog\lambda$ (corresponding to a change of representative of the conformal metric by the gauge function λ).

Weyl explained that in his new geometry a *gauge principle* has to be observed. That is, invariant geometrical concepts (e.g. vectors, tensors etc.) are those which are not only covariant under coordinate transformations but also invariant or covariant under gauge transformations. In the easiest cases gauge covariant quantities are those which are to be multiplied by a factor λ^k ("*quantity of weight k*") etc.[12]

In analogy to the characterization of curvature in Riemannian or general affinely connected spaces Weyl considered the infinitesimal monodromy of the length function transport around an infinitesimal parallelogram spanned by vectors v, w and arrived easily at the conclusion that length changes according to $\delta l = f(v, w)l$ with f derived by exterior derivation of ω

$$f = d\omega , \quad f_{ij} = \frac{\partial \omega_i}{\partial x^j} - \frac{\partial \omega_j}{\partial x^i} \tag{4}$$

For obvious reasons he called $f = d\omega$ the *length curvature* (*Streckenkrümmung*) of the underlying Weylian metric (1918a, 20; RZM [3]1919, 111) and asked whether affine connections Γ compatible with a Weylian metric (g, ω) exist. In such a case the parallel transport of vectors ξ, η by Γ has to change lengths like the length connection, i.e. for ξ', η' in P' parallelly displaced by Γ from ξ, η in P

$$\xi'^i = \xi^i + \Gamma^i{}_{jk} h^j \xi^k, \quad \text{with } h := \vec{PP'}$$

it has to hold $\quad g(\xi', \eta') = g(\xi, \eta)(1 - \omega(h))$

And in fact Weyl gave a definite and positive answer to this question.

Theorem 1 (Weyl 1918a, 16f.; RZM [3]1919, 111f.):[13] *To each Weylian metric (g, ω) there exists exactly one compatible affine connection Γ, given in local coordinates by*

$$\Gamma^i_{jk} = \tilde{\Gamma}^i_{jk} + \frac{1}{2} \left(g^{i\mu}(g_{\mu j}\omega_k + g_{\mu k}\omega_j - g_{jk}\omega_\mu) \right) \tag{5}$$

with $\tilde{\Gamma}^i_{jk}$ Riemann-Levi-Civita connection of g_{ij} .

Let us call Γ the *Weyl connection*. Weyl showed that its curvature tensor (*Weyl curvature $F^i{}_{jkl}$*) is gauge invariant, as is the Weyl-Ricci tensor arising by contraction $F_{jk} = F^i{}_{jil}$, whereas the Weyl scalar curvature $F = F^i{}_i = g^{ij}F_{jk}$ obviously is not;[14] F is a scalar function of (gauge) weight -1.

---

[12]Thus the metrical tensor is a gauge covariant quantity of weight 1; the length connection is of even more complicated gauge behaviour.

[13]Weyl called this property the *fundamental theorem of infinitesimal geometry* (RZM [3]1919, 111). See for its conceptual importance the analogue postulate in the *mathematical analysis of the space problem*, section 8.

[14]"Lifting of indices" involves the inverse of the metrical tensor.

Because of this gauge property of scalar curvature in Weylian geometry, it is always possible to choose the gauge function λ in such a manner that scalar curvature is normalized to a constant, say $F \equiv 1$.[15] Weyl called this procedure the choice of *normal gauge*. Later it was to play an important role in his physical investigations.

Of course a Weylian manifold may essentially be Riemannian. That is the case if and only if integration of the length connection does not depend on the path. Length curvature is obviously a measure of infinitesimal deviation from the Riemannian situation and Weyl showed quite easily that this infinitesimal criterion can be integrated:

Theorem 2 (Weyl 1918a, 16, 20; RZM [3]1919, 111): *A Weylian manifold* (M, g, ω) *is locally Riemannian, if and only if* $d\omega = 0$.

Weyl did not discuss the influence of the topology of the manifold on the global gauging problem.[16] Nevertheless this result shows convincingly how classical Riemannian differential geometry arises as a special case of Weyl's "purely infinitesimal" geometry. But one should observe that the Weylian viewpoint gives an additional conceptual degree of freedom even in the case of a Riemannian manifold. One is not *forced* to use the gauge with trivial length connection $\omega = 0$. The length connection could be represented just as well by an exact differential $\omega = -d \log \lambda$ for any gauge function λ, for example Weyl's "normal gauge" setting Weylian scalar curvature constant equal 1.[17]

6　Some Remarks on Weyl's Unified Field Theory

Weyl chose this "normal gauge" when he discussed physical geometry. In the period between 1917 and 1923 Weyl shared Hilbert's conviction that it should be possible to derive the laws of matter dynamics from a single Hamiltonian principle combining gravitational and electromagnetic action. Hilbert had taken up an idea of the German physicist Gustav Mie (Hilbert 1915, 1917) to explain matter in completely field theoretic terms and had tried to reformulate and adapt it to the framework of general relativity [Mehra 1973, 115ff.]. Weyl hoped to have found the clue for a radical improvement of this approach in the observation that the length connection ω and length curvature $f = d\omega$ of his new geometry formally look very much like the electromagnetic potential and the relativistic electromagnetic field tensor. He thus proposed a keen hypothesis about a possible connection between his new geometric theory and physics.

---

[15]Set $\lambda = F$; then scalar curvature is regauged by $\tilde{F} = \lambda^{-1} F = 1$.

[16]If the first Bettinumber is $\neq 0$, a linear combination of the periods of ω will yield an undetermined additive term in the global length gauge.

[17]In this case Riemannian and Ricci curvature coincide with their Weylian counterparts, whereas Riemannian scalar curvature R is only one of the gauge representatives of Weylian scalar curvature F and in general $R \neq F$.

Hypothesis (Weyl 1918b; 1919; RZM [3]1919, 243f.): *The physical world (geometry, gravitational and electromagnetic fields and all matter phenomena) can be represented mathematically by a Weylian manifold (M, g, ω) with $\dim M = 4$, of signature (3,1) such that*

- *the length connection represents the electromagnetic potentials ($\omega_i = \varphi_i$) and the length curvature ($f = d\omega$) the electromagnetic field f_{ij},*

- *the metric g_{ij} represents the potentials of the purely gravitational effects,*

- *and the Weylian connection Γ represents the guiding field of the combined actions of inertia, gravitation, and electromagnetism.*

This assumption leads to a new theory, because the geometry of the Weylian manifold couples electromagnetism and gravitation non linearly in the guiding field (formula 5). Weyl assimilated the Mie-Hilbert approach to his geometric framework and looked for a *gauge invariant* Hamiltonian of the form

$$\delta \int \mathcal{H} \, dx = 0, \tag{6}$$

variation taken over all g_{ij}, φ_i, where

$$\mathcal{H} = \mathcal{R} + \mathcal{L}, \tag{7}$$

with $\mathcal{R}$ the field action density of gravitation and $\mathcal{L}$ the field action density of electromagnetism.

Weyl hoped, like Mie and Hilbert before him, that once the "right" Hamiltonian was found it might turn out to be unnecessary to take into account additional terms for the *substance action* of matter. Thus it ought to be possible to derive those properties of matter, that traditionally are described as substance properties, as results of the field behaviour in regions of high energy concentration ("energy nods"). Insofar Weyl's conception of matter was *purely dynamistic* at that time.

Already in his first article on the subject (1918b) Weyl started the investigation of a possible specification of the Hamiltonian, which he extended in (1919a, 75ff.) and the later editions of RZM ([3]1919ff.):

$$\mathcal{H} = -\frac{1}{4}F^2 \sqrt{g} + \alpha f_{ik} f^{ik} \sqrt{g} \tag{8}$$

$$\text{with } g = \sqrt{|\det g_{ij}|}$$

Both action terms are gauge invariant and derived from basic geometric invariants.

On both levels, the general (7) and the specified (8), Weyl derived the (Euler) differential equations for stationary solutions to the variational problem. Thus he was led to general Weylian field equations (1919, 74; RZM [3]1919, 250f.; [5]1923, 313f.) and to specialized Weylian field equations (1919, 77) which he later gave in normalized coordinates derived from "natural" coordinates (with $F \equiv 1$) as (RZM ^{5}1923, 305):

$$R_{ik} - \frac{1}{2}R\,g_{ik} \;=\; T_{ik} + \epsilon^2\sqrt{g}\left((1 - 3\varphi_\mu\varphi^\mu)g_{ik} + 6\varphi_i\varphi_k\right) \tag{9}$$

$$\frac{\partial f^{ik}}{\partial x^k} + \frac{3\epsilon^2}{\alpha}\varphi^i \;=\; 0 \tag{10}$$

$$df \;=\; 0 \tag{11}$$

Here ϵ is very small, more precisely the length of the meter in terms of a cosmical absolute unit of length in Weyl's "natural" coordinates.[18] Thus the first set of equations (*Einstein-Weyl equations* – (9)) is in good agreement with the Einstein equations, and the second set (*Maxwell-Weyl equations* – (10), (11)) agrees with the source-free Maxwell equations in the vacuum, if the electromagnetic potentials φ_i are not immensely large.

Initially Weyl was very cautious with this specification. In RZM (^{3}1919, 268) he commented that this approach is "...easiest to pursue algorithmically (whereas I do not claim that it is realized in nature)." It may be surprising that he commented on the specialization of his field equations, notwithstanding the divided reception of his theory by the physicists, in the fifth edition[19] much more self-consciously as a

> "...theoretically very pleasing synthesis and interpretation of all our field theoretic knowledge ..., which is extended by consequent assumptions which are plausible in themselves" (^{5}1923, 308).

Weyl could present the approach as a "pleasing synthesis" with good reasons because

- for $\epsilon \longrightarrow 0$ equations (9) approach the Einstein equations and (10), (11) the Maxwell equations for the matter and source-free case (^{5}1923, 303),
- for $\mathcal{L} = 0$ the Einstein equations with cosmological term result (^{5}1923, 300),
- for $\varphi_i\varphi^i \ll \epsilon^{-2}$ (which is usually the case in what Weyl called the "human" range) Riemannian geometry is a sufficiently good approximation to Weylian geometry, if one takes the "natural" gauge of the Weyl metric up to constant factor,
- and — most important for Weyl — his synthesis was based on a new geometry "... perfect in itself, a necessary and closed building (...) particularly because of its (...) group theoretic foundation" (^{5}1923, 308) (see section 8).

---

[18]In "natural" coordinates scalar curvature F is normalized to $F = 1$; i.e. the unit of length is chosen by averaging the radii of sectional curvatures of the world through one point. The "natural" unit is thus of the order of magnitude of the apparent "radius of the world", ϵ its reciprocal value.

[19]In RZM ^{4}1921 the presentation of this point remained identical to ^{3}1919.

So we see that Weyl was not really discouraged by the internal difficulties of his unified field theory program in 1923, notwithstanding a growing gap between Weyl and most of the physisics involved in the debate (see below). On the other hand he became increasingly aware of the problematics of quantum phenomena. That led him to a growing scepticism with regard to a purely dynamistic conception of matter, as is shown by increasingly definite side-remarks in the last two editions of RZM.[20]

7   The Reception of Weyl's Theory by the Physicists

Weyl's bold presentation of his theory in (RZM [5]1923) is in considerable contrast to the position taken at the time by the majority of the physicists working on general relativity. Most striking in this respect was Einstein's reaction. In March 1918 Weyl had sent the manuscript of his article (1918b) to Einstein. Einstein responded immediately, although he expressed his opinion step by step with increasing precision in several successive cards and letters [Straumann 1987]. Characteristic for his response was a high estimation for Weyl's ideas as a mathematical theory.

> "Your memoir has arrived. It is a coup of genius of the first rate..."
> (Einstein to Weyl, 6. 4. 1919; Straumann 1987, 414])[21]

But he combined this positive evaluation with a definite rejection of it as a physically sound approach.

> "Although your idea is so beautiful, I have to declare frankly that, in my opinion, it is impossible that the theory corresponds to nature."
> (Einstein to Weyl, 15. 4. 1918)

Einstein's counterargument went straight to the principles of measurement which had served as the cornerstone for the erection of relativity theory. Consider two equivalent atomic clocks (radiating with equal frequencies), e.g. two hydrogen atoms, in adjacent world points. Suppose they pass along different world trajectories and meet again in adjacent world points. Then in the general case of Weylian geometry their spectral behaviour should be different, assuming that their oscillations are regulated by the Weylian length connection. Thus the radiation of atomic oscillators depends on the history of the atoms, and stable spectra cannot occur. Empirical evidence contradicts such a prediction, as atoms separate in discrete, equally radiating classes. And this observation, according to Einstein, must not be regarded as an empiricist premature refutation, because

---

[20]It would be an interesting point in itself to follow up the development of Weyl's conception of matter in the successive editions of RZM. Additional evidence mainly from archival sources, which stand in striking agreement with the findings from RZM are to be found in [Sigurdsson 1991].

[21]This, like all further citations from the Einstein-Weyl correspondence, are translated from [Straumann 1987] by the author.

"... (if) one drops the connection of the ds to the measurement of distance and time, then relativity looses all its empirical basis" (Einstein to Weyl, 15. 4. 1918).

Einstein's counterargument was printed in a short amendment to Weyl's article (1918b), and Weyl had to cope with it. Weyl was neither convinced by Einstein's argument, nor was he too much concerned about it. He thought that there were sufficient philosophico-mathematical reasons to build relativistic physics and unified field theory upon his geometry. In order to reject Einstein's argument he demanded (in *this* respect not unlike the stubborn defenders of euclidean geometry) that in a first purely theoretical step the principles of the theory had to be developed mathematically, before the physics of the measuring process can be understood and taken into account properly. Only then a discussion of observational effects would make sense. Then, so he hoped, the surprising Riemannian behaviour of real clocks and measuring rods should be explainable in the framework of his theory and allow confrontation with empirical evidence (Weyl 1918b, 41; 1919, 67).

Already several of the most important approximations in RZM (^{5}1923) led credence to this expectation (see end of section 6). But apparently that did not suffice Weyl. So he tried to shunt the mathematical difficulties between the development of the foundations of his approach and the detailed elaboration of the dynamics of the measuring process by a daring hypothesis. He wanted to secure a direct physical meaning to his "natural gauge" ($F \equiv 1$) by the

Ad hoc hypothesis (Weyl RZM ^{5}1923, 303): *Atomic clocks and measuring rods calibrate on the scalar curvature in each world point, independent of their world trajectory. Thus the value of scalar curvature measured in physical units is a constant, and physical measurement leads to the* normal gauge *of Weylian spacetime.*

This shows clearly that Einstein and Weyl followed rather different methodological approaches to the foundation of a theory of mathematical physics at that time. Einstein was still convinced that the basic principles ought to be derived by a critical reflection on established physical laws of measurement. The analysis of the physical laws had to have a direct influence on the conceptual choices implemented in the principles of the theory.[22] Weyl was, for the reasons just mentioned, not convinced of the necessity of Einstein's approach. But he reacted sensitively and openly to it, in spite of this intellectual difference. Towards the end of their correspondence of 1918 he wrote to Einstein:

---

[22]Only a few years later Einstein came to emphasize more and more how important "good" conceptual guesses were in order to create a physical theory. Methodologically we can observe a crossover with Weyl's position in the early 1920s, who came to value more and more highly the determinative power of empirical evidence for the principles of a theory of mathematical physics. It is an irony of history that Einstein's aboutturn in 1923 occurred as an indirect influence of Weylian concepts, mediated by Eddington who tried a purely affine field theory in 1921 and set Einstein on that road for a year or two [Stachel 1986; Pais 1986, 347ff.].

> "This [insistence, E.S.] irritates me of course, because experience has
> proven that one can rely on your intuition; as unconvincing your coun-
> terarguments seem to me, as I have to admit." (Weyl to Einstein 10.
> 12. 1918)

A little later in the same letter he admitted:

> "By the way, you should not believe that I was driven to introduce
> the linear differential form in addition to the quadratic one by physical
> reasons. I wanted, quite to the contrary, to get rid of this 'methodolog-
> ical inconsistency (*Inkonsequenz*)' which was a stumbling stone to me
> already much earlier. And then, to my surprise, I realized that it looks
> as if it might explain electricity. You clap your hands above your head
> and shout: You can't do physics this way!" (ibid.)

Good luck for Weyl that not all the physicists reacted like Einstein. Einstein
already forwarded a very positive first reaction of M. PLANCK to him [Straumann
1987, 417]. A. SOMMERFELD wrote enthusiastically to Weyl that there was "...
hardly doubt, that you are on the correct path and not on a wrong one" (Sommerfeld to Weyl, 3. 7. 1918; cited from [Sigurdsson 1991, 160]). Sommerfeld's reaction
was of particular importance, as he had taken responsibility for the article on relativity in the *Encyclopädie der Mathematischen Wissenschaften*. When he decided
he was too busy to write the article himself, he handed the task over to his student
W. Pauli.

PAULI studied Weyl's theory very seriously and contributed to a better understanding of the Hamiltionian principle to be used in Weylian geometry in order
to derive a good candidate for a unified field theory. He thus contributed to the
special Einstein-Weyl and Maxwell-Weyl equations mentioned above (Pauli 1919).
This support may have contributed considerably to the rising self-consciousness in
Weyl's commentary on the specified Hamiltonian principle (8), changing its presentation from a purely calculatory example to the conviction that it synthesizes
"all field theoretic knowledge". But although Pauli had an active share in the second phase of elaboration of Weyl's theory and although he contributed to a wider
reception of the latter through his lucid and precise presentation of the theory, he
also signalized a growing reserve among the physicists towards Weyl's approach
at the end of this passage of his article, where he stated:

> "... Summing up one may say that Weyl's theory has not yet brought
> us closer to the solution of the problem of matter." (Pauli 1921, 770)

This evaluation was of particular significance, as it was precisely the theory
of matter which had been Weyl's ultimate goal when he proposed his approach.

Internationally the most important early reaction was the one of A. S. ED-
DINGTON. Not unlike Schrödinger and Pauli Eddington first reacted very positively,
but after a short period of enthusiasm, with increasing distance. In his semi-popular
book on *Time, and Gravitation* Eddington, like Weyl, stressed very strongly the

physical importance of the time-like geodetical structure of space-time (Eddington 1920, 75, 138, 150, 155), and he wrote a long and physically intuitive discussion of Weyl's gauge field theory (ibid. 167ff.). Three years later, in his sophisticated and much more technical book *Mathematical Theory of Relativity* (1923) his position had changed.

Now Eddington distinguished clearly between a purely mathematical geometrization of physics, which in his opinion is a very general "graphical representation of all quantities concerned in physics" and the "*true* geometry understood by the physicist" (Eddington 1923, 198). Following Einstein in this point, Eddington considered Weyl's theory as geometry of the first (mathematical) kind only, that is as a purely conventional geometrization of physical quantities.[23] This allowed him to present Weyl's theory in considerable detail from the mathematical point of view, although at the same moment he denied its value as a hypothetical structure of physical geometry. Of course, Weyl's *ad hoc hypothesis* on the physical background for the natural gauge was an easy point to comment sarcastically upon from the point of view of physics ("... tenable, though perhaps a bit overstrained..." (ibid, 207)).

So Eddington was perhaps the last among those who had given initial support to Weyl's program of unified field theory to distance himself and to look for different approaches. Ironically Eddington himself turned to even more mathematical abstraction (affine field theories) combined with an attempt to derive field equations, which placed considerable emphasis on purely mathematical a priori arguments. And it was with these attempts at affine field theories that Eddington attracted Einstein's attention for some time.[24] In this respect Eddington's approach became more one-sided than Weyl's: Weyl had at least attempted to arrive at a balanced combination of a priori considerations and knowledge abstracted from experience.

8   Digging Deeper: Weyl's Infinitesimal Version of the Space Problem

The more Weyl sensed the reservation of physicists with respect to his theory, the more he tried to state clearly why he was convinced that his geometry was a better approach than classical Riemannian differential geometry. He took up an idea expressed in a different context already in early 1919, when he had reedited Riemann's inaugural lecture on the *Hypotheses lying at the basis of geometry*. There he had commented on Riemann's rather pragmatic remarks about the selection of the quadratic differential form for the definition of the metric (Riemann 1854, 278) by a short discussion of how he could imagine that the (square root of the)

---

[23] "Indeed it [Weyl's gauge field metric] is no longer a hypothesis, but a graphical representation of the facts, and its value lies in the light suggested by this graphical representation." (Eddington 1923, 208).

[24] See footnote 22.

quadratic differential form can be characterized conceptually among the wider class of Finsler metrics. Weyl referred to the fundamental nature of the concept of parallel displacement for the foundation of geometry and conjectured that the class of Riemannian metrics is singled out among the larger one by looking for those metrics which admit a compatible affine connection (Weyl 1919b, 27).[25] A year later Weyl tried to adapt the analogous figure to his own (Weylian) metrics. His first remarks in this respect were published in RZM ([4]1921, 124ff), soon followed by his articles (1921, 1922) and his book (1923).

Weyl argued that one ought to delimit a framework for metrical concepts in space as generally as possible, thus a *general concept of metrics*, into which physical space cannot but fit. In this framework the "*nature of metrics (Natur der Metrik)*" has to be determined, giving expression to the "*metrical essence of space (das metrische Wesen des Raumes)*" ([4]1921, 125–127; 1921, 223–225; 1923, 46). The (metrical) essence of space has, according to Weyl, to be characterized by the combination of a principle of *widest possible freedom for the choice of parameters*[26] inside the limits of the "nature of metrics" and a restrictive principle of *unique determination of a compatible affine connection* for each of the admissible choices of parameters ([4]1921, 127f.; 1921, 224; 1923, 46). In such a manner the *essence of space* understood as a counterbalancing pair of principles (freedom/restriction) ought to delimit a more specific class of metrics characterized by what Weyl called the *nature of metrics* among the general concept of metrics.[27] The determination of the parameters inside the more specific class of metrics could no longer be the task of mathematics or conceptual analysis but was, of course, to be determined by empirical investigations (physics).

Mathematically, Weyl characterized the *general concept of metrics* by an adaptation of congruence ideas to infinitesimal geometry. To speak of infinitesimal congruence in a differentiable manifold M, so went his argument, one has to give a family of groups $G_P \subset SL(T_P M)$, in local coordinates $G_P \subset SL(n, \mathbf{R})$, which are all conjugate in $GL(n, \mathbf{R})$ to one of them, say G.[28] A transformation $A \in G_P$ can then be considered to be an infinitesimal *congruence* at P; and a $U \in N :=$ Normalizer(G) as a *similarity*.

---

[25]This conjecture had long been overlooked, before it was taken up and positively answered by D. Laugwitz in 1958; cf. [Scheibe 1988, 65].

[26]Weyl repeatedly claimed that this principle of widest possible *freedom* of the choice of parameters had to replace the principle of homogenity in the classical Helmholtz-Lie analysis of the space problem in the light of relativity theory, greatest possible variability now being postulated only on a metalevel of reflection (Weyl 1921, 224; 1923, 46).

[27]Thus "essence of space" and "nature of space" are coextensive characterizations of the same restricted class in the *general concept of metrics*. They differ only methodologically. The *essence of space* contains general properties arrived at by a *phenomenological analysis* of the concept of spacetime, which is on the level of insight reached by modern relativistic physics. The *nature of space* is an equivalent characterization of the same restriction, which is mathematically more illuminating in the context of *infinitesimal geometry*, although in itself philosophically not as informative as the first characterization.

[28]$G_P \subset SL(T_P M)$ because of volume invariance.

To allow comparison of vectors in infinitesimally close points $P, P' \in M$ possible, with $\vec{PP'} = dx$, Weyl introduced what he called a *metrical connection* associating linearly to each vector $dx(= dx^i, 1 \leq i \leq n)$ an infinitesimal similarity transformation $\Lambda(dx)$. As this similarity transformation is only determined up to congruence in $T_{P'}M$, $\Lambda(dx)$ is more precisely speaking an infinitesimal transformation in $H := N/G$, that is an element in the Lie algebra h of H.[29] Evaluation of $\Lambda(dx)$ on a tangential vector $\xi \in T_P M$ gives

$$(dx, \xi) \mapsto \Lambda(dx)[\xi] \text{ in coordinates } \Lambda^i_{jk} dx^k \xi^j$$

where Λ^i_{jk} is a n^3-dimensional system of coefficients in $\mathbf{R}$. Evaluation of an infinitesimal transformation in h on a vector ξ is determined only up to infinitesimal transformations from the Liealgebra of G, called g, because $H = N/G$ (RZM [4]1921, 125–127). Let us resume this argument in the following

Definition 3: A *general Weylian congruence structure* on a differentiable manifold M consists of a smooth family of groups G_P, $(P \in M)$ with all G_P conjugate to a $G \subset SL(n, \mathbf{R})$ and a connection Λ in the principal bundle with characteristic fibre $H = \text{Normalizer}(G)/G$, called the *metrical connection* (of the congruence structure).

This corresponded to the *general concept of metrics* of Weyl's more philosophical reflections. And it enabled Weyl to formulate his two principles of the "essence of space" (freedom of parameter choice and unique determination of affine connections) mathematically as axiomatic conditions for the group G. His discussion can be condensed to

Definition 4 (RZM [4]1921, 127f.; [5]1923, 139f.): A congruence structure on a differentiable manifold with characteristic congruence group G may be called a *specialized Weylian congruence structure* if G fulfills the following conditions:

Axiom I. Each number system Λ^i_{jk} may arise as system of coefficients of the metrical connection.

Axiom II. Each metrical connection Λ contains exactly one affine connection Γ (modulo infinitesimal transformations from g).

This was, of course, a clear cut mathematical expression of the "essence of space" on which the mathematical analysis of the "nature of space" could now be built. To start with, the second axiom means that Λ can be decomposed as

$$\Lambda^i_{jk} = A^i_{jk} - \Gamma^i_{jk}$$

---

[29]Slightly modernized, the *metrical connection* Λ is a connection in the principal bundle over M with typical fibre $H = N/G$.

in exactly one way, with affine connection Γ and that infinitesimal transformation $A \in g$ depends linearly on dx. Because of the symmetry $\Gamma^i_{jk} = \Gamma^i_{kj}$ the uniqueness postulate of Axiom II leads to

$$A^i_{jk} = A^i_{kj} \Longrightarrow A^i_{jk} = 0$$

and by a counting argument for free parameters $dim\, G = \frac{n}{2}(n-1)$. As moreover $G \subset SL(n, \mathbf{R})$, Weyl rather easily arrived at

Lemma 2 (RZM ^{4}1921, 132): *For a group $G \subset SL(n, \mathbf{R})$ of a specialized Weylian congruence structure with Liealgebra g the following hold:*

1. $A \in g \implies trace\, A = 0$,

2. $dim\, G = dim\, g = \frac{n}{2}(n-1)$,

3. *if coefficients A^i_{jk} with $A^i_j \in g$ are formed such that $A^i_{jk} = A^i_{kj}$, then $A^i_{jk} = 0$.*

Now Weyl could state his *central conjecture of the space problem* (RZM ^{4}1921, 132), later proven as

Theorem 3 (Weyl 1922a, 1923): *A group G fulfilling the conditions 1) to 3) of Lemma 2 is isomorphic to a special orthogonal group (of general signature) $G \cong SO(k, n-k; \mathbf{R})$.*

This conjecture was of particular importance, as Weyl could comparably easily show

Theorem 4 (Weyl 1923, 51f.): *To each specialized Weylian congruence structure on a manifold M with group $G \cong SO(k, n-k; \mathbf{R})$ there always exists a Weylian metric on M, from which it can be derived.*

Theorems *3* and *4* finally showed that the structure of Weylian manifolds is a good characterization of the *nature of space* and thus nothing less than a proper expression of the *essence of space*.

In his report for the *Deutsche Mathematiker-Vereinigung* (1921) Weyl already spoke of the central conjecture as *theorem* and admitted that in the 4-th edition of RZM he had been

"... so keen (hatte ich die Stirn) to state the just-mentioned purely algebraic theorem about infinitesimal groups of linear transformations as a conjecture, alone on the basis of its transient meaning for the space problem" (Weyl 1921, 228).

He could already talk about the contents of the conjecture as *theorem*, because he had been able to verify it in a complicated case by case discussion published in (Weyl 1922a).[30] He simplified his proof in lectures given the same year at Barcelona and Madrid and published as *Die mathematische Analyse des Raumproblems* (Weyl 1923).

With the French translation of the 4-th edition of RZM (Weyl TEM) E. CARTAN learned of Weyl's formulation of the space problem. He modified it slightly, but not without consequences, substituting the concept of a *G-structure* on the manifold M for the *general concept of metrics*, i.e. essentially a principal G-bundle over M ($G \subset GL(n, \mathbf{R})$) together with a connection in this bundle. Such a structure can be realized locally as the orbit of a n-frame field on M under the action of G, in the prinicpal G-bundle. Then he restricted his investigation to more specialized structures by the postulate of existence and uniqueness of an affine connection in the principal bundle (corresponding to Weyl's Axioms I, II). Using his developed structural knowledge of Lie algebras, Cartan then proved that the complexification of the Liealgebra of G is conjugate to an orthogonal group of general signature over $\mathbf{C}$ (Cartan 1923).[31]

Weyl seems to have been fascinated by Cartan's methods in the theory of Lie algebras [Borel 1986, 54] and took this encounter with a new problem field as incentive to enter his next large field of activities in the 1920s. In combination with another incentive – an engagement in questions of invariant theory – Weyl's study of the space problem evolved into his most fruitful mathematical field of work, the theory of representation of Lie groups. *So he let the problem shift appearing in Cartan's work uncommented,*[32] *although it was of considerable influence on the geometric results of the analysis*: Cartan's special G-structures lead to semi-Riemannian manifolds throwing Weylian metrics back again to the fringe of oblivion.

Weyl seems to have accepted a mathematically interesting turn of his problematics, although it went counter to the philosophical and geometrical intentions of his original program. This does not mean at all that he lost confidence in his own philosophical viewpoint. Quite to the contrary, he concluded the second chapter of the 5-th edition of RZM with a remark that his investigations seemed to him to be "... a good example for the analysis of essence (Wesensanalyse) intended by Husserl's phenomenological philosophy ..." ([5]1923, 140f.). He warned philosophers, though, of trying too impatiently to grasp the "essence" by one stroke of exemplaric presentation and hinted at the long and involved development of mathematical and physical knowledge, which had been necessary before reason became

---

[30] Already in (RZM [4]1921) Weyl remarked that he had verified the conjecture for the dimensions n = 2, 3 (ibid., 132) – a first step for the case discussion in (1922a).

[31] See also [Borel 1986], who does not mention that Cartan solved a slightly different problem than Weyl's. This has been clearly pointed out by [Scheibe 1988].

[32] As Scheibe remarks [1988], this shift was taken over unnoticed by succeeding investigators like Klingenberg, Freudenthal etc.

able to see *the essence of space* as clearly as it appeared to him now. Moreover he thought the example of the space problem very illuminating for

> "... that question of phenomenology, which appears to me as the most important: to what extent the delimitation of the entities accessible to reason (die Abgrenzung der dem Bewußtsein aufgehenden Wesenheiten) expresses a structure owed to the empire of the given and to what extent pure convention is involved in it" (RZM ⁵1923, 141).

This remark leads to questions of methodology in Weyl's work on relativity and could be a starting point to explain some of his differences with Einstein.[33] But this question too is beyond the scope of this paper.

---

[33]In particular the question of genesis and structure of the guiding field (and metrics) of a matter free state of the world, which turned up in Weyl's remarks to the early cosmological debate (Weyl's discussion of the Einstein and de Sitter models) (RZM ⁵1923, 287ff.).

Literature

Binder, Christa; Hlawka, Edmund 1986. Über die Entwicklung der Theorie der Gleichverteilung in den Jahren 1909 bis 1919. *Archive for the History of Exact Sciences* **36**, 197–249.

Borel, Armand 1986. Hermann Weyl and Lie Groups. In [Chandrasekharan 1986, 53–82].

Cartan, Elie 1923. Sur un théorème fondamental de M. H. Weyl. *Journal de Mathématiques* **2**, 167–192. Ouevres Complètes **3.1**, Paris: Gauthier-Villars 1955, 633–658.

Chandrasekharan, K. (ed.) 1986. *Hermann Weyl 1885–1985*: Centenary Lectures Delivered by C. N. Yang, R. Penrose, A. Borel at the ETH Zürich. New York etc.: Springer.

Chevalley, Claude; Weyl, André 1957. Hermann Weyl (1885–1955). *L'Enseignement Mathématique* (2) **3**, 157–187.

Dieudonné, Jean 1989. *A History of Algebraic and Differential Topology, 1900–1960.* Boston - Basel: Birkhäuser.

Deppert, Wolfgang (Hrsg.) 1988. *Exact Sciences and Their Philosophical Foundations: Exakte Wissenschaften und ihre philosophische Grundlegung.* Vorträge des internationalen Hermann-Weyl-Kongresses, Kiel 1985. Frankfurt/M - Bern etc.: Peter Lang Verlag.

Einstein, Albert; Grossmann, Marcel 1913. *Entwurf einer verallgemeinerten Relativitäthstheorie und einer Theorie der Gravitation.* I. Physikalischer Teil von A. Einstein, II. Mathematischer Teil von M. Grossmann. Leipzig: Teubner. [Extended reprint in:] *Zeitschrift für Mathematik und Physik* **62** (1914), 225–261.

Einstein, Albert 1915a. Erklärung der Perihclbewegung des Merkur aus der allgemeinen Relativitätstheorie. *Sitzungsberichte Akademie der Wissenschaften Berlin, Math.-phys. Kl. 1915*, 831–839.

Einstein, Albert 1915b. Die Feldgleichungen der Gravitation. *Sitzungsberichte Akademie der Wissenschaften Berlin, Math.-phys. Kl. 1915*, 844–847.

Einstein, Albert, 1916a. Die Grundlage der allgemcinen Relativitätstheorie. *Annalen der Physik* **49**, 769–822.

Einstein, Albert 1916b. Hamiltonsches Prinzip und allgemeine Relativitätstheorie. *Sitzungsberichte Akademie der Wissenschaften Berlin, Math.-phys. Kl. 1916*, 1111–1116.

Ehlers, Jürgen 1988. Hermann Weyl's contributions to the general theory of relativity. In [Deppert 1988, 83–106].

Frei, Günther 1989. Hermann Weyl an der ETH Zürich (1913–1930). In: C. Binder (Hrsg.). II. *Österreichisches Symposium zur Geschichte der Mathematik*, Neuhofen Ybbs, 22. bis 28. Oktober 1989. Institut für Technische Mathematik, TU Wien, 10–15.

Hilbert, David 1915. Die Grundlagen der Physik. Erste Mitteilung. *Nachrichten Gesellschaft der Wissenschaften Göttingen*, 395–407. [not in *Gesammelte Abhandlungen*]

Hilbert, David 1917. Die Grundlagen der Physik. Zweite Mitteilung. *Nachrichten Gesellschaft der Wissenschaften Göttingen*, 395–407. [not in *Gesammelte Abhandlungen*]

Hilbert, David 1924. Die Grundlagen der Physik. *Mathematische Annalen* **92**, 1–32. Gesammelte mathematische Abhandlungen **3** (1935), 258–289 [*modified reprint of* (1915, 1917)].

Howard, Don; Stachel, John 1989. *Einstein and the History of General Relativity*. Einstein Studies **1**. Boston – Basel etc.: Birkhäuser.

Levi-Civita, Tullio 1917a. Nozione di parallelismo in una varietà qualunque e conseguente specificacione geometrica della curvatura Riemanniana. *Rendiconto Circulo Mathematico Palermo* **42**, 173–205.

Levi-Civita, Tullio 1917b. Statica Einsteiniana. *Rendiconti Accademia Lincei* **26**, 458ff.

Mehra, Jagdish 1973. Einstein, Hilbert, and the theory of gravitation. In: J. Mehra (ed.). *The Physicist's Conception of Nature*. Dordrecht – Boston: Reidel, 1973, 92–178.

Pais, Abraham 1982. *'Subtle is the Lord...': The Science and the Life of Albert Einstein*. Oxford: Oxford University Press.

Pais, Abraham 1986. *"Raffiniert ist der Herrgott..." Albert Einstein. Eine wissenschaftliche Biographie*. Aus dem Englischen von R. U. Sexl, H. Kühnelt, E. Streeruwitz. Braunschweig/Wiesbaden: Vieweg.

Pauli, Wolfgang 1919. Zur Theorie der Gravitation und der Elektrizität von H. Weyl. *Physikalische Zeitschrift* **20**, 457–467.

Pauli, Wolfgang 1921. Relativitätstheorie. *Encyklopädie der Mathematischen Wissenschaften* **5.2**, Leipzig: Teubner, 539–775.

Planck, Max 1899. Die Maxwellsche Theorie der Electrizität von der mathematischen Seite betrachtet. *Jahresberichte DMV* **7**, 77–89. Physikalische Abhandlungen **1**, 601–613.

Riemann, Bernhard 1854. Über die Hypothesen, welche der Geometrie zu Grunde liegen. Habilitationsvortrag Göttingen. *Göttinger Abhandlungen* **13** (1867). Werke, 272–287.

Riemann, Bernhard Werke. *Gesammelte mathematische Werke und wissenschaftlicher Nachlaß.* Leipzig 1876, 2. erweiterte Aufl. 1892. Neudruck New York: Dover 1955. Neudruck Nendeln: Sändig 1978. Erweiterter Neudruck (Hrsg. Narasimhan) Berlin etc.: Springer 1990.

Scheibe, Erhard 1988. Hermann Weyl and the nature of spacetime. In [Deppert 1988, 61–82].

Schwarzschild, Karl 1916. Über das Gravitationsfeld eines Massepunktes nach der Einsteinschen Theorie. *Sitzungsberichte Akademie der Wissenschaften Berlin* 189–196.

Sigurdsson, Skuli 1991. Hermann Weyl, Mathematics and Physics, 1900–1927. *Dissertation* Harvard University. Cambridge.

de Sitter, Willem 1916/1917. On Einstein's theory of gravitation and its astronomical consequences, 3 parts. Monthly Notices Royal Astronomical Society London **76** (1916), 699–728; **77** (1916), 155–183; **78** (1917).

de Sitter, Willem 1917. On the relativity of inertia. *Verh. Akademie Wetenschappen Amsterdam* **19**, 1217–1225.

Speiser, David 1988. "Gruppentheorie und Quantenmechanik": The book and its position in Weyl's work. In [Deppert 1988, 161–190].

Stachel, John 1986. Eddington and Einstein. In: E. Ullmann-Margalitt (ed.), *The Prism of Science*, Dordrecht: Reidel, 225–250.

Stachel, John 1989. Einstein's search for general covariance, 1912–1915. In [Howard/Stachel 1989, 63–100].

Straumann, N. 1987. Zum Ursprung der Eichtheorien bei Hermann Weyl. *Physikalische Blätter* **43** (11), 414–421.

Weyl, Hermann 1917. Zur Gravitationstheorie. *Annalen der Physik* **54**, 117–145. Werke **1**,670–698 [29].

Weyl, Hermann 1918a. Reine Infinitesimalgeometrie. *Mathematische Zeitschrift* **2**, 384–411. Werke **2**, 1–28 [30].

Weyl, Hermann 1918b. Gravitation und Elektrizität. *Sitzungsberichte Akademie der Wissenschaften Berlin*, 465–480. Werke **2**, 29–42 [31].

Weyl, Hermann 1919a. Eine neue Erweiterung der Relativitätstheorie. *Annalen der Physik* **59**, 101–133. Werke **2**, 55–87 [34].

Weyl, Hermann 1919b. Vorwort und Erläuterungen in: Riemann, Bernhard. *Über die Hypothesen welche der Goemetrie zu Grunde liegen.* Neu herausgegeben und erläutert von H. Weyl. Berlin: Springer, III–V, 24–47.

Weyl, Hermann 1921. Das Raumproblem. *Jahresberichte DMV* **30**, 92ff. Werke **2**, 212–228 [45].

Weyl, Hermann 1922a. Die Einzigartigkeit der Pythagoreischen Maßbestimmung. *Mathematische Zeitschrift* **12**, 114–146. Werke **2**, 263–295 [49].

Weyl, Hermann 1922b. Die Relativitätstheorie auf der Naturforscherversammlung. *Jahresberichte DMV* **31**, 51–63. Werke **2**, 315–127 [52].

Weyl, Hermann 1923. *Mathematische Analyse des Raumproblems.* Vorlesungen gehalten in Barcelona und Madrid. Berlin: Springer. Neudruck Darmstadt: Wissenschaftliche Buchgesellschaft, 1963.

Weyl, Hermann RZM. *Raum, Zeit, Materie.* Vorlesungen über allgemeine Relativitätstheorie. Berlin usw.: Springer 1918. ^{2}1919. Veränderte Auflagen ^{3}1919, ^{4}1920, ^{5}1923. Nachdruck Berlin etc.: Springer als ^{6}1970, ^{7}1988.

Weyl, Hermann STM. *Space, Time, Matter.* Translated from the 4th German Edition. London: Methuen 1922.

Weyl, Hermann TEM. *Temps, éspace, matière* (de la 4ème édition allemande). Paris: Blanchard 1922.

Weyl, Hermann Werke. *Gesammelte Abhandlungen.* 6 vols. Ed. K. Chandrasekharan. Berlin - Heidelberg - New York: Springer.

The Origins of Infinite Dimensional Unitary Representations of Lie Groups

Sugiura Mitsuo

1 Introduction

The systematic investigation of infinite dimensional unitary representations of Lie groups began with the works of V. Bargmann[1947] on $SL(2, \mathbf{R})$ and I. M. Gelfand–M. A. Naimark[1947] on $SL(2, \mathbf{C})$. The theory of the finite dimensional representations of Lie algebras and Lie groups had been established by E. Cartan[1913] and H. Weyl[1925/26], [1934/35]. However the finite dimensional theory did not develop spontaneously into the infinite dimensional one. The purpose of this paper is to study the circumstances.

The Cartan's theory of representations dealt with the problem to determine all possible irreducible linear (or projective) groups. It was the problem within the framework of Lie groups. In this framework, there was no room to consider infinite dimensional representations, because the general linear group $GL(V)$ are not a (finite dimensional) Lie group if V is infinite dimensional.

After 1925, the infinite dimensional representations arose from several problems both in physics and mathematics. Quantum mechanics popularized the notion of Hilbert spaces and operators on them and gave the directest motive for the introduction of the infinite dimensional representations. In mathematics the discovery of the Haar measure enabled the introduction of the regular representation for an arbitrary locally compact group and the theory of rings of operators used unitary representations of discrete groups in order to construct the factors of type II and of type III. But these were abstract examples based on the general theory of measures and did not concern the detailed structure of the groups. More detailed studies on the infinite dimensional representations of the concrete Lie groups only began with the works of E. Wigner[1939] and P. A. M. Dirac[1945] on the Lorentz groups. These two works were the origins of the decisive papers of Bargmann and Gelfand–Naimark cited above.

2   The Finite Dimensional Representations of Lie Groups

E. Cartan[1913] established the theory of finite dimensional representations of Lie algebras over C. As its title showed, Cartan's aim was to get all irreducible linear groups. Cartan[1909] proved that every irreducible linear Lie algebra $\underline{g}$ over C is either semisimple or the direct sum of a semisimple algebra and a one–dimensional center which consists of scalar transformations. Hence we can assume that the linear Lie algebra $\underline{g}$ is semisimple without loss of generality. Let $\underline{h}$ be a fixed Cartan subalgebra of $\underline{g}$ and put ℓ be the dimension of $\underline{h}$ which is called the rank of $\underline{g}$. A linear form ω on $\underline{h}$ is called a weight of $\underline{g}$ if there exists a vector $x \neq 0$ such that $Hx = \omega(H)x$ for all H in $\underline{h}$. Nonzero weights of the adjoint Lie algebra of $\underline{g}$ are called the roots of $\underline{g}$. (The roots and weights were introduced by W. Killing[1988–90].) There exist ℓ roots $\alpha_1, \cdots, \alpha_\ell$ and every root α of $\underline{g}$ can be expressed as a linear combination of $\alpha_1, \cdots, \alpha_\ell$ with the integral coefficients of the same sign. Every weight λ is a linear combination of $\alpha_1, \cdots, \alpha_\ell$ with rational coefficients: $\lambda = \sum_{i=1}^{\ell} m_i \alpha_i$. We introduce the lexicographic order with respect to $(m_1, \cdots, m_\ell)$ in the set of all weights. The maximal weight of $\underline{g}$ in this order is called highest. Two isomorphic linear semisimple Lie algebras $\underline{g}$ and $\underline{g}'$ are linearly equivalent if and only if their highest weights λ and λ' coincide with each other (i.e. $\lambda' = \lambda \circ \phi$ where ϕ is an isomorphism of $\underline{g}'$ onto $\underline{g}$). Cartan found that there exist exactly ℓ irreducible linear Lie algebras $\underline{g}_1, \cdots, \underline{g}_\ell$ which are isomorphic to $\underline{g}$ and have the highest weights $\lambda_1, \cdots, \lambda_\ell$ respectively. He showed that every irreducible linear Lie algebra isomorphic to $\underline{g}$ is linearly equivalent to one which has the highest weight $p_1\lambda_1 + \cdots + p_\ell\lambda_\ell$ for some non negative integers $p_1, \cdots, p_\ell$ (Cartan's classification theorem). Cartan's proof of classification theorem consisted of two steps. First he proved the existence of linear Lie algebras $\underline{g}'_i$s $(1 \leq i \leq \ell)$ individually by the concrete realizations for each simple Lie algebra. Secondly he proved the existence of a linear Lie algebra $\underline{g}$ with the highest weight $p_1\lambda_1 + \cdots + p_\ell\lambda_\ell$ by constructing $\underline{g}$ from $\underline{g}_1, \cdots, \underline{g}_\ell$. The process is called making the Cartan product. The Cartan product is realized on a subspace of the tensor product of p_i–copies of the representation spaces of $\underline{g}'_i$s for all i. The history of the Cartan product was studied by Th. Hawkins[1988]. Cartan's theory gave the complete invariant for irreducible linear Lie algebras, the highest weight, and established the core of the finite dimensional representation theory.

H. Weyl[1925/26] developed the theory in many ways. His main innovations were the introduction of the global Lie groups and their covering groups, the integral formula for compact semisimple Lie groups, the principle of unitary restriction and the explicit character formula. Using these tools, Weyl found that the proof of Cartan's classification theorem can be reduced to the completeness of the set of all irreducible characters of compact semisimple Lie group G in the space of all square–integrable central functions on G. This completeness is a corollary to that of all matrix coefficients of irreducible representations in the space $L^2(G)$. The latter was proved by F. Peter–H. Weyl[1927] with the aid of the eigenvalueproblem of integral operators. Thus Weyl[1934/35] was able to give a unified proof to Cartan's classification theorem. This proof was a remarkable progress from the case by case verification of Cartan.

These were the main results of the finite dimensional representation theory of Lie groups by Cartan and Weyl. It had a beautiful façade of classification theorem. But it lacked a unified method of constructing irreducible representations. Weyl[1925/26], [1939] constructed the irreducible representations of $SL(n, C)$ and $Sp(n, C)$ on the spaces of certain tensors. The Lie algebra $\underline{o}(n, C)$ requires the space of spinors besides those of tensors. The constructions were carried out for each classical Lie groups and Lie algebras separately. A unified method was given later by A. Borel and A. Weil (cf. §7).

On the other hand E. Cartan[1929] generalized Peter–Weyl theorem to the completeness of the spherical functions on the homogeneous spaces of compact Lie groups. Moreover he gave an analogue of his classification theorem for the irreducible representations realized by the spherical functions on compact Riemannian symmetric spaces G/K. He showed that the restricted highest weight of such representations make λ (=rank G/K)-dimensional lattices. He did not determine which irreducible representations are realized by spherical functions on G/K. The determination was obtained by Sugiura[1962] and Helgason [1966].

We remark that Peter–Weyl and Cartan dealt with solely the finite dimensional representations. They neither used nor defined the regular representation of a Lie group. Infinite dimensional vector spaces and function spaces such as L^2, ℓ^2 and $C([0,1])$ had been studied since the first decade of the 20th century by D. Hilbert, E. Schmidt, J. Hadamard, M. Fréchet, F. Riesz, E. Fischer and others (cf. J. Dieudonné[1981]). The notion of Banach space was defined in 1922 by S. Banach and N. Wiener. However the infinite dimensional representations did not appear immediately. As long as mathematicians considered the representations of Lie groups as the homomorphisms between two Lie groups, infinite dimensional representations were beyond their scope.

3 The Quantum Mechanics

The situation has changed completely since the rise of the quantum mechanics introduced by W. Heisenberg[1925] and E. Schrödinger[1926]. Soon the mathematical equivalence of their theories were recognized. M. Born[1926] gave the probabilistic interpretation of a normalized wave function $\psi(x, y, z, t)$ of a particle that $p_t(A) = \iiint_A |\psi(x, y, z, t)|^2 dx dy dz$ is equal to the probability of the particle to be found in a subset A of $\boldsymbol{R}^3$ at the time t. The Hilbert space of the square integrable functions arose naturally in this context. The mathematical formulation of the quantum mechanics was given by J. von Neumann[1927a], [1927b] and [1932]. In [1927a] he gave the first definition of an abstract Hilbert space. In [1932] a state of a physical system is represented by a ray in a Hilbert space and a physical quantity is represented by a self-adjoint operator.

In this trend of studies, many applications of group representations to quantum mechanics were found in the period from 1926 to 1932. But most of them concern the finite dimensional representations. We pick up two examples of the infinite dimensional representations.

Weyl[1928] introduced an infinite dimensional representation U of the rotation group $SO(3)$ on the Hilbert space $L^2(\mathbf{R}^3)$. The representation U and its restriction to the subgroup $SO(2)$ explain the azimuthal quantum number and the magnetic quantum number.

Weyl[1927] remarked that the Heisenberg commutation relation

(1) $$PQ - QP = iI$$

for two self–adjoint operators P and Q can be derived from the relation

(2) $$U_s V_t U_s^{-1} V_t^{-1} = e^{ist} I, \qquad s, t \in R$$

for the one parameter group $U_s = e^{isP}$ and $V_t = e^{itQ}$ of unitary operators. M. H. Stone[1930] proved that if U_s and V_t satisfies the Weyl relation (2), then they are unitarily equivalent to a direct sum of operators given by

$$(U_s\psi)(q) = \psi(q - s) \qquad \text{and} \qquad (V_t\psi)(q) = e^{itq}\psi(q).$$

The same results was obtained by von Neumann[1931] independently. Much later, it was found that Stone–von Neumann Theorem is essentially equivalent to give a certain unitary representations of the Heisenberg group (the group of upper triangular unipotent matrices of the degree 3). cf. A. A. Kirillov[1962], and R. How[1980].

4   The Arising of Infinite Dimensional Representations

In 1930's the infinite dimensional representations of topological groups became gradually to be the object of investigations. M. H. Stone[1932] gave the spectral decomposition of a one-parameter group of unitrary operators. The theorem gave the irreducible decomposition of a unitary representation of the additive group $\mathbf{R}$ of real numbers in an abstract form. It was used later by V. Bargmann[1947] to construct the representation of the Lie algebra from the given unitary representation of the Lorentz group.

The invariant integration on groups was first used by A. Hurwitz[1897] to prove the finite generation of invariants of the orthogonal groups. Then Weyl [1925/26] and Peter–Weyl[1927] successfully used the invariant integration on compact Lie groups.

The discovery of the Haar measure by A. Haar[1933] was the starting point of many investigations on locally compact groups. Haar[1933] defined the right regular representation of an arbitrary (separable) locally compact group. Using

Haar's work, J. von Neumann[1933] solved the Hilbert's fifth problem for compact groups.

von Neumann–F. J. Murray[1936] and von Neumann[1940] constructed the factors of type II and III using a unitary representation of a countable discrete group acting ergodically on a measure space.

In the meanwhile there appeared some works which suggested the limitation of the finite dimensional unitary representations. A topological group G is called maximally almost periodic if each two distinct points in G are separated by a finite dimensional unitary representation of G. Compact groups and locally compact abelian groups are maximally almost periodic by the works of Peter–Weyl[1927], Haar[1933] and L. S. Pontrjagin[1934], van Kampen[1935] respectively. Conversely H. Freudenthal[1936] proved that if a connected locally compact group G is maximally almost periodic then G is the direct product of a compact group and a vector group $\boldsymbol{R}^n$. Moreover von Neumann–E. Wigner[1940] proved that every non compact connected simple Lie group has no finite dimensional irreducible unitary representations except the identity representations. On the other hand, I. M. Gelfand–D. A. Raikov[1943] proved that every locally compact group G has a sufficiently many irreducible unitary representations, that is, every two distinct points of G are separated by an irreducible unitary representation of G which may be infinite dimensional. These facts clearly indicated the necessity for the infinite dimensional unitary representations.

5   The Lorentz Group

The significance of getting all irreducible unitary representations of the Lorentz group was first pointed out in a paper of E. Wigner[1939]. In order to classify all possible types of relativistic wave equations of a free particle, he desired to determine all irreducible unitary representations U of the inhomogeneous Lorentz group G. The group G is a semidirect product of the proper Lorentz group L and the vector group $\boldsymbol{R}^4$ of translations. He showed that such a representation U is determined uniquely up to the unitary equivalence by each orbit $\underline{O}$ of the Lorentz group L on $\boldsymbol{R}^4$ and an irreducible unitary representation M of the stationary subgroup L_0 of L corresponding to a point p in the orbit $\underline{O}$. The orbits are parametrized by the value of the Minkowski metric

$$P = \langle p, p \rangle = -p_1^2 - p_2^2 - p_3^2 + p_4^2$$

at the point $p \in \underline{O}$. The orbits are divided into four classes according to the value of P:

I	$P > 0$,	$L_0 = SO(3)$,	
II	$P = 0$,	$p \neq 0$,	$L_0 = M(2)$ (2–dim. Euclidean motion group)
III	$P = 0$,	$p = 0$,	$L_0 = L = SO_0(3,1)$,

IV $P < 0,$ $L_0 = SO_0(2,1).$

Except the case $L_0 = SO(3)$, the irreducible unitary representations of $L_0 = M(2), SO_0(3,1)$ and $SO_0(2,1)$ were not known at that time. Wigner left the problem of determinimg these representations.

P. A. M. Dirac[1945] found some infinite dimensional unitary irreducible representations of the Lorentz group $L = SO_0(3,1)$ and pointed out some applications to quantum mechanics. The papers of Wigner and Dirac were the origin of the works of Bargmann[1947] and Gelfand–Naimark[1947]. Also inspired by Dirac, Harish–Chandra[1947] gave an infinitesimal classification of infinite dimensional irreducible unitary representations of the Lorentz group L.

6 $SL(2, \mathrm{R})$ and $SL(2, \mathrm{C})$

The systematic investigations of global unitary representations of the Lorentz groups were achieved by Bargmann[1947] and Gelfand–Naimark[1947]. They studied $SL(2, \boldsymbol{R})$ and $SL(2, \boldsymbol{C})$ respectively which are the 2–fold covering groups of the proper Lorentz groups $SO_0(2,1)$ and $SO_0(3,1)$. The methods of the two papers were rather different from each other. But both papers obtained the following results for the group $G = SL(2, \boldsymbol{R})$ or $SL(2, \boldsymbol{C})$.

(i) The principal and complementary series of the irreducible unitary representations of G are constructed on some function spaces. The principal series of $SL(2, \boldsymbol{R})$ is divided into the continuous series and the discrete series. The latter is characterized by the square integrability of their matrix coefficients. The principal series of $SL(2, \boldsymbol{C})$ and the continuous series of $SL(2, \boldsymbol{R})$ consists of the representations induced from one dimensional unitary representations of their minimally parabolic subgroups.

(ii) Any irreducible unitary representation of G is unitarily equivalent to one of the representations described in 1. or the identity representation.

(iii) The matrix coefficients of the principal series representation constitute a complete system in some sense. Gelfand–Naimark proved an analogue of Plancherel theorem on the Fourier transforms for the group $SL(2, \boldsymbol{C})$. Similar result for $SL(2, \boldsymbol{R})$ was proved by Harish–Chandra[1952].

The works of Bargmann and Gelfand–Naimark[1947] clearly showed the fruitfulness of the studies on the unitary representations of semisimple Lie groups. Immediately after their works, the important investigations appeared one after another in this new field. Thus the theory of infinite dimensional unitary representations of Lie groups has been established.

7   Approaches from Pure Mathematics

As stated above the theory of the infinite dimensional unitary representations was initiated under the stimulus from quantum mechanics. However several investigations in pure mathematics were approaching to the infinite dimensional representations. In this last section, we study one of them, the induced representations.

The induced representation was a general method which construct a representation U^L of a finite group G from a representation L of a subgroup H. The method was invented by Frobenius[1897], [1898]. The extension of the notion to compact groups was achieved by A. Weil[1940]. Weil proved the Frobenius reciprocity law $(U^L : V) = (V|H : L)$ for any irreducible representation V of a compact group G. A special case in which L is the identity representation was proved by E. Cartan[1929]. Weil did not define the induced representation for an arbitrary locally compact group. This generalization required to overcome two difficulties. One of them arose from the fact that the factor space G/H may have no G-invariant measure and was surmounted by the discovery of the quasi–invariant measure by Dieudonné [1948]. The other was more serious. The induced representations U^L of a locally compact group is, in general, not a direct sum but a direct integral of irreducible representations. Hence the multiplicity lose their usual meaning. Before the general theory of the induced representations was established, the notion was used successfully by Bargmann and Gelfand–Naimark to construct the principal series representations. Wigner's analysis of orbits was deeply related to the induced representations. The general theory was given by G. W. Mackey[1952] who determined the irreducible unitary representations of the motion group $M(n)$ by an orbital analysis. cf. also S. Ito[1952]. After these works were done, A. Borel and A. Weil proved a theorem which was published in J. P. Serre[1954]. The Borel–Weil theorem asserts that every finite dimensional irreducible holomorphic representation of connected complex semisimple Lie group G is holomorphically induced from a 1–dimensional holomorphic representation of a Borel subgroup B of G. On the other hand, every unitary representation of G in the principal series is unitarily induced from a 1–dimensional unitary representation of B. The parallelism between the two theories is very remarkable. But the parallelism was found after the theory of the infinite dimensional representations had been established.

In his commentary in [1979] on his book [1940] written in 1936, A. Weil recalled his work and remarked as follows: "Above all I tried to open the way to a generalization of representation theory of finite groups and compact groups. However not only I did not attain the Land of Promise, the infinite dimensional representations, but also I could not catch a glimpse of the landscape. I was discouraged too early when I found that the matrix coefficients of the finite dimensional representations of a noncompact simple Lie group are not square integrable. This fact denied the hope to look for the representations in $L^2(G)$ and to go forward to the unknown world starting from the known country. Perhaps such a bold march might require the unprejudiced spirit of physicist. In fact, they were Dirac, Wigner and Bargmann who opened the way concerning the Lorentz group."

This frank remark of Weil clearly indicated the difficulty to open the way to the infinite dimensional representations. Under the circumstances the finite dimensional theory of representations could not develop spontaneously into the infinite dimensional theory but required the stimulus from the quantum mechanics.

References

[1] Bargmann, V., 1947, *Irreducible unitary representations of the Lorentz group*, Ann. Math. 48, 568–640.

[2] Born, M., 1926, *Quantenmechanik der Stossvorgänge*, Zeit. für Physik 38, 803.

[3] Cartan, E., 1909, *Les groupes de transformations continus, infinis, simples*, Ann. Ec. Norm. sup. 26, 93–161.

[4] Cartan, E., 1913, *Les groupes projectifs qui ne laissent invariante aucune multiplicité plane*, Bull. Soc. Math. France 41, 53–96.

[5] Cartan, E., 1929, *Sur les détermination d'un système orthogonal complet dans un espace de Riemann symétrique clos*, Rend. Circ. mat. Palermo, 8, 181–225.

[6] Dieudonné, J., 1948, *Sur le théorème de Lebesgue–Nykodym* (III), Ann. Univ. Grenoble 23, 24–53.

[7] Dieudonné, J., 1981, *History of Functional Analysis*, North-Holland, Amsterdam.

[8] Dirac, P. A. M., 1945, *Unitary representations of the Lorentz group*, Proc. Roy. Soc. A 183, 284–295.

[9] Freudenthal, H., 1936, *Topologischen Gruppen mit genügend vielen fastperiodischen Funktionen*, Ann. Math. 37, 57–77.

[10] Frobenius, F. G., 1897, *Über die Darstellung der endlichen Gruppen durch linearen Substitutionen*, Sitz. Preuss. Akad. Wiss. Berlin, 944–1015.

[11] Frobenius, F. G., 1898, *Über Relationen zwischen den Charakteren einer Gruppen und denen ihrer Untergruppen*, *ibid.* 501–515.

[12] Gelfand, I. M.–Naimark, M. A., 1947, *Unitary representations of the Lorentz group* (Russian), Izv. Akad. Nauk SSSR ser. Mat. 11, 414–504

[13] Gelfand, I. M.– Raikov, D. A., 1943, *Irreducible unitary representations of locally compact groups* (Russian), Mat. Sbornik 13, 301–319.

[14] Harish–Chandra, 1947, *Infinite irreducible representations of the Lorentz group*, Proc. Roy. Soc. A 189, 372–401.

[15] Harish–Chandra, 1952, *Plancherel formula for the 2×2 real unimodular group*, Proc. Nat. Acad. Sci. 38, 337–342.

[16] Haar, A., 1933, *Der Massbegriff in der Theorie der kontinuierlichen Gruppen*, Ann. Math. 34, 147–169.

[17] Hawkins, Th., 1982, *Wilhelm Killing and the structure of Lie algebras*, Arch.Hist.exact Sci. 26, 127–192

[18] Hawkins, Th., 1988, *Hesse's principle of transfer and the representation of Lie algebras*, Arch. Hist. exact Sci. 39, 41–73.

[19] Heisenberg, W., 1925, *Über quantentheoretisch Umdeutung kinematischer Beziehungen*, Zet. für Physik 33, 879–893.

[20] Helgason, S., 1966, *A duality in integral geometry on symmetric spaces*, Proc. of the U.S.–Japan Seminar in Differential Geometry, Kyoto, 1965, Nippon Hyoronsha, Tokyo.

[21] How, R., 1980, *On the role of the Heisenberg group in harmonic analysis*, Bull. A. M. S. 3, 821–843.

[22] Hurwitz, A., 1897, *Über die Erzeugung der Invarianten durch Integration*, Gött. Nachr. 71–90.

[23] Ito, S., 1952/53, *Unitary Representations of Some Linear Groups I, II*, Nagoya Math. J. 4, 1–13; 5, 79–96.

[24] Killing, W., 1888–1890, *Die Zusammensetzung der stetigen endlichen Transformationsgruppen*, Math. Ann. 31, 252–290, 33, 1–48, 34, 57–122.

[25] Kirillov, A. A., 1962, *Unitary representations of nilpotent Lie groups*, Russ. Math. Surv. 17, 53–104.

[26] Mackey, G. W., 1952, *Induced representations of locally compact groups*, Ann. Math. 55, 101–139.

[27] Mackey, G. W., 1980, *Harmonic analysis as the exploitation of symmetry — a historical survey*, Bull. A.M.S. 3, 543–698.

[28] Murray, F.J.,– von Neumann, J., 1936, *On rings of operators*, Ann. Math. 37, 116–229.

[29] Peter, F.– Weyl, H., 1927, *Die Vollständigkeit der primitiven Darstellungen einer geschlossenen kontinuierlichen Gruppen*, Math. Ann. 97, 737–755.

[30] Pontrjagin, L. S., 1934, *The theory of topological commutative groups*, Ann. Math. 35, 361–388.

[31] Schrödinger, E., 1926, *Quantesierung als Eigenwertproblem* I-V, Ann. Physik, 79;361, 489; 80;437; 81;109.

[32] Serre, J. P., 1954, *Représentations linéaires et espaces homogènes kähleriens des groupes de Lie compacts*, Sém. Bourbaki(1953/54), n° 100, Inst, H.Poincaré, Paris.

[33] Stone, M. H., 1930, *Linear transformations in Hilbert space III, operational methods and group theory*, Proc. Nat. Acad. Sci. 16, 172–175.

[34] Stone, M. H., 1932, *On one-parameter unitary groups in Hilbert space*, Ann. Math. 33, 645–648.

[35] Sugiura, M., 1962, *Representations of compact groups realized by spherical functions on symmetric spaces*, Proc. Japan Acad. 38, 111–113.

[36] van Kampen, E., 1935, *Locally bicompact abelian groups and their character groups*, Ann. Math. 36, 448–456.

[37] von Neumann, J., 1927a, *Mathematische Begründung der Quantenmechanik*, Gött. Nachr., 1–57.

[38] von Neumann, J., 1927b, *Wahrscheinlichkeittheoretischer Aufbau der Quantenmechanik*, Gött. Nachr., 245–272.

[39] von Neumann, J., 1931, *Die Eindeutigkeit der Schrödingerschen Operatoren*, Math Ann. 104, 570–578.

[40] von Neumann, J., 1932, *Mathematische Grundlagen der Quantenmechanik*, Springer, Berlin.

[41] von Neumann, J., 1933, *Die Einführung analystischer Parameter in topologischen Gruppen*, Ann. Math. 34, 170–190.

[42] von Neumann, J., 1940, *On rings of operators* III, Ann. Math. 41, 87–93.

[43] von Neumann, J.- Wigner, E., 1940, *On minimally almost periodic groups*, Ann. Math. 41, 746–750.

[44] Weil, A., 1940, *L'intégration dans les groupes topologiques et ses applications*, Hermann, Paris.

[45] Weil, A., 1979, *Collected Papers* 3 vols, Springer, Berlin.

[46] Weyl, H., 1925/26, *Theorie der Darstellung kontinuierlicher halb-einfacher Gruppen durch linearen Transformationen* I, II, III, Math. Zeit. 23, 271–309; 24, 328–395.

[47] Weyl, H., 1927, *Quantenmechanik und Gruppentheorie*, Zeit. für Physik 46, 1–46.

[48] Weyl, H., 1928, *Gruppentheorie und Quantenmechanik*, Hirzel, Leipzig.

[49] Weyl, H., 1934/35, *The structure and representations of continuous groups*, Lecture Note taken by N. Jacobson and R. Brauer, The Inst. for Advanced Study, Princeton.

[50] Weyl, H., 1939, *The Classical Groups, their Invariants and Representations*, Princeton Univ. Press, Princeton.

[51] Wigner, E., 1939, *On unitary representations of the inhomogeneous Lorentz group*, Ann. Math. 40, 149–204.

Dispelling a Myth: Questions and Answers about Bourbaki's Early Work, 1934–1944

Liliane Beaulieu

Abstract

The group of mathematicians who wrote the treatise *Éléments de mathématique* under the assumed name "N. Bourbaki" inspired and created legends and controversies. This paper retraces Bourbaki's initial goals and analyzes some of the first written outlines for the treatise. This account answers some questions about the team's original visions. It shows how Bourbaki set out to write a treatise of analysis that could serve various purposes, some of which pertained to research and others to the teaching of mathematics. Bourbaki's first outlines covered much classical analysis, including the range of topics found in the then current French *Cours d'analyse*. While Bourbaki intended to recast these subjects in a more modern setting, it was not immediately obvious to the group what particular setting would be appropriate. Other features of Bourbaki's early outlines include the scarcity of set theory, algebra and topology, along with a marked conservativeness in the treatment of these subjects. From this perspective, the paper argues that Bourbaki's treatise did not arise from a ready-made, architectural and unitary vision of mathematics which the members of the group would have shared at the outset. Rather, any coherent doctrine that may eventually be attributed to Bourbaki was the outcome of painstaking reworkings and discussions.

Introduction[1]

In 1934, a group of French mathematicians[2] began to collaborate on a treatise that has been published under the title *Éléments de mathématique* from 1939 to the present day. This treatise now stands as a monument of mathematical exposition. It covers large tracts of theories, namely: set theory, algebra, general topology, functions of a real variable, topological vector spaces, integration, Lie groups and algebras, commutative algebra, spectral theories, and differentiable

---

[1]The texts that were used as a basis for this paper are analyzed and described extensively in my thesis [Beaulieu 1989].

[2]The founding members of the group were all French, by birth or acquired citizenship. In the course of time, the team coopted mathematicians from other countries.

and analytic manifolds.[3] For mathematicians, N. Bourbaki became the proponent of an abstract and rigorous mathematical style based on a systematic recourse to axiomatics. As the team had chosen to write under a pseudonym, it shrouded itself in mock mystery. This contributed to the creation of a Bourbaki myth. The group and its work stood in the mathematical limelight, especially in the 1950s and 1960s, and inspired contradictory views among mathematicians who would strongly approved of Bourbaki's program or who either vehemently condemned it. Even laymen developed passionate opinions about this mysterious and prolific author. Bourbaki's name and its works were associated with structuralism, as well as with the "new math" or "mathématiques modernes" movements of pedagogical reforms that swept various countries, especially in the Western world.

Bourbaki's best-known piece of work, however, is not the voluminous *Éléments de mathématique* but a short article entitled "L'architecture des mathématiques" [1948].[4] It was signed "Nicolas Bourbaki" and written by a single member of the group who acted mostly alone. For outsiders at least, this article represented Bourbaki's manifesto, though it was not discussed, as such, by the members at their meetings. The power of this article rests on its use of a metaphor by which mathematics (i.e. pure mathematics) is represented as a whole, a unified piece of architecture held together by, and founded on, the "mother structures" of set theory, algebra and topology. By the time this article was written, more than a decade after Bourbaki started its collective work, most of the members of the group probably shared this architectural view of mathematics. Some evidence for this view can be found in the subtitle of the first part of *Éléments*, "Les structures fondementales de l'analyse," and in Bourbaki's insistence on structures in the treatise.[5]

This vision, however, did not prevail at the outset of Bourbaki's venture. One may say that the often quoted "Architecture" article distorts the historical perspective to some extent by projecting Bourbaki's later hallmark onto the aims and concerns of the team at its inception. My intention here is to give a voice to these aims, which differ in many ways from the image that accompanied Bourbaki's later fame.

---

[3]Each of these subjects corresponds to a volume of the *Éléments*. Some volumes were translated into English, Russian, and Japanese.

[4]It is also the first of Bourbaki's work to have been translated into English [1950]. It was later translated into Portuguese, Russian, and German.

[5]This subtitle and the title *Éléments de mathématique* were chosen, at the latest, in 1938, before the first booklet was published. Bourbaki purposely chose to use the singular "mathématique" instead of the usual plural, in order to emphasize its view of mathematics as a unified whole. By then, Henri Cartan [1943] and Jean Dieudonné [1939] had already voiced similar personal views on axiomatics and the foundations of mathematics. These articles are cited in "L'architecture," p. 40.

What Were Bourbaki's Initial Goals?

In the academic year 1934–1935, the team included the following people: Henri Cartan, Claude Chevalley, Jean Delsarte, Jean Dieudonné, Szolem Mandelbrojt, René de Possel and André Weil. Paul Dubreil and Jean Leray also took part in some meetings but they did not stay on as active members. Jean Coulomb - who was a physicist - and Charles Ehresmann joined the group in the summer and the fall of 1935, respectively.

The first purpose of this group was to change the teaching of mathematics in French universities.[6] In order to do so, writing an analysis treatise seemed a reasonable course of action since such a text would normally constitute the basis for higher mathematics instruction in France. Thus, the group set out to write, collectively, a treatise of analysis they hoped would change the teaching of "differential and integral calculus" for twenty-five years, at least. At the very first meeting, in December 1934, André Weil stated this goal explicitly:

> [...] to define for 25 years the syllabus for the certificate in differential and integral calculus by writing, collectively, a treatise on analysis.
> Of course, this treatise will be as modern as possible.[7]

Nearly all the participants in this project were then teaching at the "licence" level,[8] or had been responsible, at some point, for the course that constituted the "certificate" of differential and integral calculus.

They sought to reform the teaching of mathematics in France from within the system, by replacing the textbooks from which this course had been taught for decades in the French university system.

The texts that Bourbaki[9] set out to improve on were the then current *Cours d'analyse* of Édouard Goursat, Jacques Hadamard, or Émile Picard, to name a few. These treatises were the written texts of the courses given by these professors in a French institution of higher education, the University or the École polytechnique, for instance.The most typical and extreme case was the often-mentioned *Cours d'analyse* of Édouard Goursat, from which its author and other professors at the Sorbonne taught for over fifty years. The numerous editions and later translations of this text stand as testimony to its extensive use as a textbook and a sourcebook, in France and in other countries. The French *Cours d'analyse* constituted a tradition to which the members of Bourbaki belonged as a result of their own training. The *Cours d'analyse* also represented a historical trend into which Bourbaki wilfully inscribed its own collective venture.

---

[6] Actually, the French university system consisted of a unique state "Université" with "facultés" in different locations.

[7] Report on the first meeting, xii/10/1934, p. 1. My translation.

[8] The "licence" corresponds roughly to the undergraduate level in North America. This program included three "certificates," one of which was differential and integral calculus.

[9] Referring to the group as "Bourbaki" in 1934–1935 is a slight anachronism since the pseudonym was not adopted before the summer of 1935.

Writing an analysis textbook was not Bourbaki's only goal. At the second
meeting, Weil expressed the wish that their treatise would address a broader audience of learners, teachers, users and producers of mathematics.

> We must write a treatise which will be useful to all: to researchers
> (*bona fide* or not), "finders," aspirants to posts in public education,
> physicists, and all technicians.[10]

With this readership in mind, the treatise to be written became more of a sourcebook than a simple textbook. Thus, it could act on research as well as on teaching,
and would be used by working mathematicians as well as working scientists in other
disciplines, such as physics.

From the beginning, Bourbaki considered a wide range of topics. The outlines that the team drafted during its first year included many of the traditional
subjects of the *Cours d'analyse* such as: integration,[11] analytic functions, Fourier
series, differential equations, partial differential equations, integral equations, elliptic functions, special functions, calculus of variations. Making up individual
inventories of contents for each topic took up most of the team's early efforts.

As Weil's statement indicates, Bourbaki pursued yet another goal, that of
creating a piece of work that would be "as modern as possible." Some members of
the group had singled out Stokes' theorem as a paradigm of the kinds of results in
"differential calculus" which they wanted to "modernize."[12] But what did modern-
izing mean? What topics would the treatise cover? What point(s) of view would it
promote? The group wanted to produce both a conventional and a modern text,
but how was it to reconcile the old and the new? No ready-made doctrine solved
this quandary.

At the group's first meeting, Jean Delsarte suggested that

> [...] at the beginning of the treatise there should be an abstract and
> axiomatic presentation of some essential general notions (such as field,
> operation, set, group, etc.).[13]

As an example of the kind of presentation he had in mind, Delsarte cited Van
der Waerden's *Moderne Algebra* [1930, 1931]. His colleagues, however, resolved to
restrict the abstract presentation to a minimum. This small "abstract package,"
as they called it, would consist merely of a modicum of general elementary notions
and mathematical tools.[14]

The members of Bourbaki did not agree on what this package would contain.
Should it include algebraic notions, set theoretic ones, elements of topology? And,

---

[10]Report on the second meeting, i/14/1935, p. 1. My translation.

[11]They planned on an exposition of integration theory rather than the techniques of differential
and integral calculus, as in the more traditional *Cours d'analyse*.

[12]This is discussed in greater detail in Beaulieu 1990, 36–38.

[13]Report on the first meeting, xii/10/1934, p. 3. Quoted in Beaulieu 1989, vol. I, 151. My
translation.

[14]Report on the first meeting, xii/10/1934, p. 3.

if so, which ones? There were even disputes as to whether to include set theory and algebra at all. During the first six months of the group's existence, it drafted no specific plan for set theory, for instance. In fact, wherever set theoretical notions appeared, they were completely fused with algebraic or topological concepts. No work was done on topology either, during this period. As for algebra, it was confined to an elementary introduction of groups, rings and fields - a minimum of notions necessary to the treatment of real and complex numbers from an algebraic point of view. Linear algebra consisted mostly of matrices, determinants and their use in the solution of systems of equations. While some of the members wrote elaborate projects on integration theory, the team resolved to limit its presentation of integration to a brief exposition of Lebesgue's generalized theory (on abstract sets) and integration in topological vector spaces (introducing Radon measures).[15]

At that stage, Bourbaki explicitly refused to introduce abstract theories for their own sake and interest. Rather, it strove to devise the rudiments of a vocabulary it deemed necessary to write the treatise - and nothing more than what was needed for this task would be included in the first part of its treatise. Thus, the notions comprised by the abstract package were not yet conceived of as "mother structures," as they would later be called. No particular unifying vision of mathematics arose explicitly from these meetings nor from the inventories which the group produced at that point. No discourse on mathematics as a whole seems to have found its way into Bourbaki's early reports, although these reports sometimes recorded major discussions and decisions about which point of view the group would adopt on a given subject.[16]

To summarize, in Bourbaki's early work the goal was not to discuss mathematics as a whole. The group's main goals were to write a rather conventional and limited *Cours d'analyse* and to produce a sourcebook which would reach the widest possible mathematical readership. At the same time, Bourbaki strove to create a treatise that would be as modern as possible.

How Did the Project Change in Scope?

The changes that occurred in the scope of the project resulted from both global and local choices. The global choices involved the overall plans of the treatise as well as its general orientation. The local choices appeared in the many attempts to present the various subjects that Bourbaki decided to include. Along with these choices, the very nature of the treatise was modified.

From the traditional *Cours d'analyse*, Bourbaki had already inherited an array of subjects. In 1935, seeing that the group was considering an increasingly long list of topics, one of the participants, René de Possel, suggested that the project be called "Traité de mathématiques" (plural) rather than "Traité d'analyse." But

---

[15]See Beaulieu 1989, vol. I, 164–203.
[16]This is true, at least, for the documents I perused, which are described in Beaulieu 1989, vol. II.

his colleagues promptly rejected his suggestion and continued, for a while, to label the project an "analysis treatise," and no official title was yet chosen.[17]

A first turning point occurred in the summer of 1935 during what Bourbaki later called its "foundation meeting," which was held in Besse-en-Chandesse, a village near Clermont-Ferrand. There, Bourbaki officially adopted the idea that axiomatics would be used in the treatise, and it resolved to write an introductory text justifying this choice. At the same time, however, Bourbaki decided not to axiomatize set theory, choosing instead to include only the rudiments of a naive set theory. On the one hand, Bourbaki asserted that axiomatics provided the best tools to set forth mathematical facts. On the other hand, it insisted that axiomatization should be restricted to a bare minimum in its own treatise.[18]

At that meeting, Bourbaki created a general outline for the treatise. This consisted of an ordered list of subjects, but not yet an actual table of contents of volumes and chapters. The list contained the following: abstract sets, algebra, real and complex numbers with series, topology and existence theorems, topology of complex numbers, quadratic and hermitian forms, convex fields and orthogonal group, integration, functions of real variables, exterior multiplication, determinants, Pfaffian forms, tensors and geometry, infinite products, inequalities, general analytic functions, special analytic functions, approximations, operational calculus, general diferential equations, special differential equations, Élie Cartan's methods, integral equations, potential, elliptic equations, hyperbolic and parabolic equations, calculus of variations, special functions (Bessel, algebraic, elliptic, theta, gamma, zeta), and numerical calculus. Each member was assigned the task of reporting on a number of these topics. It is interesting to note that the list also gave an estimate of the number of pages that would be required for each topic. Together, set theory, algebra, real and complex numbers, topology, integration and functions of real variables added up to a fifth of the estimated total number of pages in the treatise. What may have been conceived as the "abstract package" remained relatively small.[19]

There was no attempt to cover all of modern mathematics. Nor does the modesty which at that time marked the particular plans of set theory, algebra, topology and integration, indicate a determination to lay down all-encompassing grounds. This idea emerged later. In 1939, the introduction to the "Fascicule de résultats" of Set Theory[20] stated that the "Fundamental Structures of Analysis," the first part of the treatise which was by then entitled *Éléments de mathématique*, was to "lay the foundations for the rest of the treatise and, even, for the whole of modern mathematics."[21] The idea of introducing some mathematical tools is clear from Bourbaki's list of 1935, as well as from the choices it was making on

---

[17] Report on the 4th meeting, ii/11/1935, p. 2.

[18] July 1935, Algebra and set theory, p. 1.

[19] July 1935, Order and Evaluation.

[20] This was a digest of notions and theorems which was the very first published booklet of the treatise.

[21] Bourbaki 1939, "Mode d'emploi de ce traité."

individual topics. Doing this for the whole of modern mathematics, however, was not yet Bourbaki's ambition in 1935.

After a first year of planning, the writing process started. The group then resolved to concentrate on the "abstract package" while developing its plans on other topics. Since the tool kit needed to be ready before other theories could be expounded, the authors devoted most of their efforts to this particular task. As a consequence of this procedure, reading and criticizing the draft texts of algebra, topology, set theory and integration occupied nearly all the meetings between 1935 and 1938. This activity forced a real shift in focus, from the traditional subjects of analysis to, mostly, an elaboration of the volumes on algebra and topology.[22] Although this was not the only factor that transformed the scope of the treatise, it was certainly an important one.

Changes in the scope of the project may be best observed in the plans and draft texts on individual subjects. Each one of these evolved in its own particular way, but a pattern appears throughout the topics that would constitute the first volumes of *Éléments*. In order to illustrate this phenomenon, I will retrace the processes through which parts of Bourbaki's algebra were produced.

The Case of Algebra

At first, Bourbaki was quite timid about algebra. As I mentioned above, its first algebra program was intended as nothing more than a minimum for doing real and complex numbers algebraically, i.e. to treat the real numbers as a commutative field and the complex numbers as an extension of the reals. The purpose of doing matrices and linear transformations was to study systems of equations. The team specifically did not want to rewrite van der Waerden's *Moderne Algebra*. For its own purpose, it was thought sufficient to introduce some elementary notions and results which could be used directly in other parts of the treatise. At the outset, Bourbaki did not wish to write a monograph on algebra.[23]

At the foundation meeting of 1935, set theory and real and complex numbers were considered as parts of algebra. Algebra proper evolved about the concept of field, with both commutativity and non-commutativity. The general outline for algebra comprised: composition laws (with fields as the central concept), polynomials, and linear algebra with matrices and tensors. Linear algebra covered essentially what is now a standard elementary presentation of finite dimensional vector spaces (linear independence, basis, dimension, matrices, etc.), but it also included tensors, defined on vector spaces;[24] this was apparently done in order to prepare the ground for a treatment of exterior differential forms. The section on real and complex numbers mixed algebraic axioms with topological notions and stressed,

---

[22]Bourbaki encountered some difficulties in dealing with set theory and integration which set back the publication of these volumes. For this reason, nothing more than a digest of results was published on set theory until the 1950s.

[23]Beaulieu 1989, vol. I, 168–178.

[24]Later, these will be defined on modules.

as the main result, simply the fact that complex numbers form a vector space over the reals. Although it had grown and broadened in scope since the first semester, the plan for algebra was still thought of in terms of what would be used directly in other parts of the treatise. Since these other parts were as yet undeveloped, the content of the volume on algebra was not very far-reaching.[25] For this reason also, Bourbaki had many discussions over algebra between 1935 and 1938, especially concerning the choice of concepts and their order of appearance in this volume.

By 1936, Bourbaki had severed real and complex numbers from the core of algebra and lodged them within topology. Thus, writing on real and complex numbers was withheld until the first chapters on topology and integration had progressed. Meanwhile, Bourbaki clearly separated set theory from algebra. By then, the algebra program admitted exterior multiplication and exterior algebras and the notion of group received a more thorough treatment than it had before. The section on tensor algebra developed into a whole new chapter. Through this process, Bourbaki created a volume of algebra which was relatively independent from other theories and which was increasing in size and complexity.[26]

This process continued in the following year, as a new chapter on natural numbers was added to algebra. Linear algebra included groups with operators, while tensors were taken out of linear algebra to form an independent chapter. Exterior algebras also constituted another chapter, together with determinants and matrices. There were, by then, seven chapters in the volume on algebra.[27]

In 1938, Bourbaki revised its plan for algebra, reshuffling the former seven chapters and condensing them into five. Field, rather than group, was still the dominant concept of the first chapter. In fact, the essence of this first chapter remained stable until its publication while Bourbaki revised the other chapters of algebra, up to the eve of their publication date in some cases.[28] Linear algebra constituted the second chapter and was supposed to be developed both from the point of view of modules and the point of view of vector spaces. The next chapter expounded on rings and quotient rings. Chapter four consisted of algebras and determinants, while matrices were the subject of the last chapter. So we see that, although in one way Bourbaki's algebra remained tied to analysis, in another way it was absorbing an increasing number of algebraic theories, such as: rings, ideals, fields, polynomials, exterior algebras, modules, tensors, and vector spaces.[29]

Despite the difficulties endured in World War II, Bourbaki published its first chapter of Algebra, entitled "Algebraic Structures" in 1942, and pursued its work

---

[25] Beaulieu 1989, vol. I, 247–254.

[26] Beaulieu 1989, vol. I, 332–337.

[27] Beaulieu 1989, vol. I, 337–339.

[28] The first editions of the different chapters of the volume on algebra are: chapter 1-Structures algébriques, 1942; chapter 2-Algèbre linéaire, 1947; chapter 3-Algèbre multilinéaire, 1948; chapters 4 and 5-Polynômes et fractions rationnelles et Corps commutatifs, 1950; chapters 6 and 7- Groupes et corps ordonnés et Modules sur les anneaux principaux, 1952; chapter 8-Modules et anneaux semi-simples, 1958; chapter 9-Formes sesquilinéaires et formes quadratiques, 1959; chapter 10 -Algèbre homologique, 1980.

[29] Beaulieu 1989, vol. I, 339–342.

on linear algebra (chapter 2) and on multilinear algebra (chapter 3). Both of these chapters boasted significant breakthroughs, mainly in presenting linear algebra from the point of view of modules and in devising streamlined definitions of tensor and tensor product of modules for multilinear algebra. In the case of multilinear algebra, Bourbaki accomplished what it considered one of its most remarkable expository successes.[30]

This history of Bourbaki's treatment of algebra illustrates well the tendency throughout its whole work to move almost unconsciously from an initial narrowly focused goal to a more ambitious global and foundational vision. Thus, the volume on algebra became an extensive monograph that presented a network of theories and provided the instruments with which Bourbaki algebraized its treatise.

A General Outline for the *Éléments de mathématique*

The different general outlines that Bourbaki drafted for the *Éléments* provide another indicator of the changing scope of the project. In 1941, Dieudonné circulated among his colleagues a general outline that featured a wide variety of subjects. Some of these had been considered by Bourbaki as early as 1935, others had not yet been worked on. Between 1935 and 1941, Bourbaki's list of subjects had grown considerably and the team had devised a classification of its material into parts, volumes and chapters, each representing a cluster of interrelated yet relatively independent theories. By 1941, Bourbaki had already published a digest of results on Set Theory (1939) and two chapters of General Topology (1940). The first chapter of Algebra and the next two chapters of Topology were forthcoming (1942).

The 1941 outline divided the *Éléments* into four parts: Fundamental Structures of Analysis (eight volumes), Functional Analysis (seven volumes), Differential Topology (two volumes), and Algebraic Analysis (eight volumes).[31]

As outlined in 1941, the first six volumes constituting the "Fundamental Structures of Analysis" bore almost the same order and titles as they did when they were later published. Between 1935 and 1941, Bourbaki had drafted at least one complete version of each of these six volumes which included: Set Theory, Algebra, General Topology, Topological Vector Spaces, Differential Calculus, and Integration. The last two volumes of this first part, "Combinatorial Topology" (as algebraic topology was sometimes called then) and Analytic Functions, were still at the planning stage.[32]

---

[30]Both of these chapters were published after 1945 although the major decisions concerning them had been drawn in 1942 and 1943. These decisions are accounted in Beaulieu 1989, vol. I, 404–414.

[31]Tribu nº6, x/15/1941, 1–2. The "Tribu" is Bourbaki's newsletter. A simplified version of this outline appears in Beaulieu 1989, vol. II, 104.

[32]The "Fundamental Structures of Analysis" finally comprised six volumes instead of eight. Other alterations in plan were that the order between topological vector spaces and differential calculus was eventually permuted, and the volume entitled "Differential Calculus" became "Functions of a Real Variable."

In this 1941 outline, we find again some standard subjects of French analysis treatises, such as: differential calculus, analytic functions, integrals, calculus of variations, differential equations, partial differential equations, elliptic differential equations, hyperbolic and parabolic differential equations. These, however, were now considered in the language of algebra and topology, two theories on which Bourbaki had worked intensively and effectively since 1935. For instance, systems of equations were dealt with from the point of view of functional spaces.

There were also topics not usually covered by conventional analysis treatises. The last two parts of the outline comprised, for instance: Lie groups, differential geometry, number theory, polynomial ideals and algebraic geometry, all of which were then areas in rapid expansion. As some members of the team were specialists in these fields, Bourbaki entertained some hope of carrying out these plans successfully. In this respect, the outline of 1941 reflected the state of Bourbaki's accomplishments as well as the scope of its ambitions.

The contents of the remaining three parts were intended to constitute the core of the treatise. As such, they would offer a justification for the contents of the "Fundamental Structures." Most of the topics included in these last parts, however, did not have even a first report written on them yet. In 1941, Dieudonné already saw the chasm widening between the "Fundamental Structures" and the rest of the project. When he tried to rally the weakening forces of a dispersed membership during the war years, he urged his colleagues to write state-of-the-art reports on some topics the group had not yet approached, even in a preliminary form. He threatened that, should these topics remain untouched, there would be such a gap between the publication date of the first volumes of the *Éléments* and that of their sequel that the "Fundamental Structures of Analysis" would seem disconnected from the subjects which justified them. Hence, the first part would appear unreasonably abstract and purposeless. Dieudonné did not succeed in getting his co-workers to write these reports. His prophecy became Bourbaki's fate and its Achilles' heel, at least for its detractors.

At the same time, Dieudonné suggested that Bourbaki concentrate its efforts on the first six volumes of "Fundamental Structures" so that they would be near ready for publication whenever the group could resume its regular activities. Indeed, the volumes on algebra, on topology, and on functions of a real variable continued to progress.

Many of the subjects that Bourbaki was contemplating in 1941, however, were finally left out of the *Éléments de mathématique*. Such was the fate of algebraic topology, algebraic geometry, number theory, differential equations,[33] partial differential equations, elliptic, hyperbolic, and parabolic differential equations, calculus of variations, algebraic differential equations, and numerical calculus. Some subjects were subsumed by other volumes or briefly examined in them. For instance, linear functionals are dealt with in the tenth chapter of General Topology

---

[33] Although some part of Bourbaki's volume on Functions of a Real Variable [1951] and of Differentiable and Analytic Manifolds [1971] deal with differential equations.

(1949), in the chapter of Topological Vector Spaces that covers Hilbert spaces (1955) and in the fourth chapter of Integration (1952).

A thorough explanation of why this overall project, as it stood in 1941, never materialized would span a larger part of Bourbaki's history than is intended here. Such an explanation would also call into play the developments of different mathematical theories, especially in the 1940s and the 1950s. Of course, the war adversely affected Bourbaki's collective work, resulting in the disappearance of the projected topics that had never been examined before. After 1945, the group resolved to prepare for publication the chapters which were closest to a final version. Thus, Bourbaki continued to neglect the subjects to which it had yet as given little or no attention. In the 1950s, Bourbaki also decided to write more specialized books on theories that interested the group by then and for which there were new specialists among the members, namely Lie groups and algebras and commutative algebra. Thus, many of Bourbaki's plans eventually fell into oblivion.

Conclusion

The picture of Bourbaki's early work that arises from this sketch is one of variety and uncertainty, of trial and error. The team proceeded to write its expository text while bearing in mind several goals. By the 1940s, the idea of an abstract treatment of major parts of mathematics triumphed over other purposes. Nevertheless, the subjects that constituted the core of the traditional French analysis treatises were still a part of Bourbaki's plans for its *Éléments de mathématique.*

In the 1930s there were currents of thought proclaiming the unification of different fields of scientific knowledge and particularly of mathematics. A form of "structuralism" also prevailed in the philosophy of mathematics at that time. But these visions were not the main motivation behind Bourbaki's decision to write its treatise. The group did not strive, originally, to implement such a view in its work. As a team, it did not seem to hold a common and unified vision of mathematics at the outset. If Bourbaki ever promoted such a doctrine, it was the result and not the spark of its own collective work.

Acknowledgments

This paper was written while I was a visiting fellow at the Office for History of Science and Technology of the University of California at Berkeley. I wish to thank the Social Sciences and Humanities Research Council of Canada for the fellowship that provided me with the freedom and the resources necessary to further my research. I am deeply indebted to Henri Cartan, Jean Dieudonné, André Weil and the late Claude Chevalley, who granted me interviews or who made their papers available to me. My thanks are also due to Joel Hillel, Gregory H. Moore, and David Thompson for their helpful comments on a previous draft.

Biblilography

BEAULIEU, Liliane

1990 Proofs in expository writing - Some examples from Bourbaki's early drafts. *Interchange*, **21**, n$^{\circ}$2, 35–45.

1989 *Bourbaki. Une histoire du groupe de mathématiciens français et de ses travaux (1934–1944)*. Doctoral dissertation, Université de Montréal. (2 vols.)

BOURBAKI, Nicolas[34]

1939 *Éléments de mathématique*, Vol. I, Théorie des ensembles, "Fascicule de résultats." Paris: Hermann.

1948 L'architecture des mathématiques, in *Les grands courants de la pensée mathématique*. François Le Lionnais (ed.), Paris: Cahiers du Sud, 35–47.

1950 The architecture of mathematics. Translation of [1948] by Arnold Dresden, *American Mathematical Monthly*, **57**, 221–232.

CARTAN, Henri

1943 Sur le fondement logique des mathématiques. *Revue scientifique*, **81**, 3–11.

DIEUDONNÉ, Jean

1939 Les méthodes axiomatiques modernes et les fondements des mathématiques. *Revue scientifique*, **77**, 224–232.

Van der WAERDEN, Bartel L.

1930 *Moderne Algebra*, vol. I. Berlin: Springer.

1931 *Moderne Algebra*, vol. II. Berlin: Springer.

---

[34] A near complete list of Bourbaki's works appears in Beaulieu 1989, Vol. II, appendix II.

Questions in the Historiography of Modern Mathematics: Documentation and the Use of Primary Sources
[Abstract]

E. Neuenschwander

Documentation and the use of primary sources play an important role in the history of mathematics, since they provide the means for any critical study in the field. As can be seen from numerous examples, historical developments are seldom as straightforward and logical as many theoretically oriented historians and philosophers of science like to pretend. The precise historical development can rarely be reconstructed by purely intellectual means, since influences from outside the field, and even mere accidents, sometimes play an important role. In order to determine the significant and decisive factors and to avoid false conclusions, it is therefore essential to undertake an extensive evaluation of primary and secondary sources.

The usefulness of the study of primary sources in the history of mathematics was demonstrated in my talk with the aid of research that I conducted into the estates of three mathematicians from three different countries, namely Joseph Liouville (1809–1882), Bernhard Riemann (1826–1866) and Felice Casorati (1835–1890). My investigations revealed the richness of the estates of mathematicians of the 19th century, as nearly all the mathematical notes of these three mathematicians have survived, and in the case of Liouville and Riemann, for example, even their school and university reports are still to be found. On the other hand, it is possible to reconstruct the discovery of mathematical theorems down to the smallest detail by the use of primary sources, as was shown by a survey of the history of the Casorati-Weierstrass-Theorem.

My talk concluded with some remarks on the present state of research into the documentation of secondary sources. This field has been greatly neglected in the past, and many older studies are now nearly forgotten, which leads to pointless duplication and false conclusions of yet another kind. It would therefore be highly desirable for all this material to be systematically documented in a mathematical-historical bibliography or database. For further information on the estates mentioned, see my publications: The Casorati-Weierstrass Theorem (Studies in the History of Complex Function Theory I), Historia Mathematica **5**

(1978), 139–166; Der Nachlaß von Casorati (1835-1890) in Pavia, *Archive for History of Exact Sciences* **19** (1978), 1–89; Interactions Among the French School, Riemann, and Weierstrass (Studies in the History of Complex Function Theory II), *Bulletin of the American Mathematical Society* **5** (1981), 87–105; Joseph Liouville (1809–1882): Correspondance inédite et documents biographiques provenant de différentes archives parisiennes, *Bollettino di Storia della Scienze Matematiche* **4** (1984), No. 2, 55-132; A Brief Report on a Number of Recently Discovered Sets of Notes on Riemann's Lectures and on the Transmission of the Riemann Nachlass, *Historia Mathematica* **15** (1988), 101–113; The Unpublished Papers of Joseph Liouville in Bordeaux, *Historia Mathematica* **16** (1989), 334-342.

List of Invited Speakers at the Tokyo History of Mathematics Symposium 1990

1. Liliane BEAULIEU, Université du Quebec à Montréal, Case postale 8888, succursale A, Montréal (Québec), H3C 3P8 Canada

2. Umberto BOTTAZZINI, University of Bologna, Department of Mathematics, Via Plutarco 12, 20145 Milan, Italy

3. Joseph W. DAUBEN, The Graduate School and University Center of the City University of New York, Ph. D. Program in History/Box 505, Graduate Center: 33 West 42 Street, New York, NY 10036-8099, USA

4. Chandler DAVIS, University of Toronto, Department of Mathematics, Toronto M5S 1A1, Canada

5. DU Shi Ran, Institute for the History of Natural Science, Academia Sinica, 137 Chao Nei Avenue, Beijing, People's Republic of China; Buddist College, 96 Hananobo-cho, Murasakino-kita, Kita-ku, Kyoto 603, Japan

6. William J. ELLISON, Laboratoire de Physique des Interactions Ondes-Matière, Ecole Nationale Supérieure de Chimie et de Physique de Bordeaux, Université de Bordeaux 1, Bordeaux, France

7. Craig G. FRASER, University of Toronto, Institute for the History and Philosophy of Science and Technology, Room 316, Victoria College, Toronto, M5S 1K7 Canada

8. Günther FREI, Mathematik, Eidgenosische Technische Hochschule Zürich, CH-8092 Zürich, Switzerland

9. Jeremy GRAY, Open University, Faculty of Mathematics, Milton Keynes MK7 6AA, England

10. HARA Kokiti, 2-5-18, Asahigaoka, Ikeda 563, Japan

11. Annick HORIUCHI, 69, rue Dunois, 75646, Paris, France

12. Christian HOUZEL, Comité national français des mathématiciens, 11, rue Pierre et Marie Curie, Institut Henri-Poincaré, 75231 Paris, France

13. Eberhard KNOBLOCH, Technische Universität Berlin, Institut für Philosophie, Wissenschafttheorie, Wissenschafts- u. Technikgeschichte, Sekr. TEL 2. Ernst-Reuter-Platz 7, D-1000 Berlin 10, Germany

14. Jesper LÜTZEN, Københavns Universitets Matematiske Institut, Universitetsparken 5, 2100 København Ø, Denmark

15. NAGAOKA Ryosuke, Daito Bunka University, Faculty of Jurisprudence, 1-9-1, Takashimadaira, Itabashi-ku, Tokyo 175, Japan

16. Erwin NEUENSCHWANDER, Universität Zürich, Mathematisches Institut, Rämistrasse 74, CH-8001 Zürich, Switzerland

17. Roshdi RASHED, REHSEIS, CNRS, 27, rue Damesme, 75013 Paris, France

18. David E. ROWE, Pace University, Department of Mathematics, 861 Bedford Road, Pleasantville, New York 10570-2799, USA

19. Erhard SCHOLZ, Bergische Universität, Gesamthochschule Wuppertal, Gauss-Strasse 20, Postfach 100127, 5600 Wuppertal 1, Germany

20. SASAKI Chikara, University of Tokyo, Department of History and Philosophy of Science, 3-8-1, Komaba, Meguro-ku, Tokyo 153, Japan

21. SUGIURA Mitsuo, Tsuda College, Department of Mathematics, 2-1-1, Tsuda-cho, Kodaira-shi, Tokyo 187, Japan

22. TAKASE Masahito, Kyushu University, Department of Mathematics, 6-10-1, Hakozaki, Higashi-ku, Fukuoka 812, Japan

List of Speakers and Titles of Their Lectures at Session B (Short Communications)

August 31

Itō Yoshihiko (Japan), On the Process of the Change of Education for Differential and Integral Curriculum in Japan

Nakane Michiyo (Japan), The Role of Mathematical Analogy Played in the Optical-Mechanical Research by W. R. Hamilton

Hoshida Haruyo (Japan), Fourier's View on the Role of Mathematical Analysis in the Physical Science

Murata Tamotsu (Japan), A Tentative Reconstruction of the Formation Process of the Xth Book of the *Elements*

September 1

Joran Friberg (Sweden), New Insights into the Nature of Late Babylonian Mathematics: Unpublished Texts from the Iraq Museum and the British Meseum

Simizu Tatsuo (Japan), Ruan Yuan and Endō Toshisada, Founders of History of Mathematics in China and Japan

Ubiratan D'Ambrosio (Brazil), The Emergence of Modern Mathematics in Brazil from the Middle of the Nineteenth Century to the First Quarter of the Twentieth Century

Hirano Yoichi (Japan), Le développement de l'algèbre en France au 19ème siècle

Science Networks • Historical Studies

Edited by
Erwin Hiebert, Harvard University, Cambridge MA, USA
Hans Wussing, Universität Leipzig, Germany
in cooperation with an international editorial board

The publications in this series are limited to the fields of mathematics, physics, astronomy, physical chemistry, and their applications. The publication languages are English preferentially, German, and in exceptional cases also French. The series is primarily designed to publish monographs. However, special editions featuring collected letters as well as thematic groupings of smaller individual works or proceedings can be taken into consideration. Annotated sources and exceptional biographies might be accepted in rare cases. The series is aimed primarily at historians of science and libraries; it should also appeal to interested specialists, students, and diploma and doctoral candidates. In cooperation with their international editorial board, the editors hope to place a unique publication at the disposal of science historians throughout the world.

SN 1   Scholz, E.: Symmetrie, Gruppe, Dualität. Zur Beziehung zwischen theoretischer Mathematik und Anwendung in Kristallographie und Baustatik des 19. Jahrhunderts, 1989 (ISBN 3-7643-1974-7)

SN 2   Grattan–Guinness, I.: Convolutions in French Mathematics, 1800–1840. From the Calculus and Mechanics to Mathematical Analysis and Mathematical Physics, Volume I: The Settings, 1990 (ISBN 3-7643-2237-3)

SN 3   Grattan–Guinness, I.: Convolutions in French Mathematics, 1800–1840. From the Calculus and Mechanics to Mathematical Analysis and Mathematical Physics, Volume II: The Turns, 1990 (ISBN 3-7643-2238-1)

SN 4   Grattan–Guinness, I.: Convolutions in French Mathematics, 1800–1840. From the Calculus and Mechanics to Mathematical Analysis and Mathematical Physics, Volume III: The Data, 1990 (ISBN 3-7643-2239-X)

SN 5   Kipnis, N.: History of the Principle of Interference of Light, 1990 (ISBN 3-7643-2316-7)

SN 6   Hentschel, K.: Interpretationen und Fehlinterpretationen der speziellen und der allgemeinen Relativitätstheorie durch Zeitgenossen Albert Einsteins, 1990 (ISBN 3-7643-2438-4)

SN 7   Medvedev, F.A.: Scenes from the History of Real Functions, 1991 (ISBN 3-7643-2572-0)

SN 8   Busard, H.L.L.., Folkerts, M. (Eds): Robert of Chester's (?) Redaction of Euclid's Elements, the so-called Adelard II Version, Volume I, 1992 (ISBN 3-7643-2658-1)

SN 9   Busard, H.L.L.., Folkerts, M. (Eds): Robert of Chester's (?) Redaction of Euclid's Elements, the so-called Adelard II Versinon, Volume II, 1992 (ISBN 3-7643-2727-8)

SN 10 Benoit, P., Chemla, K., Ritter, J.: Histoire de fractions, fractions d'histoire, 1992 · (ISBN 7643-2693-X)

SN 11 Reich, K.: Die Entwicklung des Tensorkalküls. Vom absoluten Differentialkalkül zur Relativitätstheorie, 1992 (ISBN 3-7643-2814-2)

BIRKHÄUSER

SN 12 / Science Networks • Historical Studies

G.E. Gorelik / V.Y. Frenkel

Matvei Petrovich Bronstein and the Soviet Theoretical Physics in the Thirties

1994. 300 pages. Hardcover
ISBN 3-7643-2752-9

Matvei Petrovich Bronstein with his short life and tragic death (1906-1938) may be seen as a symbol of his time and his country. One of the most remarkable features of Soviet history was the impressive advance of its physical sciences against the burtal and violent background of totalitarianism. Soviet advances in nuclear and space technology form an important part of world history. These achievements had their roots in the 30s, when Bronstein's generation entered science. Among his friends were the famous physicists Lev Landau and George Gamow.

Bronstein worked in the vast field of theoretical physics, ranging from nuclear physics to astrophysics and from relativistic quantum theory to cosmology. His pioneering work on quantizing gravitation goes beyond the history of physics, because today the quantum theory of gravitation occupies a special place in fundamental physics.

Bronstein was also a master of scientific explanation thanks to his profound knowledge, enthusiasm as a teacher and a gift for literature. This enabled him to write popular science for children, the widest and most responsive group of readers. He became a writer with the help of his wife Lidiya Chukovskaya, known now as an outstanding writer and fighter for human rights.

Bronstein's life was closely intertwined with the social, historical and scientific context of one of the most tragic and intriguing periods of Russian history.

If you would like regular title information from Birkhäuser please write to the following address for your personal copy of the *Birkhäuser Mathematics Quarterly.*

Please order through your bookseller or write to:
Birkhäuser Verlag AG
P.O. Box 133
CH-4010 Basel / Switzerland
FAX: ++41 / 61 / 271 76 66

For orders originating in the USA or Canada:
Birkhäuser
333 Meadowlands Parkway
Secaucus, NJ 07094-2491 / USA

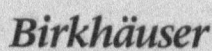

Birkhäuser Verlag AG
Basel · Boston · Berlin

BIRKHÄUSER

1993. 624 pages.
16 colour plates. Hardcover
Text English/German
ISBN 3-7643-2863-0

P.L. Butzer / D. Lohmann

Science in Western and Eastern Civilization in Carolingian Times

The intellectual crisis of the fifth to seventh centuries was followed by 'new
beginnings' in both West and East, that
become tangible by A.D. 800. Latin and
Greek works were transcribed or translated, then circulated and commented
upon, in a context of new practical needs
and scholarly interests. This interdiscipli-
nary volume includes contributions by 25
international specialists, focused on the
renewal of scientific efforts in the Latin
West, Byzantium, and the Islamic East.
It presents unpublished, original texts and
modern translations, analytical papers
on calendric computation, as well as review articles examining mathematics,
astronomy, and the applied sciences. A
comprehensive study of illuminated
Carolingian and Byzantine manuscripts
serves to illustrate the subject matter,
while a tabular overview of scientific
achievments c. A.D. 400-950 concludes
the work.

**Please order through your book-
seller or write to:**
Birkhäuser Verlag AG
P.O. Box 133
CH-4010 Basel / Switzerland
FAX: ++41 / 61 / 271 76 66

**For orders originating
in the USA or Canada:**
Birkhäuser
333 Meadowlands Parkway
Secaucus, NJ 07094-2491 / USA

Birkhäuser

Birkhäuser Verlag AG
Basel · Boston · Berlin